Gedeon Dagan

Flow and Transport in Porous Formations

With 113 Figures

Springer-Verlag Berlin Heidelberg New York
London Paris Tokyo Hong Kong

Gedeon Dagan
Tel Aviv University
Faculty of Engineering
Department of Fluid Mechanics
and Heat Transfer
P.O. Box 39040
Ramat Aviv, Israel

Softcover Edition:
ISBN 3-540-51098-2 Springer-Verlag Berlin Heidelberg New York
ISBN 0-387-51098-2 Springer-Verlag New York Berlin Heidelberg

Hardcover Edition:
ISBN 3-540-51602-6 Springer-Verlag Berlin Heidelberg New York
ISBN 0-387-51602-6 Springer-Verlag New York Berlin Heidelberg

Library of Congress Cataloging-in-Publication Data. Dagan, G. (Gedeon), 1932- . Theory of flow and transport in porous formations / G. Dagan. p. cm. ISBN 0-387-51098-2 (U.S. : alk. paper) 1. Groundwater flow–Mathematical models. 2. Porosity–Mathematical models. I. Title. TC176.D32 1989 551.49'01'5118–dc 20

Printing: Druckhaus Beltz, 6944 Hemsbach/Bergstraße; Bookbinding: J. Schäffer GmbH & Co. KG., Grünstadt
2131/3145-543210 – Printed on acid-free paper

To Ora, Sigal, Noga and Adi
A message of the spirit
Written in mathematical language
With love.

PREFACE

In the mid-seventies, a new area of research has emerged in subsurface hydrology, namely stochastic modeling of flow and transport. This development has been motivated by the recognition of the ubiquitous presence of heterogeneities in natural formations and of their effect upon transport and flow, on the one hand, and by the vast expansion of computational capability provided by electronic machines, on the other. Apart from this, one of the areas in which spatial variability of formation properties plays a cardinal role is of contaminant transport, a subject of growing interest and concern.

I have been quite fortunate to be engaged in research in this area from its inception and to witness the rapid growth of the community and of the literature on spatial variability and its impact upon subsurface hydrology. In view of this increasing interest, I decided a few years ago that it would be useful to present the subject in a systematic and comprehensive manner in order to help those who wish to engage themselves in research or application of this new field. I viewed as my primary task to analyze the large scale heterogeneity of aquifers and its effect, presuming that the reader already possesses a background in traditional hydrology. This is achieved in Parts 3, 4 and 5 of the text which incorporate the pertinent material. The term "porous formations" in the title aludes indeed to the primary preoccupation of the book with phenomena taking place in the natural environment rather than in the laboratory. However, the writing of the book has been delayed considerably for a few reasons. First, the field being in active and permanent development, I had to investigate new topics in order to fill a few gaps. The results of this activity can be found in the articles I published in this period and which are referred to in the book. Second, as I mentioned before, I regarded this book as a continuation of previously acquired knowledge from existing texts on flow through porous media. Most of these texts, however, introduce the basic concepts (Darcy's law, hydrodynamic dispersion) in a deterministic context. Thus, the reader may be left with the impression that treatment of heterogeneity in a stochastic frame is appropriate for large scales only. However, it is well known that the macroscopic laws governing flow through porous media can be derived with the aid of probability concepts also, and in my view this has definite advantages. Presenting a unified theory, based on the same basic concepts, covering flow at all scales, from the pore to the regional one, constitutes a temptation which is hard to resist by any theorist, and I am no exception. As a result, I spent a considerable time in writing Part 2, which covers classical subjects, but under a somewhat different angle. The results of this prolonged effort is a self-contained text, which may be used by those interested in becoming acquainted with the theory of groundwater flow and transport. Nevertheless, the readers with the adequate background can approach directly Parts 3, 4 and 5. Part 1, a recapitulation of a few probability concepts needed in this book, is meant to save the labor of reading extensive treatises, which cover much more material. However, it is not a substitute to a previously acquired background in basic statistics and probability theory.

It is emphasized that this is a treatise which presents the *theory* of flow and transport. It concentrates on deriving in a systematic and rational manner the basic equations governing various processes, and constitutes the starting point to various applications. Nevertheless, this is a text written for hydrologists and engineers, the ultimate goal being to apply the concepts to real life probelms. For this reason I tried to emphasize at the beginning of each new part or chapter the experimental or field findings which justify the theoretical approach. Furthermore, to facilitate the understanding of the theoretical concepts and of the accompanying equations, many examples of solutions are presented and discussed. Since their role is mainly illustrative, these solutions are either analytical or based on some rational approximations. Besides the main purpose of helping to grasp the theoretical developments, they may also serve as a benchmark for numerical solutions or ad-hoc approximations. The book does not address the growing and important field of numerical methods and of development of codes to solve complex problems. This subject deserves a separate treatment and is indeed covered by a few texts. In the same vein, the book does not attempt to treat the various practical techniques of solving field problems in order to give immediate answers to pressing questions. It provides only the theoretical background to approach the complex problems posed by real life. Finally, neither experimental techniques, laboratory or field, nor management of water resources, are subjects within the scope of this book.

The field of flow in porous media has expanded vastly in the last few years in relation to its various applications in hydrology, soil physics, reservoir engineering, chemical engineering, etc. This book treats only a limited sector of the discipline, namely saturated flow of water and transport of inert solutes. Some aspects of unsaturated flow, of transport of reactive solutes, of heat transfer and of immiscible flow are touched, but the reader is advised to approach other specialized texts for comprehensive treatments of these subjects.

Although I view the book primarily as a scientific monograph, I hope that it will also serve as an advanced text book for graduate students. For this purpose a few exercises are given at the end of most sections. These are not trivial, of the "number plugging" type exercises, but an extension of the theoretical part, and their solution requires a thorough understanding of the material.

Last not least, this book is not intended to be a compendium or a systematic literature review, but it represents a personal outlook thereby reflecting my own interests and preferences. As such it is definitely biased towards my own previous work and to the others that I considered particularly relevant. I am quite sure that by mistake or neglect some important contributions have been omitted and my apology is due to the authors who have been unintentionally overlooked. It is my belief that science is the product of a collective effort and my contribution rests on the shoulders of my predecessors and colleagues. Thus, this book expresses, directly or indirectly, the achievements of the community to which I am deeply indebted.

TABLE OF CONTENTS

PART 1 MATHEMATICAL PRELIMINARIES: ELEMENTS OF PROBABILITY THEORY AND RANDOM FUNCTIONS

PART 2 THE LABORATORY SCALE (HOMOGENEOUS MEDIA)

INTRODUCTION

Porous media and natural porous formations are heterogeneous, i.e. they display spatial variability of their geometric and hydraulic properties. Furthermore, this variability is of an irregular and complex nature. It generally defies a precise quantitative description, either because of insufficiency of information or because of the lack of interest in knowing the very minute details of the structure and flow field. Indeed, in most circumstances one is interested in the behavior of a large portion of the formation, i.e. in averaging flow or transport variables over the space. This averaging process has a smoothing effect and filters out small scale variations associated with heterogeneity. Even in those cases in which this is not achieved, because of large scale variability, determining some gross features of the processes may be quite satisfactory. The mathematical framework used to describe heterogeneity and the transition to average variables which is adopted in this book is the *statistical* one. Properties and flow and transport variables are represented with the aid of random space functions (RSF) and the actual porous formation and process are regarded as a realization of the ensemble of RSF which describes them. The main aim of this book is to derive the equations satisfied by the statistical moments (primarily the mean and the variance) of various flow and transport variables and to relate them to those characterizing the heterogeneous structure. In order to grasp the results, these equations are solved in various particular cases of interest.

The structure of the book and its partition stem from the fundamental role played by the various *length scales* which characterize the medium and the flow and transport processes. This idea, which has been explored in the article "Statistical theory of groundwater flow and transport : pore to laboratory, laboratory to formation and formation to regional scale" (Dagan, 1986), is briefly discussed herein.

A first fundamental length scale, L, is the one characterizing the extent of the flow or transport domain. Three such basic scales are selected as representative : laboratory, local (formation) and regional.

The laboratory scale pertains to samples of porous media, usually columns, which are employed in order to observe and study various processes. These are mostly one-dimensional in the average. At this scale, of the order of 10^0 meters, the observable heterogeneity is the one associated with the pore structure.

The local (formation) scale is related to the thickness of aquifers and to a similar dimension in the plane. Flow, e.g. caused by wells, and transport, e.g. a plume originating from a repository, are generally three-dimensional. At this scale, of the order of 10^1-10^2 meters, heterogeneity manifests mainly in variations of permeability in space. Such variations are associated, for instance, with layering which occurs in sedimentary formations.

Last, the largest scale considered in this book, is the regional (basin) one. The hydrologic unit is now the entire aquifer or formation, whose planar dimension, of order 10^3-10^5 meters, is much larger than the thickness. This scale is relevant, for instance, to exploitation of aquifers and determining their water balance, or to transport from non-point sources or for plumes traveling for considerable periods. At this scale the "shallow water approximation" of water waves theory is appropriate, i.e. averaging of variables over thickness and representing them as two-dimensional in the horizontal plane. A larger scale which dominates heterogeneity appears now, namely that of transmissivity and storativity.

The analysis of heterogeneity and its effect is organized in the book along these three scales.

A second fundamental length-scale, I, is that associated with heterogeneity. It is the correlation scale, i.e. roughly the distance at which properties cease to be correlated. I is determined, directly or indirectly, from measurements by using statistical inference techniques. Our basic assumptions are that $I \ll L$ and that statistical stationarity prevails. These two assumptions are necessary in order to infer the statistical structure of the RSF describing heterogeneity from a single available realization. Its validity in each particular case can be verified by using the well developed apparatus of statistics. In some circumstances these requirements can be relaxed. For instance, stationarity of the increments of the RSF of interest implies the existence of a variogram which may grow indefinitely. Nevertheless, it can be inferred from one realization for intervals much smaller than L and can be used to derive various conditional moments by "geostatistical methods".

The third fundamental length-scale, D, is the one over which properties or variables of interest are *space* averaged. This scale is related to the dimension of the measurement devices or to that of interest for management. For instance, in the case of flow and transport through a laboratory column, the velocity and concentration are averaged over the cross-section. At the local scale, piezometers or pumping wells generally serve as averaging devices, though more detailed measurements may be achieved by packers or multi-level samplers. At the regional scale D may be associated with an isolated well, at one extreme, or to the entire domain when the formation is regarded as one cell for global balances. Similarly, the scale D of a plume may vary considerably among point or non-point sources and the same is true if one is interested, for instance, in the contamination of an isolated well or in the total mass discharged into a river.

The hierarchy of the three scales, I, D and L plays a crucial role in the analysis. Thus, at the laboratory scale the ordering $I \ll D \ll L$ generally prevails. In this case, the interest resides mainly in the space average, which can be interchanged with the ensemble average under ergodic hypothesis. The main role of the theory reduces then to derive *effective* properties of the medium. The above ordering is less definite at the local scale. It may prevail in many flow problems, but not necessarily for transport processes. Finally, at the regional scale it is quite often that $D<<I$, i.e. we are concerned with "point values" of the variables of interest. Uncertainty may now be quite large, and the derivation of the higher-order statistical moments is required. Furthermore, in such cases the "geostatistical methods" like kriging or Gaussian conditioning, become of definite relevance.

The present book has much in common with the more traditional, deterministic, treatment of groundwater flow and seepage whenever the average variables are of interest (Part 2 and Chaps. 3.5, 3.6 and 5.4). However, it differs from the usual approach in those areas in which uncertainty is not negligible or in which large scale effects upon transport and the inverse problem prevail.

PART 1. MATHEMATICAL PRELIMINARIES : ELEMENTS OF PROBABILITY THEORY AND RANDOM FUNCTIONS

For easiness of reference and for readers convenience, we present here a brief recapitulation of elements of probability theory and of theory of random functions needed in this book. This subject is covered by many text books. A handy reference for introduction to probability theory is, for instance, Mood & Graybill (1963). For the theory of random functions one may consult the books by Papoulis (1965) and Yaglom (1962), or the monograph of Matheron (1965), which is close in spirit with the material presented herein. Recent texts oriented toward hydrologic applications are Bras and Rodriguez-Iturbe (1985) and Vanmarcke (1983).

1.1 RANDOM VARIABLES. STATISTICAL MOMENTS

We shall deal frequently in this book with physical entities, represented by real numbers, which are subjected to uncertainty and are regarded as random variables. If the quantity of interest, denoted generically by u, is the outcome of an experiment, then the set of possible experiments constitute an ensemble. A number P(a), called the probability of the event, can be attached to the event u = a. As a simple example let us consider a laboratory column in which we pack repeatedly the same quantity of sand by the same procedure. A fixed point of the column is within a void in some experiments, or in the solid matrix, in others. We define the random variable h as a function which takes the unit value in the first case and the zero value in the other, similarly to the outcome of the flipping of a coin. The empirical way of defining the probability P(0) of h taking the value zero, for instance, is in determining the ratio M/N for $N \gg 1$, where N is the total number of experiments and M is the number of occurrences of the event h=0. The ratio M/N fluctuates and there is no assurance that it tends to a definite limit as $N \to \infty$. There is a large body of literature discussing this difficulty and it is beyond our scope to analyze it here. We shall assume that the limit exists and it is approached with decreasing error as M and N become much larger than unity. In many cases we deal, however, with random variables which may take any value between $-\infty$ and $+\infty$ e.g., the logarithm of the permeability of the sand of the ensemble of columns defined before, measured by some standard procedure. It is convenient then to define the probability distribution function (P.d.f., for briefness) F(u) as the probability that the value of the random variable is between $-\infty$ and u (u lying between a and b meaning here $a \le u < b$). We then have by definition

$$\lim_{u \to -\infty} F(u) = 0 \quad ; \quad \lim_{u \to \infty} F(u) = 1 \tag{1.1.1}$$

whereas the probability that u lies between a and b is

$$P(a \le u < b) = F(b) - F(a) \tag{1.1.2}$$

and F(u) is seen to be a positive and monotonically increasing function.

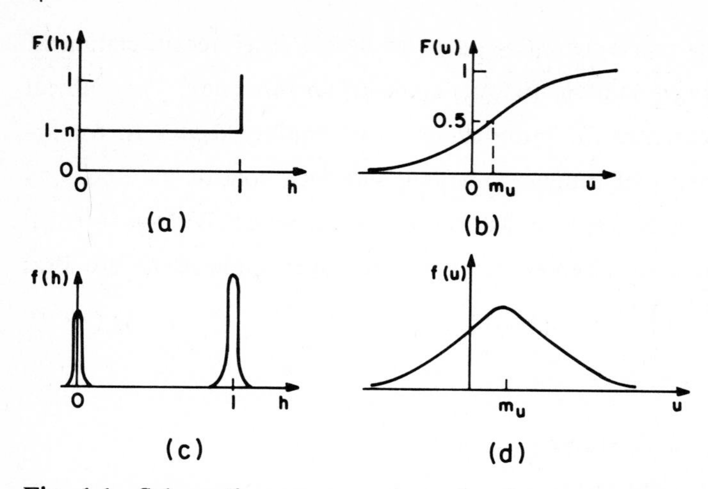

Fig. 1.1 Schematic representation of a few probability distribution functions (a and b) and associated probability density functions (c and d, respectively): (a),(c) bimodal discrete distribution; (b),(d) normal distribution.

The P.d.f. of the aforementioned discrete function h is bimodal and is represented for illustration in Fig. 1.1a. An example of continuous p.d.f. encountered frequently in this book is the normal distribution

$$F(u) = \frac{1}{\sqrt{2\pi}\,\sigma_u} \int_{-\infty}^{u} \exp[-\frac{(\lambda - m_u)^2}{2\,\sigma_u^2}]\, du = \frac{1}{2}[1 + \operatorname{erf}(\frac{u - m_u}{\sqrt{2}\sigma_u})] \tag{1.1.3}$$

which is represented schematically in Fig. 1.1b and whose important properties are well documented in the literature. The lognormal P.d.f. of the random variable v is precisely (1.1.3), with u replaced by ln(v).

The probability density function f(u) (p.d.f. in the sequel), is defined for a differentiable F(u) by

$$f(u) = \frac{dF}{du} \tag{1.1.4}$$

such that f(u)du is, at the limit du→0, the probability that u lies in the interval between u and u+du. We cannot define f(h) by ordinary functions for F(h) of Fig. 1.1a, but f(h) can be represented with the aid of the Dirac distributions (see Chap. 1.5) as

$$f(h) = (1-n)\,\delta(h) + n\,\delta(h-1) \tag{1.1.5}$$

which is represented schematically in Fig. 1.1c. By the same token, for the normal distribution (1.1.3), we obtain the Gaussian function

$$f(u) = \frac{1}{\sqrt{2\pi}\,\sigma_u} \exp[-\frac{(u-m_u)^2}{2\,\sigma_u^2}] \tag{1.1.6}$$

which is represented schematically in Fig. 1.1d.

Conversely, we also have

$$F(u) = \int_{-\infty}^{u} f(\lambda)\, d\lambda \; ; \; \int_{-\infty}^{\infty} f(\lambda)\, d\lambda = F(\infty) = 1 \tag{1.1.7}$$

The expected value (briefly, the expectation) of u is defined by

$$\langle u \rangle = E(u) = \int_{-\infty}^{\infty} u\, f(u)\, du \tag{1.1.8}$$

and $\langle u \rangle$ is also called the ensemble average or the ensemble mean of u. It can also be defined, under some limiting conditions, as

$$\langle u \rangle = \lim_{N \to \infty} \left(\frac{1}{N} \sum_{i=1}^{N} u_i \right) \tag{1.1.9}$$

where N is again the total number of experiments which have u_i as outcome. Similarly, the moment of order n is defined by

$$\langle u^n \rangle = E(u^n) = \int_{-\infty}^{\infty} u^n\, f(u)\, du \tag{1.1.10}$$

The fluctuation, or the residual, u′ is defined by $u' = u - \langle u \rangle$ and obviously $\langle u' \rangle = 0$.

The following relationships

$$\sigma_u^2 = \langle u'^2 \rangle = \langle u^2 \rangle - \langle u \rangle^2 \; ; \; s_u = \langle u'^3 \rangle \tag{1.1.11}$$

define the variance σ_u^2 and the skewness s_u, σ_u being the standard deviation and $CV_u = \sigma_u / |\langle u \rangle|$, the coefficient of variation.

For the bimodal p.d.f. of Fig. 1.1c we thus have

$$\langle h \rangle = n \; ; \; \sigma_h^2 = n(1-n) \; ; \; s_h = n - 3n^2 + 2n^3 \tag{1.1.12}$$

whereas for the normal distribution (1.1.6)

$$\langle u\rangle = m_u \quad ; \quad \langle u'^2\rangle = \sigma_u^2 \quad ; \quad \langle u'^3\rangle = 0 \tag{1.1.13}$$

and for the lognormal distribution ($u=\ln v$)

$$\langle v^n\rangle = \exp(n\, m_u + \frac{n^2}{2}\sigma_u^2) \tag{1.1.14}$$

It can be shown that under quite general conditions the set $\langle u^m\rangle$, for $m=1,2,...$, exhausts the statistical information about the random variable u and the p.d.f. of u can be constructed from a complete knowledge of its moments.

The expectation, as defined by (1.1.8) or (1.1.9), is also known as the arithmetic average u_A, a term inspired by (1.1.9). In a similar fashion, for positive random variables, the geometric ensemble mean is defined by $u_G=(\Pi\, u_1u_2...u_N)^{1/N}$ for $N\to\infty$, whereas the harmonic mean is given by $u_H=(\langle u^{-1}\rangle)^{-1}$. These average values have simple expressions for log-normal distributions (Eq. 1.1.6 with $u=\ln v$)

$$v_A = \exp(m_u + \frac{1}{2}\sigma_u^2) \; ; \; v_G = \exp(m_u) \; ; \; v_H = \exp(m_u - \frac{1}{2}\sigma_u^2) \tag{1.1.15}$$

1.2 JOINT PROBABILITY DISTRIBUTIONS. CONDITIONAL PROBABILITY. MULTIVARIATE NORMAL DISTRIBUTIONS

In the present book we shall encounter sets of random variables which are statistically dependent. For example, the water heads ϕ_1, ϕ_2,...., ϕ_N, in an aquifer at a few different points of coordinates $\mathbf{x}_1$, $\mathbf{x}_2$,...., $\mathbf{x}_N$, respectively, are viewed as statistically dependent random variables. Similarly, the pressure at a point $\mathbf{x}_1$ and the fluid velocity at a point $\mathbf{x}_2$ of a porous medium are regarded as statistically dependent random variables.

With the generic notation u_1, u_2,..., u_N for a set of N random variables, the probabilistic information about them is exhausted by the joint probability distribution function $F(u_1,u_2,...,u_N)$, which is defined as the probability that u_1 lies between $-\infty$ and u_1, while u_2 lies between $-\infty$ and u_2, etc. We have obviously

$$\lim_{u_i\to-\infty} F(u_1, u_2,...,u_N) = 0 \qquad \text{(for any } i=1,2,...N) \tag{1.2.1}$$

$$\lim_{u_i\to\infty} F(u_1,u_2,...,u_N)=1 \qquad \text{(for all } i \text{ simultaneously)} \tag{1.2.2}$$

The joint probability density function associated with F is

$$f(u_1,u_2,...u_N) = \frac{\partial^N F}{\partial u_1\, \partial u_2 ... \partial u_N} \tag{1.2.3}$$

such that $f(u_1,u_2,...,u_N)\, du_1\, du_2...du_N$ is the probability of u_1 lying in the interval u_1,u_1+du_1, u_2 in the interval u_2,u_2+du_2, etc. Thus, Eqs. (1.2.1) and (1.2.3) lead to

$$F(u_1,u_2,...,u_N) = \int_{-\infty}^{u_1}\int_{-\infty}^{u_2}...\int_{-\infty}^{u_N} f(u_1,u_2,...,u_N)\, du_1\, du_2...du_N \tag{1.2.4}$$

The marginal probability distribution function $F(u_1,u_2,...,u_{N-1}) = \lim F(u_1,u_2,...,u_N)$ for $u_N \to \infty$ is defined as the probability that u_1 lies between $-\infty$ and u_1, u_2 between $-\infty$ and $u_2,...,u_{N-1}$ between $-\infty$ and u_{N-1}, whereas u_N may be anywhere in the interval $-\infty,\infty$ and it provides no information about u_N. Various such marginal p.d.f. can be defined by letting some of the u_i tend to ∞ in $F(u_1,u_2,...,u_N)$.

The marginal joint probability density functions are defined by integration, e.g.

$$f(u_1,u_2,...,u_{N-1}) = \int_{-\infty}^{\infty} f(u_1,u_2,...,u_N)\, du_N \tag{1.2.5}$$

$$f(u_1,u_2,...,u_M) = \int_{-\infty}^{\infty}...\int_{-\infty}^{\infty} f(u_1,u_2,...u_M,u_{M+1},...,u_N)\, du_{M+1}\, ...du_N \tag{1.2.6}$$

In the case of two variables, $f(u_1,u_2)$ can be represented graphically in the space of Cartesian coordinates u_1,u_2,f as a bell shaped surface in the half-space $f>0$. Then, the marginal p.d.f. $f(u_1)$, for instance, is the area between $f(u_1,u_2)$ and $f=0$ in the plane u_1=const.

Various moments are generated with the aid of the joint p.d.f. Thus, the expected value of u_1 and its fluctuation are given by

$$\langle u_1\rangle = \int_{-\infty}^{\infty}...\int_{-\infty}^{\infty} u_1 f(u_1,u_2,...,u_N)\, du_1...du_N = \int_{-\infty}^{\infty} u_1 f(u_1)\, du_1$$

$$u_1' = u_1 - \langle u_1\rangle \tag{1.2.7}$$

The second-order moment, the covariance, is defined by

$$C_{uv} = \langle u'v' \rangle = \int_{-\infty}^{\infty}\int_{-\infty}^{\infty} (u-\langle u \rangle)\,(v-\langle v \rangle)\, f(u,v)\;du\,dv \tag{1.2.8}$$

whereas the correlation coefficient is given by

$$\rho_{uv} = \frac{C_{uv}}{\left[\langle u'^2 \rangle \langle v'^2 \rangle\right]^{1/2}} \tag{1.2.9}$$

In the case of independent random variables, the joint p.d.f. degenerates into

$$f(u_1,u_2,\ldots,u_N) = f(u_1)\, f(u_2) \ldots f(u_N) \tag{1.2.10}$$

and it is seen by (1.2.8) and (1.2.9) that $C_{uv}=0$ and $\rho_{uv}=0$ in this case. However, the converse is not true, i.e. the cancellation of the correlation coefficient does not necessarily imply that the variables are independent.

The conditional probability $f(u|v)du$ for the case of two variables, is defined as the probability of u lying between u and u+du for v fixed. From this definition it follows that

$$f(u|v) = \frac{f(u,v)}{f(v)} \tag{1.2.11}$$

where f(v) is the marginal p.d.f. of v, such that $\int f(u|v)du=1$. By the same token $f(v|u)=f(u,v)/f(u)$ and in the general case

$$f(u_1,u_2,\ldots,u_M|u_{M+1},u_{M+2},\ldots,u_N) = \frac{f(u_1,u_2,\ldots,u_M,u_{M+1},\ldots u_N)}{f(u_{M+1},u_{M+2},\ldots,u_N)} \tag{1.2.12}$$

where $f(u_1,u_2,\ldots,u_M|u_{M+1},u_{M+2},\ldots,u_N)$ is the joint p.d.f. of u_1, u_2,..., u_M for the fixed values $u_{M+1},u_{M+2},\ldots,u_N$. The fundamental Eq. (1.2.12) is known as Bayes relationship. For instance, the conditional ensemble means and variances are given by

$$\langle u_1|u_2,u_3,\ldots,u_N \rangle = \int_{-\infty}^{\infty} u_1\, f(u_1|u_2,u_3,\ldots,u_N)\, du_1 \tag{1.2.13}$$

$$C_u(\mathbf{r},\mathbf{J}) = C_{uT}(r)\, J^2 + [C_{uL}(r) - C_{uT}(r)] \frac{(\mathbf{r}.\mathbf{J})^2}{r^2} \tag{1.3.7}$$

such that C_u depends on the two isotropic function C_{uT} and C_{uL}, which can be called the transverse (normal to **J**) and longitudinal (parallel to **J**) covariances, respectively. More general expressions of this type for vectors, tensors, etc. can be obtained by using the method of Robertson (1940). The covariance $C_u(\mathbf{r})$ satisfies the following relationships

$$C_u(\mathbf{r}) = C_u(-\mathbf{r}) \quad ; \quad C(0)>0 \quad ; \quad C(\mathbf{r}) \leq C(0) \tag{1.3.8}$$

and additional conditions related to the positiveness of the variance of a linear combination of the random variables u_i.

Of special interest here are random functions whose joint p.d.f. for any set of points $\mathbf{x}_i$ are multivariate normal. We shall assume that the mean $m_u(\mathbf{x})$ is not necessarily constant, whereas the covariance $C_u(\mathbf{r})$ is stationary. In this case the elements of the covariance matrix (1.2.16) of $u(\mathbf{x}_i)$ for N fixed points (i=1,2,...,N) are given by

$$\sigma_{ij} = C_u(\mathbf{x}_i,\mathbf{x}_j) = C_u(\mathbf{x}_i-\mathbf{x}_j) \tag{1.3.9}$$

and consequently the entire statistical structure of $u(\mathbf{x}_i)$ (i=1,2,...,N) is exhausted by $m_u(\mathbf{x})$ and $C_u(\mathbf{r})$. Furthermore, the random function $u'(\mathbf{x})=u(\mathbf{x})-m_u(\mathbf{x})$ is stationary in the strict sense and all the statistical moments of $u(\mathbf{x}_i)$ can be easily calculated with the aid of Eqs. (1.2.15) and (1.3.9). Assume now that $u(\mathbf{x})$ is conditioned by its fixed values at a set of given points $\mathbf{x}_j$ (j=M+1,M+2,...,N) (see Chap. 1.2). The statistical information about $u(\mathbf{x})$ for any set of points $\mathbf{x}_i$ (i=1,2,...,M) is expressed with the aid of the conditional joint p.d.f. (1.2.12) of Chap. (1.2).

Thus, the ensemble mean at an arbitrary point $\mathbf{x}_i$ is, according to (1.2.20),

$$\langle u(\mathbf{x}_i|\mathbf{x}_{M+1},\mathbf{x}_{M+2},...,\mathbf{x}_N)\rangle = m_u(\mathbf{x}_i) + \sum_{j=M+1}^{N} \lambda_{ij}\, [u(\mathbf{x}_j)-m_u(\mathbf{x}_j)] \tag{1.3.10}$$

whereas the conditional covariance for two arbitrary points is given by (1.2.23) or (1.2.24) as follows

$$C_u(\mathbf{x}_i,\mathbf{x}_j|\mathbf{x}_{M+1},\mathbf{x}_{M+2},...,\mathbf{x}_N) = C_u(\mathbf{x}_i-\mathbf{x}_j) - \sum_{k=M+1}^{N} \lambda_{ik}\, C_u(\mathbf{x}_k-\mathbf{x}_j) =$$

$$C_u(\mathbf{x}_i-\mathbf{x}_j) - \sum_{k=M+1}^{N} \sum_{\ell=1}^{N} \lambda_{ik}\lambda_{j\ell}\, C_u(\mathbf{x}_k-\mathbf{x}_\ell)$$

with λ_{ij} solutions of the linear system (1.2.21), i.e. (1.3.11)

$$\sum_{j=M+1}^{N} \lambda_{ij}\, C_u(\mathbf{x}_j-\mathbf{x}_k) = C_u(\mathbf{x}_i-\mathbf{x}_k) \qquad (i,k=M+1,...,N) \tag{1.3.12}$$

The set of random functions $u(\mathbf{x})$ conditioned at $\mathbf{x}_j$ ($j=M+1,...,N$) constitute a subset of the ensemble of random functions which are free to fluctuate at $\mathbf{x}_j$. All the statistical information about the unconditioned set is exhausted by $m_u(\mathbf{x})$ and $C_u(\mathbf{r})$, whereas that related to the conditioned subset requires the additional data $u(\mathbf{x}_j) = u_j$ ($j=M+1,...,N$). It is pointed out that the conditioned random function is no more stationary: even if $m_u(\mathbf{x})$ is constant, the conditional average (1.3.10) depends on $\mathbf{x}_i$. Similarly, the conditional covariance (1.3.11) depends on $\mathbf{x}_i$ and $\mathbf{x}_j$ rather than $\mathbf{x}_i-\mathbf{x}_j$. The conditional variance, obtained from (1.3.11) by setting $\mathbf{x}_i=\mathbf{x}_j$, is no more constant.

It is emphasized that the conditional expected value of u (1.3.10) depends on the values $u_{M+1},...,u_N$ as well as on the position of $\mathbf{x}_i$ relative to $\mathbf{x}_{M+1},...,\mathbf{x}_N$. In contrast, the conditional covariance (1.3.11) depends only on the relative position, and the same is true for the conditional variance, which is obtained for $\mathbf{x}_i=\mathbf{x}_j$ in (1.3.11). Thus, further ensemble averaging of (1.3.10) for all possible values of $u_{M+1},...,u_N$, yields the unconditional mean $\langle u\rangle$. In contrast, the conditional covariance (1.3.11) and variance, remain unchanged under this additional averaging.

The theory of conditional probability is one of the most powerful tools employed by stochastic modeling of flow through porous formations, a tool which is used seldom in similar areas of the statistical theories of continua.

1.4 DIFFERENTIATION AND INTEGRATION OF RANDOM FUNCTIONS. MICROSCALE AND INTEGRAL SCALE

First, we shall recall briefly the notions of convergence and continuity of random functions. Among the various types of stochastic convergence, we shall adopt here the one of mean square convergence. Thus, a series of random variables u_n converges in the mean square to the random variable u if

$$\lim_{n\to\infty} \langle (u_n - u)^2\rangle = 0 \tag{1.4.1}$$

If this condition is satisfied, we also have

$$\langle \lim_{n\to\infty} u_n\rangle = \lim_{n\to\infty} \langle u_n\rangle = \langle u\rangle \tag{1.4.2}$$

i.e. the limit and the expectation signs commute.

A random function $u(\mathbf{x})$ is called continuous in the mean square at a point $\mathbf{x}$ if

$$\lim_{|\mathbf{r}|\to 0} \langle [u(\mathbf{x}+\mathbf{r})-u(\mathbf{x})]^2 \rangle = 0 \tag{1.4.3}$$

This condition may be rewritten as follows

$$\lim_{|\mathbf{r}|\to 0} C_u(\mathbf{x},\mathbf{x}+\mathbf{r}) = C_u(\mathbf{x},\mathbf{x}) = \sigma_u^2(\mathbf{x}) \tag{1.4.4}$$

In particular, in the case of second-order stationary random functions, if $u(\mathbf{x})$ is continuous in the mean square at a point $\mathbf{x}$, it is continuous at any point $\mathbf{x}+\mathbf{r}$. The necessary and sufficient equivalent condition is, according to Eq. (1.4.4),

$$\lim_{|\mathbf{r}|\to 0} C_u(\mathbf{r}) = C_u(0) = \sigma_u^2 \tag{1.4.5}$$

and $C_u(\mathbf{r})$ is necessarily continuous for any $\mathbf{r}$.

A random function $u(x)$ is differentiable in the m.s. at x and has the random function du/dx as derivative at x if

$$\lim_{r\to 0} \left[\frac{u(x+r)-u(x)}{r} - \frac{du(x)}{dx}\right]^2 = 0 \tag{1.4.6}$$

and this definition is easily extended to partial differentiation. The necessary and sufficient condition for du/dx (1.4.6) to exist is that $C_u(x,x+r)$ has partial derivatives $\partial C_u/\partial x$, $\partial C_u/\partial r$ and $\partial^2 C_u/\partial x\, \partial r$ for $r\to 0$. If du/dx exists at any point, the above condition extends to the entire plane x,r over which C_u is defined. Furthermore, we then have

$$\langle \frac{du}{dx} \rangle = \frac{d\langle u \rangle}{dx} \quad ; \quad \langle u'(x_1) \frac{du'(x_2)}{dx_2} \rangle = \frac{\partial C_u(x_1,x_2)}{\partial x_2}$$

$$\langle \frac{du'(x_1)}{dx_1} \frac{du'(x_2)}{dx_2} \rangle = \frac{\partial^2 C_u(x_1,x_2)}{\partial x_1\, \partial x_2} \tag{1.4.7}$$

In the particular case of second order stationarity we have, with $x_1=x$, $x_2=x+r$ in (1.4.7)

$$\langle \frac{du}{dx} \rangle = 0 \quad ; \quad \langle u(x) \frac{du(x+r)}{dx} \rangle = \frac{dC_u(r)}{dr} \tag{1.4.8}$$

$$\langle \frac{du(x)}{dx} \frac{du(x+r)}{dx} \rangle = - \frac{d^2C_u(r)}{dr^2}$$

Hence, differentiable random functions are characterized by twice differentiable covariance functions. Furthermore, the variance of the derivative is given by

$$<(\frac{du}{dx})^2> = - \frac{d^2C_u(r)}{dr^2}\bigg|_{r=0} \tag{1.4.9}$$

and the existence of the second derivative of $C_u(r)$ at the origin ensures its existence for any r. It is seen that in the neighbourhood of the origin $C_u(r)$ has the following Taylor expansion

$$C_u(r) = \sigma_u^2 \left[1 - \frac{r^2}{i_u^2} + O(r^4)\right] \tag{1.4.10}$$

since $dC_u/dr = (1/2) \langle d(u'^2)/dx \rangle = 0$ by (1.4.8). The constant i_u , which has the dimension of a length, is called the microscale in the literature on turbulence. Intuitively speaking, it represents the distance over which the values of u are fully correlated. By (1.4.9) and (1.4.10) we can also write

$$<(\frac{du}{dx})^2> = \frac{2\sigma_u^2}{i_u^2} \tag{1.4.11}$$

The same line of reasoning can be extended to higher-order differentiation or to functions of a few variables. Thus, with $u(\mathbf{x})=u(x_1,x_2,x_3)$, we have for instance

$$\langle \frac{\partial u}{\partial x_i} \rangle = \frac{\partial \langle u \rangle}{\partial x_i} \;;\; \langle \frac{\partial u'(\mathbf{x}_I)}{\partial x_{Ii}} \frac{\partial u'(\mathbf{x}_{II})}{\partial x_{IIj}} \rangle = \frac{\partial^2 C_u(\mathbf{x}_I,\mathbf{x}_{II})}{\partial x_{Ii} \partial x_{IIj}} \;; \ldots \quad (i,j = 1,2,3) \tag{1.4.12}$$

whereas in the case of stationary random functions (1.4.12) yields

$$\langle \frac{\partial u'(\mathbf{x})}{\partial x_i} \frac{\partial u'(\mathbf{x}+\mathbf{r})}{\partial x_j} \rangle = - \frac{\partial^2 C_u(\mathbf{r})}{\partial r_i \partial r_j} \;;\; \langle \frac{\partial^2 u'(\mathbf{x})}{\partial x_i \partial x_j} \frac{\partial^2 u'(\mathbf{x}+\mathbf{r})}{\partial x_i \partial x_j} \rangle = \frac{\partial^4 C_u(\mathbf{r})}{\partial^2 r_i \partial^2 r_j} \;; \ldots \quad (i,j = 1,2,3) \tag{1.4.13}$$

In the case of isotropy (Chap. 1.3) C_u is a function of $r=(r_1^2+r_2^2+r_3^2)^{1/2}$ solely, and the appropriate formulae are obtained from (1.4.12) by chain differentiation, i.e.

$$\frac{\partial C_u(r)}{\partial r_i} = \frac{r_i}{r} \frac{dC_u(r)}{dr} \;;\; \frac{\partial^2 C_u(r)}{\partial r_i \partial r_j} = - \frac{r_i r_j}{r^3} \frac{dC_u}{dr} + \frac{r_i r_j}{r^2} \frac{d^2C_u}{dr^2} \quad (i,j=1,2,3) \tag{1.4.14}$$

In the two-dimensional space, these formulae can be rewritten in a convenient form for the type of computations needed in this book by using the polar coordinate system $r=(r_x^2+r_y^2)^{1/2}$, $\theta=\tan^{-1}(r_y/r_x)$. Hence, $r_1=r_x=r\cos\theta$ and $r_2=r_y=r\sin\theta$ have to be substituted in (1.4.14).

Additional useful relationships for the case of isotropic covariance, based on (1.4.14), are

$$\langle\nabla u(\mathbf{x}).\nabla u(\mathbf{x}+\mathbf{r})\rangle = -\nabla^2 C_u = -\frac{1}{r}\frac{d}{dr}\left(r\frac{dC_u}{dr}\right) \; ;$$

$$\langle\nabla^2 u(\mathbf{x})\;\nabla^2 u(\mathbf{x}+\mathbf{r})\rangle = \nabla^4 C_u(r) \tag{1.4.15}$$

It is seen that for twice differentiable random functions the covariance is four times differentiable. Similar, but more involved, formulae could be derived for the axisymmetric covariance function (1.3.7).

The definition of a stochastic integral in the mean square follows the same line. With $d\mathbf{x}$ a volume, area or line element, the integral (in the ordinary, Riemann, sense) is written as usual

$$U = \int_V a(\mathbf{x})\, u(\mathbf{x})\, d\mathbf{x} \tag{1.4.16}$$

In Eq. (1.4.16) u is a random function, V is the integration domain and $a(\mathbf{x})$ is a deterministic real, bounded and piecewise continuous function. U is defined with the aid of a sequence as follows

$$U_n = \sum_{i=1}^{n} a(\mathbf{x}_i)\, u(\mathbf{x}_i)\, \Delta\mathbf{x}_i \tag{1.4.17}$$

where $\Delta\mathbf{x}_i$ is an elementary volume of the partition of V and $\mathbf{x}_i$ is in the interior of $\Delta\mathbf{x}_i$. If for $n\rightarrow\infty$ (with the measure of the largest $\Delta\mathbf{x}_i$ tending to zero) the random variable U_n converges in the mean square to U, i.e.

$$\lim_{n\rightarrow\infty} \langle(U_n-U)^2\rangle = 0 \qquad (\max \Delta\mathbf{x}_i\rightarrow 0) \tag{1.4.18}$$

then U is the stochastic integral of $a(\mathbf{x})u(\mathbf{x})$ over V. The necessary and sufficient condition for (1.4.18) to exist is that

$$\langle U'^2\rangle = \int_V\int_V a(\mathbf{x}')a(\mathbf{x}'')\, C_u(\mathbf{x}',\mathbf{x}'')\, d\mathbf{x}'\, d\mathbf{x}'' \tag{1.4.19}$$

exists and in this case (1.4.19) represents precisely the variance of U, whereas the expected value is given by

$$\langle U \rangle = \int a(\mathbf{x}) \langle u(\mathbf{x}) \rangle \, d\mathbf{x} \tag{1.4.20}$$

For second-order stationary functions (1.4.19) becomes

$$\sigma_U^2 = \int_V a(\mathbf{x}') \, d\mathbf{x}' \int_V a(\mathbf{x}'') \, C_u(\mathbf{x}'-\mathbf{x}'') \, d\mathbf{x}'' \tag{1.4.21}$$

If we define the step function

$$\Omega(\mathbf{x}) = 1 \quad (\text{for } \mathbf{x} \in V) \quad ; \quad \Omega(\mathbf{x}) = 0 \quad (\text{for } \mathbf{x} \text{ outside } V) \tag{1.4.22}$$

then U (1.4.16) can be rewritten as follows

$$U = \int a(\mathbf{x}) \, u(\mathbf{x}) \, \Omega(\mathbf{x}) \, d\mathbf{x} \tag{1.4.23}$$

where the integration domain is extended to infinity. Following Matheron (1965, Chap. V) we also define the useful function

$$H(\mathbf{r}) = \int \Omega(\mathbf{x}) \, \Omega(\mathbf{x}+\mathbf{r}) \, d\mathbf{x} \tag{1.4.24}$$

which represents simply the volume of the intersection of the domain V with its translation by $-\mathbf{r}$. Then, for a=1, i.e. for the integral of u over V, we have in (1.4.21), by (1.4.23) and (1.4.24),

$$\sigma_U^2 = \int C_u(\mathbf{r}) \, H(\mathbf{r}) \, d\mathbf{r} \tag{1.4.25}$$

and a similar representation for any $a(\mathbf{x})$ if Ω is multiplied by $a(\mathbf{x})$ in the first equation of (1.1.73). We shall discuss in some detail the properties of U in Chap. 1.9.

In many applications considered in this book the autocovariance of the random variable u tends to zero as the distance between the two points tends to infinity. For second-order stationarity and for sufficiently rapid decay at infinity the integrals of type

$$I_{u,x} = \frac{1}{\sigma_U^2} \int_0^\infty C_u(r_x,0,0)\, dr_x = \int_0^\infty \rho_u(r_x,0,0) dr_x \;;$$

$$I_{u,y} = \int_0^\infty \rho_u(0,r_y,0)\, dr_y \;;\; I_{u,z} = \int_0^\infty \rho_u\,(0,0,r_z)\, dr_z \;; \qquad (1.4.26)$$

$$I_{u,xy} = \left[\frac{4}{\pi}\int_0^\infty\int_0^\infty \rho_u(r_x,r_y,0)\, dr_x\, dr_y\right]^{1/2} \ldots ;\; I_u = \left[\frac{6}{\pi}\int_0^\infty\int_0^\infty\int_0^\infty \rho_u(r)\, dr_x\, dr_y\, dr_z\right]^{1/3}$$

are called integral scales of u(x,y,z). For isotropic covariance (1.3.6) the integral scales reduce to

$$I_{u,x} = I_{u,y} = I_{u,z} = \int_0^\infty \rho_u(r)\, dr \;;\; I_{u,xy} = I_{u,yz} = I_{u,zx} = \left[2\int_0^\infty \rho_u(r)\, r\, dr\right]^{1/2}$$

$$I_u = \left[3\int_0^\infty \rho_u(r)\, r^2\, dr\right]^{1/3} \qquad (1.4.27)$$

Intuitively speaking $I_{u,x}$ is a measure of the distance between two points on the x axis beyond which u(x) and $u(x+r_x)$ cease to be correlated. Similarly, $I_{u,xy}$ is the radius of a circle in which u are correlated in the x,y plane, whereas I_u is the corresponding radius of a sphere.

The extensive use of the integral scale of various variables in this book is related to the frequent encounter of integrals of random functions which lead to expressions of type

$$V(\mathbf{x}) = \int a(\mathbf{x}')\, C_u(\mathbf{x},\mathbf{x}')\, d\mathbf{x}' \qquad (1.4.28)$$

over the entire space. If $a(\mathbf{x})$ changes little over a distance of order I_u, it may be taken out of the integral in (1.4.28), leading for stationary and isotropic u, to

$$V(\mathbf{x}) \simeq a(\mathbf{x}) \int C_u(r)\, dr = a(\mathbf{x})\, (\tfrac{4\pi}{3}\sigma_u^2\, I_u^3) \qquad (1.4.29)$$

Eq. (1.4.29) could be formally obtained from (1.4.28) by replacing $C_u(r)$ by a Dirac distribution

$$C_u(r) = \frac{4\pi}{3} I_u^3 \sigma_u^2 \delta(r) \qquad (1.4.30)$$

such that Eq. (1.4.27) is satisfied.

A more rigorous derivation of (1.4.30) can be carried out by expanding first a($\mathbf{x}$) in a Taylor expansion in the neighborhood of $\mathbf{x}=\mathbf{x}'$. The next term in (1.4.29) is then given by $\nabla^2 a(\mathbf{x})$ $[2\pi \int C_u(r) r^4 dr]$.

A stochastic process which has (1.4.30) as its covariance is known as "white-noise". Approximation (1.4.30), which leads to (1.4.29), will be shown to be quite useful in applications.

A particular case of covariance of an one-dimensional weakly stationary process u(x) is the one for which $I_u=0$. This may happen in the degenerate case of complete lack of correlation between events at any two points and such a process will be called completely random or entirely uncorrelated. Another case is that of a "hole function", i.e. an autocorrelation function $\rho_u(r)$ which is negative for some interval of r so that its positive and negative contributions cancel out exactly in the process of integration in (1.4.26). The derivatives of a stationary and differentiable random function have this property, as one can see immediately from Eq. (1.4.8). A periodic random function will also generally display a covariance function which changes sign.

The normal (or Gaussian) random functions, defined in Chap. 1.3 as multivariate normal for any set of points, have some important properties which stem from the linearity of the differentiation and integration operation. Thus, if u($\mathbf{x}$) is Gaussian, not necessarily stationary, its derivative is also Gaussian with the expectation and covariance given by (1.4.7). Similarly U (1.4.16) is univariate normal with mean given by (1.4.20) and with variance given by (1.4.21).

1.5 DIFFERENTIATION OF RANDOM DISCONTINUOUS FUNCTIONS

In this book we shall encounter random functions which are discontinuous. The typical, and most simple example is that of a porous medium of rigid random structure for which we defined in Chap. 1.1 the basic index function

$$h(\mathbf{x}) = 1 \quad (\mathbf{x} \in V_v) \; ; \; h(\mathbf{x}) = 0 \quad (\mathbf{x} \in V_s) \qquad (1.5.1)$$

where V_v and V_s are the void and solid domains, respectively. For instance, if we figure out a straight line (the x axis) crossing the medium, a realization of h(x) is similar to the sketch of Fig. 1.2a. The expected value or ensemble average of h($\mathbf{x}$), denoted by $n(\mathbf{x})$, is the probabilistic definition of the porosity, which is assumed to be a continuous, differentiable function.

The derivative of h(x) is a generalized function, also known as a distribution or a Dirac measure. A complete, or even summary, discussion of the properties of distributions and mathematical operations with them is beyond the scope of this book. Such a discussion can be found for instance in the book by Matheron (1965), but its reading requires an extensive mathematical background. The simple-minded but illuminating approach, adopted for instance in Lighthill (1958), is to regard $\partial h/\partial x$ as the limit of a sequence of well-behaved functions (of bounded support and infinitely differentiable), for the measure of their support tending to zero.

The derivative $\partial h/\partial x$ in the realization of Fig. 1.2a is represented in Fig. 1.2b as a sequence of Dirac distributions of alternating signs. Our discussion here is limited to stating in a heuristic manner the relationship between the mean and covariance of the derivatives of a discontinuous function and those of the function itself. To simplify matters we consider first h(x), i.e. a one-dimensional variation, as achieved for instance for a layered medium, made up from parallel solid slabs of random thickness and spacing (Fig. 1.2c). To avoid the delicate process of averaging generalized functions, we rewrite the derivative as a limit as follows

$$\frac{dh(x)}{dx} = \lim_{\Delta x \to 0} \frac{1}{\Delta x} \int_{x-\Delta x/2}^{x+\Delta x/2} \frac{dh(x')}{dx'} dx' \tag{1.5.2}$$

and although this limit becomes unbounded at a point of discontinuity of h as $\Delta x \to 0$, the expectation value has a well defined limit. Indeed,

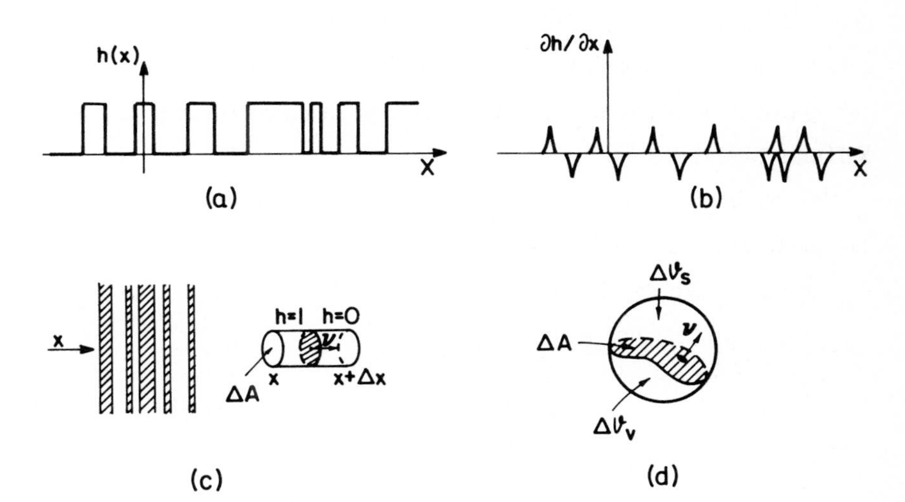

Fig. 1.2 The discontinuous h(**x**) function: (a) intersect along the x axis in a given realization; (b) the derivative of h of (a); (c) definition sketch of a layered medium and of the elementary volume ΔV and of the void-solid interface ΔA and (d) generalization for a three-dimensional structure.

$$\langle \frac{1}{\Delta x} \int_{x-\Delta x/2}^{x+\Delta x/2} \frac{dh}{dx'} dx' \rangle = \langle \frac{h(x+\Delta x/2) - h(x-\Delta x/2)}{\Delta x} \rangle = \frac{n(x+\Delta x/2)-n(x-\Delta x/2)}{\Delta x} \qquad (1.5.3)$$

Thus, for $\Delta x \rightarrow 0$ we obtain from (1.5.2) and (1.5.3)

$$\langle \frac{dh}{dx} \rangle = \lim_{\Delta x \rightarrow 0} \frac{1}{\Delta x} \int_{x-\Delta x/2}^{x+\Delta x/2} \langle \frac{dh}{dx'} \rangle \, dx' = \frac{dn(x)}{dx} \qquad (1.5.4)$$

precisely like in the case of ordinary functions. This result can be explained by observing that the difference $h(x+\Delta x/2)-h(x-\Delta x/2)$ is equal to zero whenever the void-solid interface is outside the interval Δx and is equal to +1 or -1 when the interface between the void and the solid crosses Δx. The average in Eq. (1.5.4) can be interpreted therefore as the density of excess of interfaces, i.e. the difference between the number of solid-void interfaces and that of void-solid interfaces, per unit length. In the stationary case there is mutual cancellation as the two numbers are equal and consequently $dn/dx=0$.

Toward generalization for the three-dimensional case, Eq. (1.5.4) can be rewritten in the following form, after multiplying it by ΔA,

$$\langle \frac{dh}{dx} \rangle = - \lim_{\Delta V \rightarrow 0} \left(\frac{1}{\Delta V} \langle \Delta A \boldsymbol{\nu}.\mathbf{i} \rangle \right) \qquad (1.5.5)$$

where $\mathbf{i}$ is a unit vector in the positive x direction, ΔA (Fig. 1.2c) is the cross-sectional area of a small cylinder, $\Delta V = \Delta A \; \Delta x$ is an infinitesimal volume and $\boldsymbol{\nu}$ is a unit vector normal to ΔA, selected as positive when it points out from the void to the solid. Hence, the ensemble average of the gradient of h is equal to the density of oriented interfacial areas, an interpretation suggested by Saffman (1971).

We shall consider now the covariance $C_h(r)$ for the same one-dimensional h(x) (Fig. 1.2c), and in particular its derivative at the origin, i.e. lim dC/dr for $r \rightarrow 0$. By the definition of $C_h = \langle [h(x+r)-n][h(x)-n]$, we get

$$\lim_{r \rightarrow 0} \frac{dC_h}{dr} = \lim_{\Delta x \rightarrow 0} \langle \frac{[h(x+\Delta x)-h(x)] \; h(x)}{\Delta x} \rangle \qquad (1.5.6)$$

Referring again to Fig. 1.2c, it is seen again that the product $h(x+\Delta x)-h(x)$ is different from zero and equal to ±1 only when the void-solid interface is within the volume ΔV. Multiplication by h(x), however, cancels the contribution of solid-void interfaces, which is equal to half of the total number of interfaces for a stationary h(x), and leaves only the contribution -1 corresponding to void-solid interfaces. Hence, we may write

$$\lim_{r\to 0} \frac{dC_h}{dr} = -\frac{1}{2} \lim_{\Delta V\to 0} \langle \frac{\Delta A}{\Delta V} |\boldsymbol{\nu}.\mathbf{i}| \rangle \tag{1.5.7}$$

where $|\boldsymbol{\nu}.\mathbf{i}|=1$ has been written toward generalizing (1.5.6) in the sequel for a three-dimensional medium. The expression $\sigma = \lim\langle \Delta A/\Delta V\rangle$ for $\Delta V\to 0$ is the definition of the specific area, i.e. the total area of void-solid interfaces per unit volume of medium.

The generalization of these results for a three-dimensional medium, i.e. for h(x,y,z), is straightforward. Thus, (1.5.5) can be written as follows

$$\langle \nabla h(\mathbf{x})\rangle = \nabla n(\mathbf{x}) = -\lim_{\Delta V\to 0} \left(\frac{1}{\Delta V}\langle \int_{\Delta A} \boldsymbol{\nu}\, dA\rangle\right) \tag{1.5.8}$$

where ΔV is an arbitrary volume, ΔA is the void-solid interface within ΔV and $\boldsymbol{\nu}$ is a unit vector normal to dA pointing from the void toward the solid (Fig. 1.2d). In words, the ensemble average of the gradient of h is equal to the density of oriented interfacial areas. In the stationary case, for constant n, the density is zero due to mutual cancellation.

We consider now the derivative of the covariance at the origin, generalization of (1.5.7). In the three-dimensional case, it is not only the event of crossing by an interface which is random, but also the direction of the unit vector $\boldsymbol{\nu}$. The simplest case is of an isotropic medium, for which the probability of the unit vector $\boldsymbol{\nu}$ to be in any direction is the same. Hence, with $\boldsymbol{\nu}.\mathbf{i} = \cos\theta$, the averaging in (1.5.7) yields

$$\langle |\boldsymbol{\nu}.\mathbf{i}| \rangle = \int_0^{\pi/2} \cos\theta \sin\theta \, d\theta = \frac{1}{2} \tag{1.5.9}$$

Thus, we arrive at the final important result

$$\lim_{r\to 0} \frac{dC}{dr} = -\frac{1}{4} \lim_{\Delta V\to 0} \langle \frac{1}{\Delta V} \int_{\Delta A} dA\rangle = -\frac{1}{4}\sigma \tag{1.5.10}$$

where $\sigma = \langle dA/dV\rangle$ is the specific area, the important geometrical quantity mentioned above. Hence, (1.5.10) can be stated in words as follows: the slope of the isotropic covariance of h near the origin tends to minus one-forth of the specific area. This result has been derived in slightly different ways by Debye et al, (1957), and extended by Matheron (1967) for anisotropic media as well. Hence, unlike the case of continuous functions (Chap. 1.4), C_h has a discontinuous derivative at the origin. Similar relationships can be derived for moments of higher derivatives of h, leading eventually to generalized functions for the variances.

Let us consider now a continuous random function $u(\mathbf{x})$. The product $v(\mathbf{x})=u(\mathbf{x})h(\mathbf{x})$ is a discontinuous random function which is represented schematically, in a given realization, in Fig. 1.3. The ensemble average $\langle v(\mathbf{x})\rangle=\langle uh\rangle$ is a continuous function over the space and is defined everywhere. Of fundamental interest is the expected value of the derivative of u in the void space. By using chain differentiation we have

$$\langle h(\mathbf{x})\ \nabla u(\mathbf{x})\rangle = \nabla\langle u(\mathbf{x})\ h(\mathbf{x})\rangle - \langle u(\mathbf{x})\ \nabla h(\mathbf{x})\rangle \qquad (1.5.11)$$

which can be rewritten with the aid of (1.5.8) as follows

$$\langle h\ \nabla u\rangle = \nabla\langle v\rangle + \lim_{\Delta V\to 0} \left[\frac{1}{\Delta V} \left\langle \int_{\Delta A} u\boldsymbol{\nu}\ dA\right\rangle\right] \qquad (1.5.12)$$

Eq. (1.5.12), which has been derived in a different way by Saffman (1971), can be stated in words as follows: the ensemble average of the derivative of u over the void space is equal to the derivative of the average in the void, plus the areal density of u over oriented interfaces between void and solid. It is emphasized that u in the integral of (1.5.12) can also be regarded as the jump of the value of u across the void-solid interface. For continuous functions, i.e. u=0 on the interface, the jump is zero and the last term in (1.5.12) vanishes and the differentiation and expectation signs are commutative (Chap. 1.4). Furthermore, in the stationary case both terms in the r.h.s. of (1.5.12) vanish.

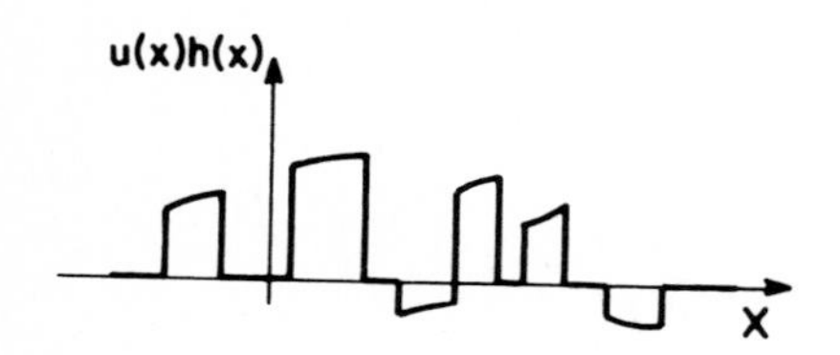

Fig. 1.3 Sketch of the discontinuous function uh.

1.6 SPECTRAL METHODS

Spectral methods are closely associated with stationary processes and dynamical systems for which the energy density can be conveniently defined. To illustrate this idea in a simple-minded way, let us consider the one-dimensional random function, defined over the interval $|x|<L$ by the Fourier sum

$$u(x) = \sum_{n=-N}^{n=N} a_n \cos\left[n\left(\frac{\pi x}{L}+b_n\right)\right] \tag{1.6.1}$$

where a_n and b_n are random variables of zero mean. It is easier, however, to operate with the complex random function

$$u(x) = \sum_{n=-N}^{n=N} a_n \exp\left(\frac{in\pi x}{L}\right) \tag{1.6.2}$$

where here i is the imaginary unit and a_n are complex random numbers of zero mean, such that (1.6.1) is the real part of (1.6.2). In a dynamical system, the energy is proportional to the square of the "amplitudes" and the spectrum is defined as follows

$$S_n = \langle a_n a_n^* \rangle \tag{1.6.3}$$

where a_n^* is the complex conjugate of a_n. The energy is equal to ΣS_n, such that S_n can be called spectral density.

The autocovariance of u is defined with the aid of (1.6.2) as follows

$$C_N(x_1,x_2) = \langle u(x_1)u^*(x_2)\rangle = \sum_{n=-N}^{m=N} \langle a_n a_m^* \rangle \exp\left(i\,\frac{n\pi x_1 - m\pi x_2}{L}\right) \tag{1.6.4}$$

For a stationary process $C_N(x_1,x_2)$ is a function of x_1-x_2 and this happens only if in (1.6.4) the following orthogonality relationships hold

$$\langle a_n a_m^* \rangle = 0 \qquad \text{for} \qquad m \neq n \tag{1.6.5}$$

Then C_N (1.6.4) becomes

$$C_N(x_2-x_1) = \sum_{n=-N}^{n=N} \langle a_n a_n^* \rangle \exp[i\frac{\pi n(x_1-x_2)}{L}] = \sum_{n=-N}^{n=N} S_n \exp[-i\frac{\pi n(x_2-x_1)}{L}] \quad (1.6.6)$$

It is seen, therefore, that S_n are the Fourier coefficients of C_N and we can write

$$S_n = \frac{1}{2L}\int_{-L}^{L} C_N(r)\exp(i\pi r/L)\,dr \;\; ; \;\; r = x_2-x_1 \quad (1.6.7)$$

In the general case in which $L\to\infty$ for u(x) defined over an unbounded domain, the spectral density of the stochastic process u(x) is defined by the Fourier integral transform

$$S(k) = \frac{1}{\sqrt{2\pi}}\int_{-\infty}^{\infty} C(r)\,e^{ikr}\,dr \quad (1.6.8)$$

which is generalized for the space with m = 1,2 or 3 dimensions as follows

$$S(\mathbf{k}) = \frac{1}{(2\pi)^{m/2}}\int_{-\infty}^{\infty} C(\mathbf{r})\,e^{i\mathbf{k}.\mathbf{r}}\,d\mathbf{r} \quad (1.6.9)$$

with **k** being the wave-number vector and **r** the coordinate vector. By inverting (1.6.9) we also have

$$C(\mathbf{r}) = \frac{1}{(2\pi)^{m/2}}\int_{-\infty}^{\infty} S(\mathbf{k})\,e^{-i\mathbf{k}.\mathbf{r}}\,d\mathbf{k} \quad (1.6.10)$$

and in particular

$$\sigma^2 = \frac{1}{(2\pi)^{m/2}}\int_{-\infty}^{\infty} S(\mathbf{k})\,d\mathbf{k} \quad (1.6.11)$$

These definitions of S (1.6.8, 1.6.9) are different from the usual ones, in which the factor $(2\pi)^{-m}$ rather than $(2\pi)^{-m/2}$ appears in front of the integrals. The advantage of the present convention, following the definition of the integral FT (Fourier transform) in Carrier et al (1966), is in the symmetry between the FT (1.6.9) and its inverse (1.6.10).

In the isotropic case and for m=3, Eqs. (1.6.9) - (1.6.11) simplify to

$$S(k) = \frac{1}{k\sqrt{2\pi}} \int_0^\infty C(r) \sin(kr)\, r\, dr \quad ; \quad C(r) = \frac{1}{r\sqrt{2\pi}} \int_0^\infty S(k) \sin(kr)\, dk$$

$$\sigma^2 = \sqrt{2\pi} \int_0^\infty S(k)\, k^2\, dk \tag{1.6.12}$$

where $k=|\mathbf{k}|$ and $r=|\mathbf{r}|$. The relationship between S (1.6.8) and u(x), however, is not as simple as (1.6.7) for the finite case. Indeed, the orthogonality requirement (1.6.5) leads to a representation of a_n as a set of Dirac distributions which become more and more dense as $N \to \infty$. This difficulty can be circumvented by replacing the ordinary Fourier integral transform by a Fourier-Stieltjes integral such that (1.6.1) becomes at the continuous limit

$$u(\mathbf{x}) = \frac{1}{(2\pi)^{m/2}} \int_{-\infty}^{\infty} e^{-i\mathbf{k}.\mathbf{x}}\, dA(\mathbf{k}) \tag{1.6.13}$$

and the increments of the complex random distribution function $A(\mathbf{k})$ satisfy the following relationships, equivalent to (1.6.5),

$$\langle dA(\mathbf{k})\, dA^*(\mathbf{k}')\rangle = 0 \qquad (\text{for } \mathbf{k} \neq \mathbf{k}')$$

$$\langle dA(\mathbf{k})\, dA^*(\mathbf{k})\rangle = S(\mathbf{k})\, d\mathbf{k} \qquad (\text{for } \mathbf{k}=\mathbf{k}') \tag{1.6.14}$$

The "representation theorem" known also as Wiener-Hinchin theorem, ensures that for u stationary Eqs. (1.6.13) and (1.6.14) are satisfied.

Spectral representations are used in the present book when dealing with linear differential equations satisfied by stationary random functions. Thus, differentiation in (1.6.13) yields

$$\frac{\partial u}{\partial x} = -i\, k_x \int_{-\infty}^{\infty} e^{-i\mathbf{k}.\mathbf{x}}\, dA(\mathbf{k}) \quad ; \quad \frac{\partial^2 u}{\partial x^2} = -k_x{}^2 \int_{-\infty}^{\infty} e^{-i\mathbf{k}.\mathbf{x}}\, dA(\mathbf{k}) \; ; \; \ldots \tag{1.6.15}$$

such that substitution of (1.6.15) in a linear differential operator leads to relationships between the complex amplitudes $dA(\mathbf{k})$. Furthermore, like in the case of Fourier transforms, we obtain by using the orthogonality relationships (1.6.14), algebraic relationships between amplitudes and spectra, rather than between differentials.

The use of the not so familiar Fourier-Stieltjes integral (1.6.13) can be circumvented in a few ways. The simplest one is to employ Fourier transforms in a finite domain, precisely like (1.6.2), which apply to random functions, and to let $L \to \infty$ only in the final expressions of the various statistical moments.

Another possible approach, which is adopted here, is to generalize the Fourier transform to include the Dirac distributions and its derivatives. This approach is described in a comprehensive manner by Lighthill (1958). Although this generalization is not sufficient for representing some random functions of interest here, all the transform results can be converted into the Fourier-Stieltjes framework and the final results for the various statistical moments are the same. This is the line followed by Dagan (1985b), based on Carrier et al (1966), and herewith a few useful relationships.

First, the FT and its inverse, are given by

$$\hat{u}(\mathbf{k}) = \frac{1}{(2\pi)^{m/2}} \int u(\mathbf{x})\, e^{i\mathbf{k}.\mathbf{x}}\, d\mathbf{x} \;\; ; \;\; u(\mathbf{x}) = \frac{1}{(2\pi)^{m/2}} \int \hat{u}(\mathbf{k})\, e^{-i\mathbf{k}.\mathbf{x}}\, d\mathbf{x} \qquad (1.6.16)$$

where, again m=1,2,3 stands for the number of dimensions of the space, integration is for the entire space, **k** is the wave-number vector of dimension m and i is the imaginary unit. The particular transform of u=1 is given by (Carrier et al, 1966)

$$\int e^{-i\mathbf{k}.\mathbf{x}.}\, d\mathbf{x} = \int e^{i\mathbf{k}.\mathbf{x}}\, d\mathbf{x} = (2\pi)^m\, \delta(\mathbf{k}) \qquad (1.6.17)$$

where δ is the Dirac distribution

We shall make use of the following useful relationships for derivatives and gradients

$$\mathrm{FT}\,[\nabla u(\mathbf{x})] = -i\mathbf{k}\, \hat{u}(\mathbf{k})$$

$$\mathrm{FT}\,[u_1(\mathbf{x})\, u_2(\mathbf{x})] = (2\pi)^{-m/2} \int \hat{u}_1(\mathbf{k}_1)\, \hat{u}_2(\mathbf{k} - \mathbf{k}_1)\, d\mathbf{k}_1 \qquad (1.6.18)$$

as well as of the Faltung theorems

$$\mathrm{FT}\,[\int u_1(\mathbf{x}')\, u_2(\mathbf{x}' + \mathbf{x})\, d\mathbf{x}'] = (2\pi)^{m/2}\, \hat{u}_1^*(\mathbf{k})\, \hat{u}_2(\mathbf{k})$$

$$\mathrm{FT}\,[\int\int u_1(\mathbf{x}')\, u_2(\mathbf{x}'')\, u_3(\mathbf{x}+\mathbf{x}'+\mathbf{x}'')\, d\mathbf{x}'\, d\mathbf{x}''] = (2\pi)^m\, \hat{u}_1^*(\mathbf{k})\, \hat{u}_2^*(\mathbf{k})\, \hat{u}_3(\mathbf{k}) \qquad (1.6.19)$$

We consider now the statistical moment (n-points correlation) of the functions $u_1(\mathbf{x}_1)$, $u_2(\mathbf{x}_2)$,..., $u_N(\mathbf{x}_N)$ of zero mean. The FT with respect to $\mathbf{x}_1$, $\mathbf{x}_2$,..., $\mathbf{x}_N$ is given by definition as follows

$$\mathrm{FT}\,[\langle u_1(\mathbf{x}_1)\, u_2(\mathbf{x}_2)...u_n(\mathbf{x}_N)\rangle] = \langle[\hat{u}_1(\mathbf{k}_1)\, \hat{u}_2(\mathbf{k}_2)...\hat{u}_n(\mathbf{k}_N)]\rangle \qquad (1.6.20)$$

If the random process represented by these functions is stationary, the moment (1.6.19) is invariant to a translation in space, i.e. it is a function of the N-1 variables $\mathbf{r}_2 = \mathbf{x}_2-\mathbf{x}_1$, $\mathbf{r}_3 = \mathbf{x}_3-\mathbf{x}_1$,..., $\mathbf{r}_N = \mathbf{x}_N-\mathbf{x}_1$. We then have the important representation

$$\langle\left[\hat{u}_1(\mathbf{k}_1)\, \hat{u}_2(\mathbf{k}_2)...\hat{u}_n(\mathbf{k}_N)\right]\rangle = (2\pi)^{m/2}\, \delta(\mathbf{k}_1+\mathbf{k}_2+...+\mathbf{k}_N)\, \hat{C}(\mathbf{k}_2,\mathbf{k}_3,...,\mathbf{k}_N) \qquad (1.6.21)$$

where $\hat{C}$ is the FT of the covariance

$$C(\mathbf{r}_2,\mathbf{r}_3,...,\mathbf{r}_N) = \langle u_1(\mathbf{x}_1)\, u_2(\mathbf{x}_2)...u_n(\mathbf{x}_N)\rangle \qquad (1.6.22)$$

1.7 RANDOM FUNCTIONS OF STATIONARY INCREMENTS

In many applications of this book we shall encounter random variables which are not stationary in the ordinary sense and which may have unbounded variance. As a simple and frequent example consider the case of the random function equal to the indefinite integral of a one-dimensional stationary process

$$U(x) = \int_0^x u(x')\, dx' \qquad (1.7.1)$$

We obtain by ensemble averaging

$$\langle U(x)\rangle = \langle u\rangle\, x \qquad (1.7.2)$$

such that $\langle U\rangle$ is not constant. Furthermore, to calculate the covariance we observe that H(r) (1.4.24) is given, for the domain defined as the interval x=0 to x, by

$$H(r) = x-r \quad (r<x) \quad ; \quad H(r) = 0 \quad (r>x) \qquad (1.7.3)$$

The variance of U becomes by (1.4.25)

$$\sigma_U^2(x) = x \int_0^x C_u(r)\, dr - \int_0^r r\, C_u(r)\, dr \tag{1.7.4}$$

For sufficiently large x, for which the contribution of C_u and rC_u in the two integrals of (1.7.4) become negligible, Eq. (1.7.4) yields by the definition of I_u (1.4.27)

$$\sigma_U^2(x) \rightarrow \sigma_u^2\, I_u\, x - \text{const} \tag{1.7.5}$$

such that the variance increases indefinitely with x. This example suggests the extension of the definition of stationarity to a wider class of random functions. Let us consider the increment

$$\epsilon_{ij} = u(\mathbf{x}_j) - u(\mathbf{x}_i) \tag{1.7.6}$$

i.e. the difference between the values of the random function u at two points. The random variable ϵ_{ij} is defined statistically by the P.d.f. $F(\epsilon_{ij}, \epsilon_{km}, ...)$ for arbitrary pairs of points $(\mathbf{x}_i, \mathbf{x}_j)$, $(\mathbf{x}_k, \mathbf{x}_m)$,... . The random function $u(\mathbf{x})$ is of stationary increments (or in the terminology of Matheron, 1965, it is a scheme with an intrinsic dispersion law, or briefly an intrinsic scheme), if F is invariant to a translation in space of the points $\mathbf{x}_i, \mathbf{x}_j, \mathbf{x}_k, \mathbf{x}_m$,... by the same amount $\mathbf{r}$. Then, the spatial variability of $u(\mathbf{x})$ has an intrinsic character which depends only on the relative position of the points $\mathbf{x}_i$, $\mathbf{x}_j$,... but not on the location of their center of gravity.

The expected value of $\epsilon(\mathbf{r})$, where $\mathbf{r} = \mathbf{x}_i - \mathbf{x}_j$ is the separation vector between two points, can be shown to be generally a linear function of $\mathbf{r}$ for the stationary case, i.e.

$$\langle \epsilon(\mathbf{r}) \rangle = \mathbf{J}.\mathbf{r} \tag{1.7.7}$$

where $\mathbf{J}$ is a constant vector (Matheron 1965). Hence, $u(\mathbf{x})$ has a linear trend and to simplify matters we shall consider further only the residual $u'(\mathbf{x}) = u(\mathbf{x}) - \mathbf{J}.\mathbf{x} + \text{const}$. Then $\epsilon(\mathbf{r}) = u'(\mathbf{x}+\mathbf{r}) - u'(\mathbf{x})$ is of zero mean, i.e. $\langle \epsilon(\mathbf{r}) \rangle = 0$, and we discuss only such increments. It is emphasized that in the case $\mathbf{J}=0$, i.e. if u has a constant mean $\langle u \rangle$, the latter is filtered out from the increment, i.e.

$$\epsilon(\mathbf{r}) = u(\mathbf{x}+\mathbf{r}) - u(\mathbf{x}) = u'(\mathbf{x}+\mathbf{r}) - u'(\mathbf{x}) \quad ; \quad \langle \epsilon \rangle = 0 \tag{1.7.8}$$

The second order moment of ϵ is defined with the aid of the variogram

$$\gamma_u(\mathbf{x}_i, \mathbf{x}_j) = \frac{1}{2} \langle \epsilon_{ij}^2 \rangle = \frac{1}{2} \langle [u'(\mathbf{x}_i) - u'(\mathbf{x}_j)]^2 \rangle \tag{1.7.9}$$

γ (1.7.9) is defined as semi-variogram in many texts, but for the sake of brevity we shall adhere here to the term variogram. For a process of stationary increments γ_u is a function of $\mathbf{r}_{ij} = \mathbf{x}_i - \mathbf{x}_j$ (Matheron, 1965). All second order moments of ϵ can be evaluated with the aid of γ. Thus, it is easy to show that

$$\langle \epsilon(\mathbf{r}_{ij})\, \epsilon(\mathbf{r}_{km}) \rangle = \gamma(\mathbf{r}_{ij}) + \gamma(\mathbf{r}_{jk}) - \gamma(\mathbf{r}_{ik}) - \gamma(\mathbf{r}_{jm}) \tag{1.7.10}$$

In the case of random stationary functions γ_u is simply related to the covariance C_u (1.3.6) by its definition of (1.7.9), by

$$\gamma_u(\mathbf{r}) = \sigma_u^2 - C_u(\mathbf{r}) \quad ; \quad \gamma_u(\infty) = C_u(0) = \sigma_u^2 \; C_u(\mathbf{r}) = \gamma_u(\infty) - \gamma_u(\mathbf{r}) \tag{1.7.11}$$

and there is no advantage in using γ_u rather than C_u. The case of interest, however, is that in which $\gamma(\mathbf{r})$ increases indefinitely with $(|\mathbf{r}|)$, such that $u(\mathbf{x})$ has an unbounded variance and $C_u(\mathbf{r})$ cannot be defined at all. Even when $\gamma(\mathbf{r})$ becomes practically constant for $|\mathbf{r}|>\ell$, where ℓ is called the "range" in the geostatistical literature, the covariance is still hard to define if the extent of the spatial domain over which $u(\mathbf{x})$ is defined is not much larger than ℓ (this point is retaken in Chap. 1.10). Thus, the variogram can be used conveniently for characterizing the local structure of $u(\mathbf{x})$ when σ_u^2 is not accessible.

The variogram has, by definition, the following properties

$$\gamma(\mathbf{r}) = \gamma(-\mathbf{r}) > 0 \quad ; \quad \gamma(0) = 0 \tag{1.7.12}$$

Sometimes, the drop of $\gamma(\mathbf{r})$ to zero when $\mathbf{r}\to 0$ is quite abrupt and it is represented there, for the sake of convenience, as discontinuous, i.e.

$$\lim_{r\to 0} \gamma(\mathbf{r}) = w \tag{1.7.13}$$

and this is called a "nugget" effect. A typical case of a nugget effect is that resulting from uncorrelated random errors and then w in (1.7.13) is precisely their variance.

The behaviour of $\gamma(\mathbf{r})$ for $|\mathbf{r}|\to\infty$ is not arbitrary, but is subjected to the constraints imposed by the requirement that any linear combination of increments $u(\mathbf{x}_i) - u(\mathbf{x}_j)$ has a positive variance. Matheron (1965) shows in particular that if $\gamma(\mathbf{r}) \simeq r^a$ for $r\to\infty$, the inequality

$$a < 2 \tag{1.7.14}$$

must be satisfied and that $\gamma \simeq \ln r$ for $r\to\infty$ is also an admissible variogram.

The behaviour of anisotropic variograms is the same as that of anisotropic covariances (Chap. 1.3). In particular, if $\gamma_u(\mathbf{r},\mathbf{J})$ is quadratic in $\mathbf{J}$, we have the decomposition, similar to (1.3.7),

$$\gamma_u(\mathbf{r},\mathbf{J}) = \gamma_{uT}(r)\, J^2 + [\gamma_{uL}(r) - \gamma_{uT}(r)]\frac{(\mathbf{r}.\mathbf{J})^2}{r^2} \tag{1.7.15}$$

such that γ_u depends on the two isotropic, the transverse and the longitudinal, variograms $\gamma_{uT}(r)$ and $\gamma_{uL}(r)$, respectively.

The statistical moments of the derivatives of a random function can be expressed with the aid of the variogram precisely like in Chap. 1.4 by replacing everywhere the derivatives of $C_u(\mathbf{x}_1,\mathbf{x}_2)$ by minus the derivative of $\gamma_u(\mathbf{x}_1,\mathbf{x}_2)$. For instance, in (1.4.8), we get

$$\langle \frac{du(x)}{dx}\frac{du(x+r)}{dx} \rangle = \frac{d^2\gamma_u(r)}{dr^2} \tag{1.7.16}$$

From (1.7.16) we immediately see that the derivative of a differentiable random function of stationary increments is a stationary function in the ordinary sense. Its variance is given by

$$<(\frac{du}{dx})^2> = \left.\frac{d^2\gamma_u(r)}{dr^2}\right|_{r=0} \tag{1.7.17}$$

and the variogram of a differentiable function is twice differentiable at the origin, whereas the variogram of a continuous function is differentiable once. In contrast, a nugget effect (1.7.13) leads to a white noise component of the derivative.

By condition (1.7.14) it is seen that $d^2\gamma_u/dr^2 \to 0$ for $r\to\infty$ and henceforth the covariance (1.7.16) tends also to zero for $r\to\infty$. Furthermore, we can calculate the integral scale (defined in Eqs. 1.4.27) of du/dx with the aid of the variogram $\gamma_u(r)$.

In the case of a random function of two variables, and isotropic variogram, calculations similar to those leading to (1.4.15) yield

$$\langle \nabla u(\mathbf{x}) \,.\, \nabla u(\mathbf{x}+\mathbf{r}) \rangle = \frac{1}{r}\frac{d}{dr}(r\frac{d\gamma_u}{dr}) \tag{1.7.18}$$

and the integral scale of the derivative is finite if $\gamma_u \simeq \ln r$ for $r\to\infty$. By a similar calculation in the three-dimensional space it can be shown that the integral scale is finite only if γ_u tends to a constant value like r^{-1} for $r\to\infty$.

1.8 CONDITIONAL GAUSSIAN PROBABILITY AND INTERPOLATION BY KRIGING

In Chap. 1.3 we have discussed the properties of a random space function which is Gaussian, i.e. its joint p.d.f. for N arbitrary points is multivariate normal. In particular, it was shown that for conditioning of the values of $u(\mathbf{x})$ at N - M points, the joint p.d.f. of the values of u at the remaining M points is multivariate normal, with conditional expectation value and covariance given by Eqs. (1.3.10) and (1.3.11), respectively.

We wish to extend here these results to functions of multivariate normal p.d.f. of stationary increments, in view of their importance in problems of groundwater flow, by following the procedure of Dagan (1982a).

Toward this aim, we shall rewrite the system of linear equations rendering the coefficients λ_{ij} (1.3.12) in terms of the variogram rather than the covariance function, by using (1.7.11), as follows

$$\sum_{j=M+1}^{N} \lambda_{ij}\, \gamma_u(\mathbf{x}_j-\mathbf{x}_k) = \gamma_u(\mathbf{x}_i-\mathbf{x}_k) - \sigma_u^2\,(1-\sum_{j=M+1}^{N} \lambda_{ij}) \qquad (k=M+1,...,N) \tag{1.8.1}$$

which is valid at present for ordinary stationarity.

The linear system (1.8.1) can be rewritten in a symmetrical form by introducing the additional unknown μ_i as follows

$$\sum_{j=M+1}^{N} \lambda_{ij}\, \gamma_u(\mathbf{x}_j-\mathbf{x}_k) = \gamma_u(\mathbf{x}_i-\mathbf{x}_k) + \mu_i \qquad (k=M+1,...,N) \tag{1.8.2}$$

$$\mu_i = \sigma_u^2 \sum_{j=M+1}^{N} (1-\lambda_{ij}) \tag{1.8.3}$$

We extend these results to functions of stationary increments by letting $\sigma_u^2 \to \infty$ in (1.8.1). For μ_i to remain finite, we must have in (1.8.3)

$$\sum_{j=M+1}^{N} \lambda_{ij} = 1 \tag{1.8.4}$$

Hence, by this simple transformation, we have arrived at the system of N - M+1 equations (1.8.2, 1.8.4) for the unknown coefficients λ_{ij} and μ_i, which pertain to the point of coordinates $\mathbf{x}_i$ and which are expressed with the aid of the variogram γ_u solely.

The conditional mean (1.3.10) can be rewritten, for functions of residuals of stationary increments, as follows

$$\langle u(\mathbf{x}_i | \mathbf{x}_{M+1}, \mathbf{x}_{M+2}, \ldots, \mathbf{x}_N) \rangle = \sum_{j=M+1}^{N} \lambda_{ij} \, [u(\mathbf{x}_j) - m_u(\mathbf{x}_j] + m_u(\mathbf{x}_i) \tag{1.8.5}$$

the only difference being that λ_{ij} are now solutions of (1.8.2, 1.8.4) rather than (1.8.1).

It is pointed out that the conditional expected value (1.8.5) does not depend anymore on the unconditional mean m_u if the latter is constant. Indeed, by (1.8.4) we have in this case in (1.8.5)

$$\langle u(\mathbf{x}_i | \mathbf{x}_{M+1}, \mathbf{x}_{M+2}, \ldots, \mathbf{x}_N) \rangle = \sum_{j=M+1}^{N} \lambda_{ij} \, u(\mathbf{x}_j) \tag{1.8.6}$$

By the same substitution of (1.7.11) into (1.3.12), and by taking subsequently the limit $\sigma_u{}^2 \rightarrow 0$, the conditional covariance (1.3.11) becomes

$$C_u(\mathbf{x}_i, \mathbf{x}_j | \mathbf{x}_{M+1}, \mathbf{x}_{M+2}, \ldots, \mathbf{x}_N) = \sum_{k=M+1}^{N} \lambda_{ik} \, \gamma(\mathbf{x}_k - \mathbf{x}_j) - \gamma(\mathbf{x}_i - \mathbf{x}_k) + \mu_i \tag{1.8.7}$$

with (1.8.3) taken into account. Again, the conditional covariance (1.8.7) depends only on the variogram $\gamma_u(\mathbf{r})$ and not on the values u_j. Thus, the entire statistical structure of the random normal variable $u(\mathbf{x} | \mathbf{x}_{M+1}, \mathbf{x}_{M+2}, \ldots, \mathbf{x}_N)$ is determined by this procedure for functions of stationary increments as well.

The notion of increment and variogram can be generalized conveniently (Matheron, 1965). Thus, the difference $u(\mathbf{x}_i)$-$u(\mathbf{x}_j)$ discussed so far, is called an increment of zero order. More generally, the linear combination $\Sigma_{i=1}^{n} \varsigma_i u(\mathbf{x}_i)$ is called a generalized increment of order k in the three-dimensional space if, and only if

$$\sum_{i=1}^{n} \varsigma_i \, x_i^{m_x} \, y_i^{m_y} \, z_i^{m_z} = 0 \tag{1.8.8}$$

for all integers m_x, m_y, m_z > 0 such that $m_x + m_y + m_z \leq k$, and a similar definition applies to the plane or to the line. For instance, the zero order increment imposes by (1.8.8) the constraint

$$\sum_{i=1}^{n} \varsigma_i = 0 \tag{1.8.9}$$

which is obviously satisfied for n=2, $\zeta_1=1$ and $\zeta_2=-1$. By the same token a first-order generalized increment requires

$$\sum_{i=1}^{n} \zeta_i = 0 \ ; \ \sum_{i=1}^{n} \zeta_i x_i = 0 \ ; \ \sum_{i=1}^{n} \zeta_i y_i = 0 \ ; \ \sum_{i=1}^{n} \zeta_i z_i = 0 \qquad (1.8.10)$$

and ζ_i can be determined for arbitrary points $\mathbf{x}_i$ up to a constant if n = 5 (or n = 4 in the two-dimensional case and n = 3 on the line).

The fundamental property of a generalized increment is that it filters out a polynomial trend or order k.

An intrinsic random function u(**x**) of order k is defined as the one which has weakly stationary generalized increments, i.e. for which the sum $\Sigma_{i=1}^{n}\zeta_i\ u(\mathbf{x}_i+\mathbf{r})$ has a mean and a variance which do not depend on **x**. The generalized covariance $K_u(\mathbf{r})$ is defined by

$$\langle[\sum_{i=1}^{n} \zeta_i u(\mathbf{x}_i)]^2\rangle = \sum_{i=1}^{n}\sum_{j=1}^{n} \zeta_i\zeta_j\ K_u(\mathbf{x}_i-\mathbf{x}_j) \qquad (1.8.11)$$

for any ζ_i satisfying (1.8.8). For n = 2 and with $\zeta_1=1$, $\zeta_2=-1$ one obtains $K_u(\mathbf{r}) = \gamma_u(\mathbf{r})$ for K(0)=0. The generalized covariance is defined up to an even polynomial of degree $\leq$2k. Furthermore, two random functions which differ by a polynomial of order k with deterministic or random coefficients, do have the same generalized increments and covariances.

Similar results to those obtained for Gaussian conditioning of functions of stationary increments can be achieved by a different approach developed by the geostatistical school, namely by kriging (e.g., Journel and Huijbregts, 1978). In view of its wide application, we shall describe here briefly the kriging method.

Let u(**x**) be a random space function, a regionalized variable in the geostatistical terminology. Assume that u_j are the measured values of u at the points $\mathbf{x}_j$ (j=M+1,M+2,...,N). The problem is to determine u(**x**) at any point in the domain of existence of u by interpolation. It is assumed that u is given by linear interpolation, i.e.

$$u(\mathbf{x}) = \sum_{j=M+1}^{N} \lambda_j u_j \qquad (1.8.12)$$

where $\lambda_j(\mathbf{x})$ are unknown coefficients. However, u is regarded as a realization of a random space function of, say, constant mean and stationary increments and (1.8.12) is a stochastic interpolator. The coefficients λ_j are determined with the aid of two conditions: first, for $\langle u\rangle=\langle u_j\rangle=m_u$=const, the expected value of u (1.8.12) is equal to m_u, i.e. the estimator (1.8.12) is unbiased. Taking the ensemble mean of (1.8.12) gives, therefore,

$$\sum_{j=M+1}^{N} \lambda_j = 1 \tag{1.8.13}$$

Second, the variance of estimation of u, the kriging variance, is minimal, i.e.

$$\sigma_u^2(\mathbf{x}) = \sum_{j=M+1}^{N} \sum_{k=M+1}^{N} \lambda_j \lambda_k C_u(\mathbf{x}_j-\mathbf{x}_k) = \sigma_u^2 - \sum_{j=M+1}^{N} \sum_{k=M+1}^{N} \lambda_j \lambda_k \gamma_u(\mathbf{x}_j-\mathbf{x}_k) \tag{1.8.14}$$

has to be minimized. Differentiation of (1.8.14) with respect to λ_j, with constraint (1.8.13) taken into account, leads to precisely the linear system (1.8.2) for the coefficients λ_j and the Lagrangean multiplier μ. Furthermore, the kriging variance is identical to the conditional variance (1.8.7) for $\mathbf{x}_i=\mathbf{x}_j$. Hence, for functions of stationary increments the kriging estimate and the kriging variance are equal to the conditional Gaussian expected value and variance, respectively. In spite of this striking similarity, it is pointed out that kriging does not imply necessarily that u is Gaussian, but only weakly stationary.

In the case in which a trend is present, procedures similar to those leading to the generalized covariance have been developed and details may be found, for instance, in Journel and Huijbregts (1978).

1.9 SPATIAL AVERAGES OF RANDOM FUNCTIONS

So far we have considered point value random functions u(**x**). In many applications we are interested, however, in the properties of the weighed mean of the function over a portion of the space. Such a mean is also called the regularized function $\bar{u}(\mathbf{x})$ and is defined by

$$\bar{u}_a(\mathbf{x}) = \int u(\mathbf{x}')\, a(\mathbf{x}-\mathbf{x}')\, d\mathbf{x}' \tag{1.9.1}$$

In (1.9.1) integration is carried out over the entire space and a(**x**) is a deterministic weight function satisfying the constraint

$$\int a(\mathbf{x})\, d\mathbf{x}=1 \tag{1.9.2}$$

Of particular interest here is the case of ordinary space average, obtained as a particular case of (1.9.2), for

$$a(\mathbf{x}) = \frac{\Omega(\mathbf{x})}{V} \; ; \; \bar{u}(\mathbf{x}) = \frac{1}{V}\int u(\mathbf{x}'+\mathbf{x})\, \Omega(\mathbf{x}')\, d\mathbf{x}' \tag{1.9.3}$$

where the function $\Omega(\mathbf{x})$, employed already in Chap. 1.5, is equal to unity for $\mathbf{x}$ within the fixed domain V , whose centroid is at the origin and which may be a closed volume,area or a line segment, and it is zero outside it. Then $\bar{u}$ (1.9.3) is defined as the spatial mean of u over V. Our purpose is to examine briefly the relationship between the statistical moments of $\bar{u}$ and those of u for stationary random functions first. The ensemble average of (1.9.1) yields by (1.9.2)

$$\langle \bar{u}(\mathbf{x})\rangle = \int \langle u\rangle\, h(\mathbf{x}')\, d\mathbf{x}' = \langle u\rangle \tag{1.9.4}$$

since $\langle u\rangle$ is constant. Hence, the ensemble average of a stationary random function and of its space average are identical. Next, we consider the covariance of the regularized u, i.e.

$$C_{\bar{u}}(\mathbf{r}) = \langle[\bar{u}(\mathbf{x}+\mathbf{r}) - \langle u\rangle][\bar{u}(\mathbf{x}) - \langle u\rangle]\rangle \tag{1.9.5}$$

Substituting (1.9.1) into (1.9.5) yields immediately

$$C_{\bar{u}}(\mathbf{r}) = \iint C_{u}(\mathbf{x}'-\mathbf{x}''+\mathbf{r})\, a(\mathbf{x}')\, a(\mathbf{x}'')\, d\mathbf{x}'\, d\mathbf{x}'' \tag{1.9.6}$$

and in particular for (1.9.3)

$$C_{\bar{u}}(\mathbf{r}) = \frac{1}{V^2}\iint C_{u}(\mathbf{x}'-\mathbf{x}''+\mathbf{r})\, \Omega(\mathbf{x}')\, \Omega(\mathbf{x}'')\, d\mathbf{x}'\, d\mathbf{x}'' \tag{1.9.7}$$

Following Matheron (1965), we define again the function H(r) (the transitive variogram associated with h(x), in his terminology) by Eq. (1.4.24), i.e.

$$H(\mathbf{r}) = \int \Omega(\mathbf{x})\, \Omega(\mathbf{x}+\mathbf{r})\, d\mathbf{x} \quad \text{i.e.} \quad \int H(\mathbf{r})\, d\mathbf{r} = V^2 \tag{1.9.8}$$

$H(\mathbf{r})$ is the volume (area, length) of the intersection between V whose centroid is at the origin, and its translation by $-\mathbf{r}$. By a change of variable in (1.9.7) we immediately have

$$C_{\bar{u}}(\mathbf{r}) = \frac{1}{V^2} \int C_u(\mathbf{x}')\, H(\mathbf{x}'-\mathbf{r})\, d\mathbf{x}' \tag{1.9.9}$$

The transformation of (1.9.7) into (1.9.8) is also known as the Cauchy algorithm. In particular, the variance results immediately for $\mathbf{r}=0$ in (1.9.9), i.e.

$$\sigma^2_{\bar{u}} = \frac{1}{V^2} \int C_u(\mathbf{r})\, H(\mathbf{r})\, d\mathbf{r} \tag{1.9.10}$$

which is of fundamental interest in many applications.

The function $H(\mathbf{r})$ can be readily determined for given shapes of V. The simplest shapes are a segment of length d on the line, a circle of diameter d in the plane and a sphere of diameter d in space. With $\mathbf{r}$ a one-, two- or three-dimensional coordinate, and with $V = d$, $V = \pi d^2/4$ and $V = \pi d^3/6$, respectively, we get by simple geometrical relationships for these examples

$$\begin{aligned} \frac{H(r)}{V} &= 1 - \frac{r}{d} \quad \text{for } r<d \; ; \; H(r) = 0 \quad \text{for } r \geq d \\ \frac{H(r)}{V} &= \frac{2}{\pi}\left[\cos^{-1}\left(\frac{r}{d}\right) - \frac{r}{d}\left(1-\frac{r^2}{d^2}\right)\right] \quad \text{for } r<d \; ; \; H(r) = 0 \text{ for } r \geq d \\ \frac{H(r)}{V} &= 1 - \frac{3}{2}\frac{r}{d} + \frac{1}{2}\frac{r^3}{d^3} \quad \text{for } r<d \; ; \; H(r)=0 \text{ for } r \geq d \end{aligned} \tag{1.9.11}$$

respectively, with $r=|\mathbf{r}|$.

Returning now to the general expression (1.9.10), two length scales are present: I_u, the integral scale of u and d, the diameter of the averaging volume V. There are two limits of interest, depending on the ratio between these two scales. Thus, for $d \ll I_u$, $C_u(\mathbf{r})$ is approximately equal to σ_u^2 in the range $r<d$ for which H is different from zero, and then (1.9.10) and (1.9.8) yield

$$\sigma^2_{\bar{u}} = \frac{\sigma_u^2}{V^2} \int H(\mathbf{r})\, d\mathbf{r} + O\left(\frac{d}{I_u}\right) = \sigma_u^2\left[1 + O\left(\frac{d}{I_u}\right)\right] \tag{1.9.12}$$

and, as expected, the variance of the space average is approximately equal to the variance of the point-value function. However, if the covariance C_u is discontinuous at the origin, i.e. a completely random or "nugget" effect is present, the latter is wiped out in the process of space average, and the variance of $\bar{u}$ is equal to the value of C_u near the origin.

At the other extreme, for $d \gg I_u$, $H(\mathbf{r})$, by its definition, is approximately equal to V over the entire range for which $C_u \neq 0$. Then (1.9.10) yields, by the definition of the integral scale (1.4.27)

$$\sigma^2_{\bar{u}} = \sigma^2_u \frac{I_{u,x}}{d} [1 + O(\frac{I_{u,x}}{d})]$$
$$\sigma^2_{\bar{u}} = \sigma^2_u \frac{I^2_{u,xy}}{d^2} [1 + O(\frac{I_{u,xy}}{d^2})] \qquad (1.9.13)$$
$$\sigma^2_{\bar{u}} = \sigma^2_u \frac{I_u^3}{d^3} [1 + O(\frac{I_u}{d})]$$

for one-, two- or three-dimensional space averages. The important result is that the variance of the space average tends to zero like the ratio $(I_{u,m})^m/d^m$, where m is the space dimension. This procedure of variance reduction will be exploited frequently in the sequel.

The covariance of the regularized u (1.9.1) has also a few properties of interest. First, $\bar{u}_a$ is more regular than u, provided that a(x) is sufficiently smooth, and this is the reason for calling $\bar{u}_a$ a regularization of u. In the case of the space average (1.9.2), $H(\mathbf{r})$ is continuous. Hence, the same is true for $C_{\bar{u}}$ (1.9.9), even if C_u is discontinuous. In particular, a process of finite variance and zero integral scale (i.e. completely random, see Chap. 1.4) leads to $C_{\bar{u}}=0$ and $\sigma^2_{\bar{u}}=0$, i.e. space averaging wipes out such a process. A white-noise process, defined by Eq. (1.4.30), yields in (1.9.9) the covariance function

$$C_{\bar{u}}(\mathbf{r}) = \sigma^2_u \left[\frac{I_u}{d}\right]^3 \frac{H(\mathbf{r})}{V} \qquad (1.9.14)$$

which is continuous. Higher degree of regularization can be achieved by using differentiable functions a, but we do not pursue here the investigation of such weight functions.

The integral scale of $\bar{u}$ is easily derived from (1.9.9) and (1.9.8) by integrating over $\mathbf{r}$ as follows

$$\int C_{\bar{u}}(\mathbf{r})\, d\mathbf{r} = \frac{1}{V^2} \int C_u(\mathbf{x}')\, d\mathbf{x}' \int H(\mathbf{x}'+\mathbf{r})\, d\mathbf{r} \qquad (1.9.15)$$

Consequently, by using the definition (1.1.76) of I_u we can write

$$I_{\bar{u}} = [\sigma^2_u/\sigma^2_{\bar{u}}]^{1/m}\, I_u \qquad (1.9.16)$$

where m=1,2 or 3 is the space dimension. Since $\sigma^2_{\bar{u}}$ was shown to vary between σ^2_u (1.9.12) to $\sigma^2_u(I_u/d)^m$ (1.9.13), it is seen that $I_{\bar{u}}$ changes between I_u (for $d \ll I_u$) and d (for $d \gg I_u$). Hence, the effect of regularization is to decrease the variance and to increase the correlation scale.

The results obtained so far can be easily extended to the variogram of $\bar{u}$ (Matheron, 1965). Indeed, Eq. (1.9.9) yields

$$\gamma_{\bar{u}}(\mathbf{r}) = \frac{1}{V^2} \left[\int \gamma_u(\mathbf{x}') \, H(\mathbf{x}'+\mathbf{r}) \, d\mathbf{x}' - \int \gamma_u(\mathbf{x}') \, H(\mathbf{x}') \, d\mathbf{x}' \right] \tag{1.9.17}$$

Hence, the variogram of the space average is obtained from that of u by a formula similar to (1.9.9). It is emphasized that similarly to $C_{\bar{u}}$, a nugget effect which exists in $\gamma_u(\mathbf{x}')$ is wiped out in $\gamma_{\bar{u}}(\mathbf{r})$, the latter being continuous at the origin. On the other hand, the behaviour of $\gamma_{\bar{u}}$ for $r \gg d$ is similar to that of γ_u. If γ_u has a sill, the effect of the regularization is similar to that obtained for the covariance, namely the range of $\gamma_{\bar{u}}$ is larger than that of γ_u, whereas the sill is smaller.

A similar procedure can be applied to the analysis of the space average of higher moments, e.g. the covariance or the variogram. A detailed discussion of the topic may be found in Vanmarcke (1983).

1.10 THE ERGODIC HYPOTHESIS

In applications related to flow through porous formations one encounters a single realization of the medium and the concept of ensemble is a rather abstract one. The ensemble reflects merely the uncertainty in the depiction of the spatial structure of the given formation, rather than a set of similar existing formations.

This being the case, the statistical characterization of the random structure has to be based on the information contained in the given realization. Hence, if we limit the information to second-order moments, the expected value and the covariance have to be derived from space averages rather than ensemble averages. This is possible only if some type of stationarity prevails and the ergodic hypothesis is satisfied, the rigorous enunciation of the latter being beyond our scope. Intuitively speaking, the ergodic hypothesis for a system implies that all states of the ensemble are available in each realization. Since generally a single realization is present, there is no way to validate rigorously the hypothesis. The pragmatic approach adopted here, following the one prevailing in statistical continuum theories, is to presume ergodicity, to derive moments of interest by space averaging and to check subsequently the validity of the ergodic assumption by the methods of Chap. 1.9.

2.1. *INTRODUCTION.*

A porous medium is a two-phase material in which the solid matrix constitutes one phase and the interconnected void (pores) constitutes the other. The solid matrix is either rigid or it undergoes small deformations. With a proper definition of the pore scale d, the theory of flow through porous media is concerned with the behaviour of samples of typical length scales L much larger than d. In the present part we consider the scale pertaining to laboratory samples, which are obtained for instance by extracting cores from natural formations or by packing laboratory columns. These serve as experimental support to validate the basic laws governing the flow, to be applied subsequently to large scale formations.

A distinctive property of a natural porous medium is the irregular distribution (shape, size) of its pores (the irregularity of the structure is vividly illustrated by Fig. 2.2.1). The dimension L of laboratory samples is of the order of tens to thousands d, and such samples are generally homogeneous, in the sense that the irregular pore structure reproduces itself in the various portions of the sample. This is in contrast with the much larger, local or regional, scales which are considered in the remaining parts of this book.

A typical laboratory experiment consists in creating flow of a fluid, which may also carry a solute or heat, through a porous column. In such experiments, quantities of interest, like rates of

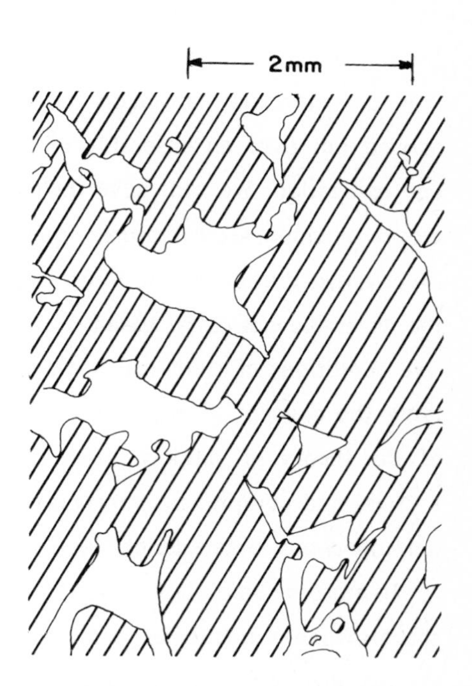

Fig. 2.2.1 A cross-section through a sample of pervious rock (chalk of permeability k=4000 md) based on a photograph from Zinszner & Meynot (1982), with the solid matrix represented by the hatched area.

flow, pressure, concentration etc., are measured over areas which cross many pores. Such space-averaged quantities, called macroscopic, are the ones of interest in applications. Unlike the microscopic, at pore scale, variables, the macroscopic quantities change in a regular manner through space and with time. The main aim of the theory of flow through porous media, and particularly of this part, is to derive the laws governing the macroscopic variables.

A few lines of attack have been used in the past in order to achieve this goal. The one followed here is to start with the known microscopic equations of transfer obeyed by the fluid as a continuum, and to arrive at the macroscopic ones by averaging over volumes or areas containing many pores. Two basic averaging approaches have been employed in the literature: spatial and statistical. In the first one, a macroscopic variable is defined as an appropriate mean over a sufficiently large representative elementary volume (r.e.v.) and is attached to its centroid. It is taken for granted in this approach that the length scale of the r.e.v. is much larger than the pore scale d, but considerably smaller than the length scale of the flow domain. Furthermore, it is also assumed that the results concerning the macroscopic variables are independent of the size of the r.e.v.

The statistical approach, to be adopted here, is related to the uncertainty of the spatial distribution of microscopic quantities. This concept may be illustrated in a simple-minded way as follows: if a column is filled repetitively with a granular material under the same laboratory conditions, the microscopic pore structure is never reproduced from one experiment to the other. In other words, the minute details of the settling of the grains in various fillings are subjected to uncertainty, although the process is repeated under seemingly identical conditions. The rational framework to deal with uncertainty is by regarding the microscopic variables as random rather than deterministic. The main conceptual difficulty stems from the fact that statistical averaging is carried out over an ensemble of realizations, whereas in practice one generally encounters only one realization. We can, nevertheless, regard the actual sample as one of the possible realizations of media of same gross features, realizing that we do not need, nor are we interested, to know the precise distribution of the microscopic quantities. The inference of the statistical information about the ensemble has to be based, however, on the unique sample. This is possible only under restrictive conditions of statistical homogeneity (stationarity) which are similar in essence to those underlying the concept of representative elementary volume. Nevertheless, the well developed apparatus of the theory of random functions (see Part. 1) enables us to derive the macroscopic laws in a rational and comprehensive manner, along the lines of the kinetic theory of gases, of the theory of heterogeneous materials, of the theory of diffusion and of other many branches of physics.

It must be said, however, that as long as we limit ourselves to deriving relationships between space averaged quantities, with no special concern about their fluctuations, the results obtained by the two approaches are essentially the same.

2.2 GEOMETRY OF POROUS MEDIA AND SPACE AVERAGING

2.2.1 Geometry of porous media

A porous medium is defined as a two-phase material, the solid matrix and the interconnected pore space. To illustrate the complexity of the pore structure, a cross-section of a core of natural chalk stone based on a photograph from Zinszner & Meynot (1982), has been reproduced in Fig. 2.2.1.

Detailed descriptions of porous materials, as related to the natural formations they belong to, or to laboratory packing conditions, may be found, for instance, in Dullien (1979). This subject, however, is beyond the scope of the present book. Instead, a few basic geometrical relationships related to the pore structure will be developed succinctly.

The geometry of the porous medium can be described with the aid of the fundamental function (Chap. 1.5) $h(\mathbf{x},t)$, which is defined as follows

$$h(\mathbf{x},t) = 1 \quad \text{for } \mathbf{x} \in V_V \;\; ; \;\; h(\mathbf{x},t) = 0 \quad \text{for } \mathbf{x} \in V_S \tag{2.2.1}$$

where V_V and V_S are the void and solid domains, respectively. Here $\mathbf{x}$ is the position vector in space, with origin attached to a fixed frame (e.g. the column containing the porous body) and t is the time. We shall use for the Cartesian components of $\mathbf{x}$ the tensorial notation x_i (i=1,2,3) or alternatively x,y,z.

In the statistical approach $h(\mathbf{x})$, called the characteristic function by Matheron (1967), is regarded as a random function of space. It is defined by its various statistical moments

$$P(\mathbf{x}_1,\mathbf{x}_2,...,\mathbf{x}_N) = <h(\mathbf{x}_1)\; h\mathbf{x}_2)...h(\mathbf{x}_N)> \tag{2.2.2}$$

where $\langle\;\rangle$ stands for an ensemble average (see Chap. 1.1) and P is the probability of the points of coordinates $\mathbf{x}_1,\mathbf{x}_2,...,\mathbf{x}_N$ to belong to the void space. The set of functions P for all values of the integer N and systems of points constitutes the "spatial law" of the random structure, in Matheron's (1967) terminology, and characterizes it completely. As mentioned in Chap. 2.1, one may visualize the ensemble as an infinite number of columns packed under the same conditions. At the fixed point $\mathbf{x}$ the function h takes the values 1 or 0, depending on whether $\mathbf{x}$ lies in V_V or V_S, respectively, in the particular realization.

A few properties of the first moments of h (2.2.2) have been discussed by Debye et al (1957) and Matheron (1967). Suppressing the dependence on t for the time being, the average

$$n(\mathbf{x}) = \langle h(\mathbf{x})\rangle \tag{2.2.3}$$

is the statistical definition of the porosity, which in words is the limit of the ratio between the number of realizations in which $\mathbf{x}$ is in V_v and the total number N, for N→∞. The function n is generally continuous in $\mathbf{x}$, a possible exception being the boundary of the porous sample, where n may be viewed as discontinuous.

The two-points covariance for $\mathbf{x}_1=\mathbf{x}$ and $\mathbf{x}_2=\mathbf{x}+\mathbf{r}$, takes the extreme values

$$C_h(\mathbf{x},\mathbf{r}) = \langle h'(\mathbf{x})\, h'(\mathbf{x}+\mathbf{r})\rangle \to n(\mathbf{x})\,[\,1 - n(\mathbf{x})]\quad \text{for}\quad |\mathbf{r}|\to 0$$
$$C_h(\mathbf{x},\mathbf{r}) \to 0 \quad \text{for} \quad |\mathbf{r}|\to\infty \tag{2.2.4}$$

where $h' = h - \langle h\rangle$ is the fluctuation.

The second relationship in (2.2.4) is underlain by the assumption that the events $\mathbf{x}\in V_v$ become completely uncorrelated as the two points depart. This excludes an interesting class of artificial porous media, namely periodic media. These are not considered, because they are not encountered in natural porous formations, which are our concern here.

The autocorrelation is defined as usual (Chap. 1.3) by

$$\rho_h(\mathbf{x},\mathbf{x}+\mathbf{r}) = \frac{\langle h'(\mathbf{x})\, h'(\mathbf{x}+\mathbf{r})\rangle}{n(\mathbf{x})[\,1 - n(\mathbf{x})]} \tag{2.2.5}$$

One can continue in a similar manner and analyze the higher-order moments for N=3,4,... in (2.2.2). Each higher order moment provides additional information on the geometry of the structure and contains the previous moments as particular cases.

At this point the usual classification of porous media can be introduced. A medium is homogeneous if

$$\frac{\partial n}{\partial x_i} = 0 \;;\quad \frac{\partial \rho_h}{\partial x_i} = 0 \quad \text{i.e.} \quad n = \text{const} \;;\; \rho_h = \rho_h(\mathbf{r}) \tag{2.2.6}$$

and, in a similar fashion, if higher order moments in (2.2.2) are invariants to a translation in space of the points $\mathbf{x}_1,\mathbf{x}_2,...$ and are, therefore, functions of the relative positions $\mathbf{r}_{12}=\mathbf{x}_1-\mathbf{x}_2$, $\mathbf{r}_{13}=\mathbf{x}_1-\mathbf{x}_3,...$.

$$C_h(\mathbf{x},\mathbf{r}) \equiv C_h(\mathbf{x},r) \tag{2.2.7}$$

where $r=|\mathbf{r}|$, and if higher order moments depend on r_{12},r_{13}, $\mathbf{r}_{12}\cdot\mathbf{r}_{13}$,...i.e., they are invariant to a rigid rotation of the configuration points. Partial degrees of isotropy, like axisymmetric, can be defined, but in the context of natural formations anisotropy is encountered mainly at the local scale (Part. 3).

For a homogeneous and isotropic medium, the autocorrelation (2.2.5) becomes

$$\rho_h(r) = \frac{C_h(r)}{\sigma_h^{\,2}} = \frac{\langle h(\mathbf{x})\, h(\mathbf{x}+\mathbf{r})\rangle - n^2}{n\,(1-n)} \tag{2.2.8}$$

Integral scales are defined like in Chap. 1.4, e.g. the linear integral scale

$$d = \int_0^\infty \rho_h(r)\, dr \tag{2.2.9}$$

assuming, of course, that ρ_h decays sufficiently fast with r. The length d is a measure of the distance of correlation for the events "point $\mathbf{x}$ and point $\mathbf{x}$ +r" being both in V_V, and can serve as a convenient definition of the pore scale.

We turn now to the gradient ∇h which is zero in both V_V and V_S and is singular at the interface A between the two phases. Its properties have been discussed in detail in Chap. 1.5 and it is reminded that

$$\langle \nabla h(\mathbf{x})\rangle = \nabla n(\mathbf{x}) \tag{2.2.10}$$

and also

$$\langle \nabla h\rangle = -\lim_{\Delta V \to 0} \left(\frac{1}{\Delta V} \left\langle \int_{\Delta A} \boldsymbol{\nu}\, dA \right\rangle\right) \tag{2.2.11}$$

where ΔV is an infinitesimal volume, ΔA is the portion of interface crossing ΔV and $\boldsymbol{\nu}$ is a unit vector normal to A (Fig. 1.2). The right-hand side of (2.2.11) can be interpreted as the ensemble average of the density of oriented void-solid interface elements. In the case of homogeneous media $\nabla n=0$ and the average density is zero, which could be expected in view of the mutual cancellation of $\boldsymbol{\nu}$ for elements of various orientations. We have also shown in Chap. 1.5 that the specific area σ, defined by

$$\sigma = \lim_{\Delta V \to 0} \left(\frac{1}{\Delta V} \langle \Delta A\rangle\right) \tag{2.2.12}$$

is related to the autocorrelation, for an isotropic medium, by

$$\sigma = -\,4\, n\,(1-n)\, \frac{d\rho_h}{dr} \qquad \text{(for r=0)} \tag{2.2.13}$$

Eq. (2.2.13) has been established by Debye et al (1957) and extended to anisotropic media by Matheron (1967).

An explicit and simple equation for $\rho_h(r)$ can be established for a perfectly random medium, in the terminology of Debye et al (1957), or in a more rigorous manner for a medium constructed according to a Boolean scheme (Matheron, 1967) or a Poisson medium (Beran, 1968). This is a medium which can be built by placing a set of given arbitrary grains (or pores) at random in the space, while keeping the porosity constant. Consider now three points along a line: $\mathbf{x}_1=\mathbf{x}$, $\mathbf{x}_2=\mathbf{x}+\mathbf{i}r$ and $\mathbf{x}_3=\mathbf{x}+\mathbf{i}(r+dr)$, where $\mathbf{i}$ is an arbitrary unit vector. In a homogeneous and isotropic medium the probability of $\mathbf{x}_1$ and $\mathbf{x}_3$ to belong both to the void space is $P(\mathbf{x}_1,\mathbf{x}_3) = C_h(r+dr)$, while for $\mathbf{x}_2$ and $\mathbf{x}_3$ it is $C_h(dr)$. We may write now for this type of porous media

$$P(\mathbf{x}_1,\mathbf{x}_3) = P(\mathbf{x}_1,\mathbf{x}_2)\,\frac{P(\mathbf{x}_2,\mathbf{x}_3)}{n} + [\,n - P(\mathbf{x}_1,\mathbf{x}_2)]\,\frac{n - P(\mathbf{x}_2,\mathbf{x}_3)}{1 - n} \tag{2.2.14}$$

as the probability $P(\mathbf{x}_1,\mathbf{x}_3)$ is the result of two sets of events: both pairs $\mathbf{x}_1,\mathbf{x}_2$ and $\mathbf{x}_2,\mathbf{x}_3$ belong to the void space, or $\mathbf{x}_1$ belongs to V_V, $\mathbf{x}_2$ belongs to V_S and $\mathbf{x}_3$ belongs to V_V. Eq. (2.2.14) reflects the complete randomness of the medium by the independence of these sets of events. Furthermore, $P(\mathbf{x}_1,\mathbf{x}_2) = C_h(r)$, while the probability of $\mathbf{x}_2,\mathbf{x}_3$ of being in V_V conditioned on $\mathbf{x}_2$ belonging to V_V is $C_h(dr)/n$, and their joint probability as independent events is given by their product. With a similar reasoning for the other set, we can rewrite (2.2.14) after letting $r\to 0$, as follows

$$n\,(1-n)\,\frac{dC_h(r)}{dr} = \frac{dC_h(0)}{dr}\,[C_h(r) - n^2] \tag{2.2.15}$$

Substitution of (2.2.8) and (2.2.13) in (2.2.15) yields

$$\frac{d\rho(r)}{dr} = -\,\frac{\sigma}{4\,n\,(1-n)}\,\rho(r) \tag{2.2.16}$$

and integration leads to the final expression of the autocorrelation

$$\rho(r) = \exp(-\frac{r}{d}) \quad ; \quad d = \frac{4\,n\,(1-n)}{\sigma} \tag{2.2.17}$$

This result has been obtained in a slightly different manner by Debye et al (1957), who were concerned with the scattering of electromagnetic radiation due to the contrast in the dielectric constants of the two phases. They have related the scattering intensity to d and have confirmed experimentally the validity of (2.2.17) for a few materials.

The linear integral scale for (2.2.17) is precisely d and it is related in a simple manner to the specific area σ by (2.2.17). For example, in the case of a porous medium built by placing at random disjoint spherical particles of radii R, of probability density distribution f(R), we have by the definition of the specific area and by simple geometrical calculations

$$\sigma = \frac{3(1-n)\displaystyle\int_0^{\infty} R^2 f(R)\, dR}{\displaystyle\int_0^{\infty} R^3 f(R)\, dR} \tag{2.2.18}$$

For uniform spheres of radius R, i.e. $f(R) = \delta(R)$, (2.2.18) yields

$$\sigma = \frac{3(1-n)}{R} \quad ; \quad d = \frac{4\, n\, R}{3} \tag{2.2.19}$$

It is well known that for the tightest packing of spheres $n = 0.2595$, but such an arrangement is periodic, rather than completely random.

A more realistic distribution of the grain size is lognormal (see Eq. 1.1.14), for which (2.2.18) yields

$$\sigma = \frac{3(1-n)}{\langle R\rangle \exp(2\sigma_{\ln r}^{2})} = \frac{3(1-n)}{\langle R\rangle}\left[\frac{\langle R\rangle^2}{\langle R^2\rangle}\right]^2 \tag{2.2.20}$$

Hence, d (2.2.17) is biased towards the grains of radii larger than $\langle R\rangle$, as one would expect. At any rate, Eqs. (2.2.17), (2.2.19) and (2.2.20) provide the relationships between the correlation scale d and the geometrical properties of the particles making up the medium. We shall make extensive use of d in the following section.

In the case of spherical pores imbedded in a solid matrix, $(1-n)$ has to be replaced by n in (2.2.18) and in the ensuing relationships.

Finally, we shall recall a few current values of n for natural media from Raudkivi and Callender (1976): fine to medium gravel (2 mm < d < 20 mm) $n = 0.30$-0.40, fine sand (0.006 mm < d < 0.2 mm) $n = 0.44$-0.49, medium sand (0.2 mm < d < 0.6 mm) $n = 0.41$-0.48, coarse sand (0.6 mm < d < 2 mm) $n = 0.39$-0.41 and loam $n = 0.35$-0.50, where d is particle representative diameters. These values of the porosity are indicative of the range of interest in this book.

2.2.2 Space averages and macroscopic variables

In this section we shall examine the connection between the ensemble averages defined in the previous section and the usual definition of macroscopic variables by space averaging. Thus, for a homogeneous medium the space average definition of the porosity is

$$\bar{n} = \frac{1}{V} \int_V \mathrm{h}(\mathbf{x}')\, dV = \frac{V_\mathrm{v}}{V} \tag{2.2.21}$$

where V is the averaging volume whose centroid is at $\mathbf{x}$. $\bar{n}$ can be defined as well with the aid of areal or line averages as follows

$$\bar{n} = \frac{1}{\mathrm{A}} \int_\mathrm{A} \mathrm{h}(\mathbf{x}')\, d\mathrm{A} \quad ; \quad \bar{n} = \frac{1}{\mathrm{L}} \int_\mathrm{L} \mathrm{h}(\mathbf{x}')\, d\mathbf{x}' \tag{2.2.22}$$

whenever the medium is homogeneous in the plane or along the line on which integration is carried out. This extension of the space average is quite useful as it allows to define $\bar{n}$ in most conceivable cases. For instance, at the planar boundary between a porous medium and a solid or the free space, Eq. (2.2.22) defines $\bar{n}$ for A parallel to the boundary.

In the statistical context $\bar{n}$ is a random variable. Its ensemble mean can be written, according to Eq. (2.2.3), as follows

$$\langle \bar{n} \rangle = \frac{1}{V} \int_V \langle \mathrm{h}(\mathbf{x}') \rangle\, d\mathbf{x}' = n \tag{2.2.23}$$

i.e. the expectation value of the space average is equal to the ensemble average.

The variance of $\bar{n}$ is given by Eq. (1.9.10) as follows

$$\sigma^2_{\bar{n}} = \langle \bar{n} - \langle \bar{n} \rangle \rangle^2 = \frac{1}{V^2} \int C_\mathrm{h}(\mathbf{r})\, \mathrm{H}(\mathbf{r})\, d\mathbf{r} = \frac{\sigma^2_\mathrm{h}}{V^2} \int \rho_\mathrm{h}(\mathbf{r})\, \mathrm{H}(\mathbf{r})\, d\mathbf{r} \tag{2.2.24}$$

where it is reminded that $\mathrm{H}(\mathbf{r})$ (1.9.8) is a function equal to unity in the joint volume defined by V and its translation by $-\mathbf{r}$ and equal to zero outside it, while $d\mathbf{r}$ is a volume element. Explicit expressions for the cases of a sphere, circle or segment are given in Eq. (1.9.11). For instance, for space averaging over o sphere of diameter D, we obtain for an isotropic medium, by using (1.9.11) and (2.2.8)

$$\frac{\sigma^2_{\bar{n}}}{\sigma^2_\mathrm{h}} = \frac{24}{d^3} \int_0^\mathrm{D} \rho_\mathrm{h}(\mathrm{r}) \left(1 - \frac{3}{2}\frac{\mathrm{r}}{\mathrm{D}} + \frac{1}{2}\frac{\mathrm{r}^3}{\mathrm{D}^3}\right) \mathrm{r}^2\, d\mathrm{r} \tag{2.2.25}$$

The ratio (2.2.25) represents the variance reduction of the space averaged variable with respect of that of the point variable. This ratio depends on the autocorrelation ρ_h and the diameter D of the averaging sphere. In order to grasp their effect we adopt the autocorrelation (2.2.17) for a completely random medium, to get explicitly in (2.2.25)

$$\frac{\sigma^2_{\bar{n}}}{\sigma^2_h} = 24\left[\frac{2}{a^3} - \frac{9}{a^4} + \frac{60}{a^6} - 3\left(\frac{1}{a^3} + \frac{7}{a^4} + \frac{20}{a^5} + \frac{20}{a^6}\right)e^{-a}\right] \tag{2.2.26}$$

where $a=D/d$ is the ratio between the diameter of the averaging sphere and the integral scale (pore scale) and $\sigma^2_h = n(1-n)$ (see Eq. 1.1.12). Eq. (2.2.26) is represented graphically in Fig. 2.2.2, to illustrate the rapid drop of $\sigma^2_{\bar{n}}$ as $a=D/d$ increases. It is seen that for $D/d \to \infty$, $\sigma^2_{\bar{n}} \to 0$ and we can assume that the requirements of the ergodic hypothesis are satisfied (Chap. 1.10), i.e. the space average $\bar{n}$ and the ensemble mean n are interchangeable. This result provides the theoretical foundation, in the statistical framework, to the definition of macroscopic variables by space averaging.

In practice we shall adopt the fundamental relationship $\bar{n}=n$ even if $a=D/d$ is finite, provided that $\sigma_{\bar{n}}/n$ is sufficiently small compared to unity. A more general analysis can be carried out in this case for an arbitrary covariance function C_h, not necessarily the exponential (2.2.17), provided that it has an integral scale. It has been shown in Chap. 1.9 (Eq. 1.9.13) that

$$\frac{\sigma^2_{\bar{n}}}{\sigma^2_h} = \left(\frac{I_h}{D}\right)^m + O\left(\frac{I_h^{2m}}{D^{2m}}\right) \tag{2.2.27}$$

where m is the number of dimensions of the space, D is the diameter and I_h is the integral scale (spatial, planar or linear) of ρ_h.

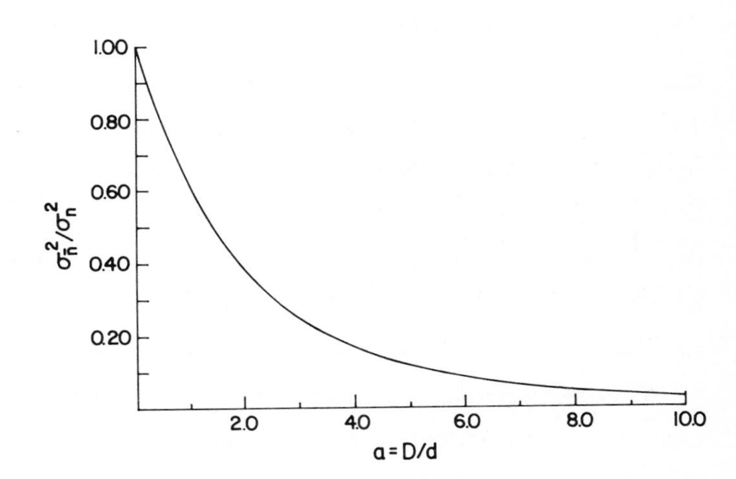

Fig. 2.2.2 Graphical representation of the dependence of the porosity variance ratio upon the ratio between the diameter of the averaging sphere and the pore scale (Eq. 2.2.26).

Returning for the purpose of illustration to the completely random media (2.2.7), we have for the three integral scales (see Eq. 1.4.2)

$$I_{h,x} = d \;\; ; \;\; I^2_{h,xy} = 18\, d^2 \;\; ; \;\; I_h^{\,3} = 48\, d^3 \qquad (2.2.28)$$

If we require the ratio σ_n / n to be smaller than a given arbitrary small number ϵ as a prerequisite of ergodicity (Chap. 1.10), we arrive from (2.2.27) at different conditions for the extent of the averaging domain, depending on whether it is a sphere, circle or line. To illustrate this point we substitute (2.2.28) into (2.2.27) to obtain for a completely random medium the requirements

$$\frac{D}{d} > \frac{1-n}{n\,\epsilon^2} \;\; ; \;\; \frac{D}{d} > \left[\frac{18(1-n)}{n\,\epsilon^2}\right]^{1/2} \;\; ; \;\; \frac{D}{d} > \left[\frac{48(1-n)}{n\,\epsilon^2}\right]^{1/3} \qquad (2.2.29)$$

for space averaging on a line, circle or sphere, respectively. We substitute, for further illustration, the relationship between the integral scale d or the related specific area σ (2.2.19) and R, for uniform spheres of radius R, in (2.2.29), and also take n=0.5 and ϵ=0.01, to obtain

$$\frac{D}{R} > 6666 \;\; ; \;\; \frac{D}{R} > 283 \;\; ; \;\; \frac{D}{R} > 52 \qquad (2.2.30)$$

It is seen that the extent of the averaging domain required in order to ensure exchange of space and ensemble averages decreases drastically with the number of dimensions of the space. The advantage of defining macroscopic variables as volume, rather than areal or linear, averages is obvious.

This analysis can be generalized in two respects. First, for a flow variable w which is defined in the pore space, we can arrive at an equation similar to (2.2.24) provided that $\sigma^2_{\bar{n}}$, σ^2_h and ρ_h are replaced by $\sigma^2_{\bar{w}}$, σ^2_w and ρ_w, respectively. If we assume that the autocorrelation ρ_w is related mainly to the randomness of the medium, it is reasonable to take it approximately equal to ρ_h. Then, the requirements for the size of the averaging domain are similar to (2.2.29), provided that we replace $(1-n)/n$ by $\sigma^2_w / \langle w \rangle^2$. The present analysis provides a rational foundation to the usual definition of macroscopic variables and arrives at quantitative criteria for the size of the averaging domain, which ensures the independence of the space average from this size. The same criteria may be adopted for defining the "representative elementary volume" (for a qualitative definition of the r.e.v. in a deterministic context, see, for instance, Bear, 1972).

A second generalization is achieved by the relaxation of the requirement of spatial homogeneity (see Chap. 1.10). Indeed, if porosity, for instance, is slowly varying in space, with the scale of its change defined by (1.10.1)

$$\ell = \frac{|\nabla n|}{|\nabla^2 n|} \qquad (2.2.31)$$

the space average still satisfies approximately the ergodic requirements if the ordering

$$d << D << \ell \qquad (2.2.32)$$

is obeyed.

We shall assume in the present part of the book that all flow variables satisfy the conditions (2.2.32) everywhere, with the possible exception of the neighborhood of the boundaries (this topic is discussed in Chap. 2.13). Thus, we will exchange freely ensemble and space averages, both serving as definitions of macroscopic variables.

Exercises

2.2.1 A porous medium is created by mixing at random solid cylinders of different radii and parallel axes. Analyze the covariance of the characteristic function and its relationship with the specific perimeter.

2.2.2 Another peculiar porous medium is the one created from platelike particles of negligible thickness. Establish the relationship between the specific area and a correlation function defined as the probability of two points to be joined by a line which lies entirely in the void space, for a homogeneous medium.

2.2.3 It is customary in soil mechanics to represent the granulometry of sandy soils, as a percentage of material which passes through standard sieves, as a function of $\log_{10}$ of the size. Derive a relationship between the integral scale d of a completely random medium of spherical particles and between the customary values d_{50} and the ratio d_{90}/d_{10}. The latter are sieve sizes corresponding to 50%, 90% and 10% of the material. Assume that the grain size distribution is lognormal.

2.2.4 Derive equations for the diameter of the averaging sphere, circle or the length of the segment, for completely random media made up of spheres of lognormal radii distribution, required in oder to reduce the porosity variance by a given amount. How is the size of the averaging domain, relative to that corresponding to a uniform medium, increasing with the variance of the log of the radii?

2.3. *THE MICROSCOPIC EQUATIONS OF FLOW AND TRANSPORT*

This book is concerned with flow of water through the void space of porous media. The solid matrix is assumed to be fixed, except for a special class of unsteady flows in which it undergoes small elastic deformations (Chap. 2.6). We also limit the scope of the book to flow through saturated media, i.e. water is filling the void space, except for possible pockets of air which are not connected, (we shall touch briefly the unsaturated flow problem in relationship to the boundary condition at a free-surface in Chap. 2.13). Besides, we are going to study two transport phenomena, namely heat or solute transfer, both by diffusion and by convection by water. These are the main processes of interest in groundwater flow to be considered here. Other important transfer phenomena are treated by various disciplines, like soil physics and chemistry, reservoir engineering, chemical engineering and soil mechanics.

The boundary between the solid matrix and the water filled pore domain is assumed to be well defined and we do not consider here clayey or similar soils. Nevertheless, many of the results obtained here can be shown to be valid in such cases on empirical or phenomenological grounds.

Our basic approach is to regard the fluid and other transferable quantities as continua, i.e. the pore scale is assumed to be much larger than the molecular scale. This is not necessarily the case for flow of water at very low saturation or for rarefied gases, which are not of our concern. Hence, we regard water at the microscopic level as a Newtonian fluid, whereas heat and solute diffusion obey the usual flux-forces relationships. Thus, the microscopic transfer laws are well defined and can be found in various text books (e.g., Bird et al, 1960, and Batchelor, 1967). We shall enumerate these laws in the sequel for convenience of reference in the following sections.

As a matter of principle, we can separate the equations of flow into three groups : equations of state, of conservation and constitutive equations. These combine in sets of partial differential equations which have to be supplemented by appropriate initial and boundary conditions.

The equations of state relate the physical properties (specific mass, temperature and concentration) as follows

$$\underline{\rho}(c,T_f) = \underline{\rho}(C_0,T_{f0}) + (C-C_0)\frac{\partial\underline{\rho}}{\partial C_0} + (T_f-T_{f0})\frac{\partial\underline{\rho}}{\partial T_{f0}} \tag{2.3.1}$$

where $\underline{\rho}$ is the specific mass of the fluid (water), T_f is the temperature of the fluid and C is the solute concentration (mass per unit volume). T_{f0} and C_0 are reference values and the coefficients $\partial\underline{\rho}/\partial C_0$ and $\partial\underline{\rho}/\partial T_{f0}$ can be taken as constant in usual ranges of concentrations and temperatures, e.g. for NaCl in water $\partial\underline{\rho}/\partial C_0 \simeq 0.73$ and for water $\partial\underline{\rho}/\partial T_{f0} \simeq 2\text{x}10^{-4}$ gm/cm³.degC. It must be said that in all problems of interest the variations of $\underline{\rho}$ with temperature or concentration are negligible at the pore scale, due to the smoothing effect of molecular diffusion. They might have a significant role in some applications at the formation scale. We shall neglect, therefore, the variation of ρ within elementary volumes.

Similarly, the density of the solid matrix ρ_s can be taken as constant at the pore scale in all problems of interest in this book.

We turn now to the equations of conservation, starting with the mass of fluid

$$\frac{D\rho}{Dt} + \rho\ \nabla.\boldsymbol{u} \equiv \frac{\partial \rho}{\partial t} + \nabla.(\rho\, \boldsymbol{u}) = 0 \tag{2.3.2}$$

with $\boldsymbol{u}$ the fluid velocity and with $D/Dt = \partial/\partial t + \boldsymbol{u}.\nabla$, the material derivative operator. In most circumstances and unless specified, ρ may be taken as constant and Eq. (2.3.2) reduces to the continuity equation for an incompressible fluid

$$\nabla.\boldsymbol{u} = 0 \tag{2.3.3}$$

The equation of conservation of energy in the fluid is given by

$$\rho\, c_f \frac{\partial T_f}{\partial t} = - \nabla.q_{hft} \tag{2.3.4}$$

where c_f is the heat capacity of the fluid (which is practically constant at ordinary temperature, e.g. for water $c_f \simeq 4.2$ joule gm/degC), and q_{hft} is the total heat flux in the fluid phase. In Eq. (2.3.4) q_{hft} is made up from the molecular heat conduction flux q_{hf} and the convective flux $\rho\, c_f\, \boldsymbol{u}$. Heat generation by viscous dissipation or by other mechanisms have been neglected. In a similar manner we have for the stationary solid phase

$$\rho_s\, c_s \frac{\partial T_s}{\partial t} = - \nabla.q_{hs} \tag{2.3.5}$$

where c_s is the heat capacity of the solid, T_s is its temperature and q_{hs} is the heat flux in the solid phase.

The equation of conservation of solute mass is similarly given for an incompressible fluid by

$$\frac{\partial C}{\partial t} = - \nabla.q_{mt} + S \quad ; \quad q_{mt} = q_m + C\, \boldsymbol{u} \tag{2.3.6}$$

where q_{mt} is the total solute flux, q_m is the molecular diffusive flux and S is a source term, e.g. radioactive decay.

Finally, the equation of conservation of momentum is given by

$$\frac{Du}{Dt} = \nabla.\underline{\sigma}_f - g \, \underline{\rho} \, \nabla z \quad ; \quad \underline{\sigma}_{f,ij} = - p \, \delta_{ij} + \tau_{ij} \quad (i,j=1,2,3) \tag{2.3.7}$$

where D/Dt is the acceleration of a fluid particle, $\underline{\sigma}_f$, i.e. $\underline{\sigma}_{f,ij}$ in the tensorial notation, is the total stress in the fluid, p is the pressure in the fluid, τ_{ij} is the viscous stress tensor, δ_{ij} is the Kronecker delta, g is the acceleration of gravity and z is a vertical coordinate pointing upwards.

The constitutive equations relate the various molecular fluxes to the gradients. Thus, Fourier law for heat conduction is given by

$$q_{hf} = - K_{hf} \, \nabla T_f \tag{2.3.8}$$

$$q_{hs} = - K_{hs} \, \nabla T_s \tag{2.3.9}$$

where K_{hf} and K_{hs} are the coefficients of heat conduction of the fluid and solid phases, respectively.

Similarly, Fick's law for solute diffusion is given by

$$q_m = - D_m \, \nabla C \tag{2.3.10}$$

with D_m being the coefficient of molecular diffusion. In both Eqs. (2.3.9) and (2.3.10), mixed terms, representing the Soret and Dufour effects, which are exceedingly small, have been neglected.

Last, the constitutive equations for stress and deformation for the liquid phase are as follows

$$(1/\underline{\rho}) \, (d\underline{\rho}/dp) = \beta \tag{2.3.11}$$

$$\underline{\tau}_{ij} = \mu \, (\partial u_i/\partial x_j + \partial u_j/\partial x_i) \quad (i,j=1,2,3) \tag{2.3.12}$$

where β is the coefficient of compressibility (the inverse of the elasticity modulus) and μ is the coefficient of dynamic viscosity, both being assumed to be constant. Because of its low compressibility, the change of ρ of water due to pressure variations is generally negligible, except for the case of unsteady flow in confined aquifers, which will be treated separately at macroscopic scale in Chap. 2.11. Otherwise, we shall assume that $\underline{\rho}$ is constant.

Substitution of (2.3.8), (2.3.9) and (2.3.10) into (2.3.4), (2.3.5) and (2.3.7), respectively, leads to the equations of heat transport in the fluid and solid phases and of solute transport in the fluid, as follows

$$\underline{\rho}\ c_f \left(\frac{\partial T_f}{\partial t} + \boldsymbol{u}.\nabla T_f\right) = K_{hf}\ \nabla^2 T_f \tag{2.3.13}$$

$$\underline{\rho}_s\ c_s\ \frac{\partial T_s}{\partial t} = K_{hs}\ \nabla^2 T_s \tag{2.3.14}$$

$$\frac{\partial C}{\partial t} + \boldsymbol{u}.\nabla C = D_m\ \nabla^2 C + S \tag{2.3.15}$$

Similarly, substitution of (2.3.12) into the momentum equation (2.3.7) yields the Navier - Stokes equations of viscous flow. It is well known that for sufficiently small Reynolds numbers, with Re = $ud\underline{\rho}/\mu$, the inertial terms of the left-hand side of (2.3.7) are negligible compared to the remaining terms of the right-hand side. We shall assume that this is indeed the case for the flows considered in this book, although retention of the inertial terms may be warranted in some applications, like flow through gravel, through fissures of large openings or for propagation of acoustic waves through porous media. With this important simplification and by making use of the continuity equation (2.3.3), we obtain from (2.3.12) and (2.3.7)

$$\mu\ \nabla^2 \boldsymbol{u} = \nabla p + \underline{\rho}\ g\ \nabla z \tag{2.3.16}$$

known as the equation of slow viscous flow (Batchelor, 1967). Division by $\underline{\rho}$ in (2.3.16) yields the usual form

$$\nabla\left(\int \frac{dp}{\underline{\rho}} + g\ z\right) = \nu\ \nabla^2 \boldsymbol{u} \tag{2.3.17}$$

where ν is the coefficient of kinematical viscosity. Further division by g and for constant $\underline{\rho}$ permits one to rewrite (2.3.17) as follows

$$\nabla\underline{\phi} = \frac{\nu}{g}\ \nabla^2 \boldsymbol{u}\ ;\quad \underline{\phi} = \frac{p}{g\underline{\rho}} + z \tag{2.3.18}$$

The entity $\underline{\phi}$ is known as the piezometric or static head and will be called the head for briefness. Multiplication of Eq. (2.3.16), for a flow field $\boldsymbol{u}'$, by the velocity vector of a different flow field $\boldsymbol{u}''$, allows us to rewrite (2.3.16) as follows

$$\sum_{i=1}^{3}\left(u''_i\ \frac{\partial p'}{\partial x_i} + \underline{\rho}\ g\ u'_i\ \frac{\partial z}{\partial x_i}\right) = \mu \sum_{i=1}^{3}\sum_{k=1}^{3}\left[\frac{\partial}{\partial x_k}\left(u'_i\ \frac{\partial u''_i}{\partial x_k}\right) - \frac{\partial u'_i}{\partial x_k}\ \frac{\partial u''_i}{\partial x_k}\right] \tag{2.3.19}$$

a form which will be shown to be useful. If we take $\boldsymbol{u}'=\boldsymbol{u}''$ and sum up in (2.3.19), which is equivalent to scalar multiplication of (2.3.16) by the velocity vector $\boldsymbol{u}$, we obtain the mechanical energy equation

$$\sum_{i=1}^{3} \frac{\partial}{\partial x_i}(p\, u_i + \underline{\rho}\, g\, z\, u_i) = \mu \sum_{i=1}^{3} \sum_{k=1}^{3} \left[\frac{\partial}{\partial x_k}\left(u_i \frac{\partial u_i}{\partial x_k}\right) - \frac{\partial u_i}{\partial x_k} \frac{\partial u_i}{\partial x_k} \right] \tag{2.3.20}$$

The left-hand side of Eq. (2.3.20) represents the rate of work per unit volume of fluid of the pressure and of the gravity, whereas the right-hand side is the rate of work of the viscous stress.

The last set of relationships to be considered here are the boundary conditions between the fluid and the solid phase. Thus, the heat flux and the temperature obey the continuity conditions

$$\boldsymbol{q}_{hf} . \boldsymbol{\nu} = \boldsymbol{q}_{hs} . \boldsymbol{\nu} \tag{2.3.21}$$

$$T_f = T_s \tag{2.3.22}$$

where $\boldsymbol{\nu}$ is a unit vector normal to the fluid-solid interface. As for the solute flux, we shall generally assume that the solid is impervious to the solute. A more general boundary condition is the one in which adsorption may be present. In such a case there is a net flux of solute from the fluid to the solid, i.e.

$$\boldsymbol{q}_m . \boldsymbol{\nu} = f(C, C_s) \tag{2.3.23}$$

where C_s is defined as the mass of adsorbed solute per unit volume of solid. Obviously, for f=0, we obtain the usual condition of impervious and neutral solid phase.

The connection between the motion of the two phases is embedded in the usual condition of velocity continuity for a viscous fluid

$$\boldsymbol{u}.\boldsymbol{\nu} = \boldsymbol{u}_s . \boldsymbol{\nu} \qquad \text{(at void-solid interface)} \tag{2.3.24}$$

and that of continuity of the total stress, i.e. $\underline{\boldsymbol{\sigma}}_f$ is transmitted to the solid at the interface. In the usual case of a stationary solid matrix, Eq. (2.3.29) becomes

$$\boldsymbol{u} = 0 \qquad \text{(at void-solid interface)} \tag{2.3.25}$$

In the case in which fixed meniscea between water and air are present in the pores, the pertinent relationships at the interface are

$$\boldsymbol{u}.\boldsymbol{\nu} = 0 \quad ; \quad p - p_{air} = 2\sigma / r_c \quad ; \quad \tau_{ij} = 0 \tag{2.3.26}$$

where p_{air} is the pressure in the air (assumed to be constant), σ stands here for the coefficient of surface tension between water and air and r_c is the radius of curvature of the meniscus. Furthermore, the meniscus intersects the solid at a wetting angle which is assumed to be constant.

The preceding set of equations of flow and transport exhaust the processes of interest in this book. They will serve as the starting point for deriving the macroscopic equations in Chap. 2.5.

Additional constitutive relationships of less generality, relating to the deformation of the solid matrix to the stress acting on it, will be discussed in 2.11.

Exercises.

2.3.1 The problem of uniform flow past a body can be solved approximately with the aid of the equation of slow viscous flow (2.3.16), for sufficiently small Reynolds numbers. The drag is then proportional to the body translational velocity and to the dynamic viscosity. Equivalently, the drag coefficient is proportionally to the inverse of the Re number. By analyzing the graphs which relate the drag coefficients of a sphere and a cylinder (e.g. in Batchelor, 1967) to Re, establish the range of Re numbers for which the slow flow approximation is valid.

2.3.2. In the case of a radioactive tracer carried by a fluid, S (2.3.15) is proportional to $\exp(-\lambda t)$, where λ is a decay time related to the "half-life". Derive a transformed C which satisfies an equation similar to (2.3.15), but without S.

2.4 AVERAGING OF DERIVATIVES OF MICROSCOPIC VARIABLES

In the process of deriving macroscopic equations we have to ensemble average various derivatives of the microscopic variables in the conservation and constitutive equations of Chap. 2.3. As a preparatory step we shall derive here the relationships between the average of the derivative of a scalar, vector or tensorial function denoted by w, $\mathbf{w}$ and w_{ij}, respectively, and the derivative of its average. We shall rely here, of course, on the results presented in Chap. 1.5 of the mathematical preliminaries.

The function $w(\mathbf{x},t)$, representing a microscopic variable in the porous medium, is assumed to be continuous in the spatial variables $\mathbf{x}$ and continuous and differentiable in the time t. w is regarded as a random function of $\mathbf{x}$, which depends deterministically on time.

A phase conditional average of w in the fluid, which fills the void space, is defined as follows

$$w_f(\mathbf{x},t) = \frac{\langle w(\mathbf{x},t)\, h(\mathbf{x},t)\rangle}{\langle h(\mathbf{x},t)\rangle} = \frac{\langle wh\rangle}{n} \tag{2.4.1}$$

Conditional averaging over the solid phase yields

$$w_s(\mathbf{x},t) = \frac{\langle w(1-h)\rangle}{1-n} \tag{2.4.2}$$

where it is reminded that n is the porosity (Chap. 2.1). w_f and w_s are called phase averages and they also can be defined in words as density of w per volume of fluid or solid, respectively. The *medium* averages of a function defined in the fluid or solid phase, respectively, are given by

$$w = \langle wh\rangle = n\, w_f \quad ; \quad w = \langle w(1-h)\rangle = (1-n)\, w_s \tag{2.4.3}$$

and they represent densities of w in the fluid or the solid phases, respectively, per total volume of medium. If w is defined everywhere, then

$$w(\mathbf{x},t) = \langle w(\mathbf{x},t)\rangle = n\, w_f + (1-n)\, w_s \tag{2.4.4}$$

is the medium average. Alternatively, the space average definitions of w_f, w_s and w are the usual ones (see Chap. 1.9)

$$\overline{w}_f(\mathbf{x},t) = \frac{1}{V_f}\int_{V_f} w(\mathbf{x}',t)\, d\mathbf{x}' \quad ; \quad \overline{w}_s(\mathbf{x},t) = \frac{1}{V_s}\int_{V_s} w(\mathbf{x}',t)\, d\mathbf{x}'$$

$$\overline{w}(\mathbf{x},t) = \frac{1}{V}\int_{V} w(\mathbf{x}',t)\, d\mathbf{x}' \tag{2.4.5}$$

where $\mathbf{x}$ is the coordinate of the centroid of V. The two definitions of averages, ensemble and spatial, are exchangeable under the conditions discussed in Chap. 2.2. and we shall adhere here to the more general procedure of ensemble averaging.

We proceed now with the averaging of the derivatives of w. For a function defined in the fluid phase we have, by the ready made formula (1.5.12)

$$\langle(\nabla w)\, h\rangle = \nabla\langle w\, h\rangle + \lim_{\Delta V\to 0}\left[\frac{1}{\Delta V}\left\langle\int_{\Delta A} w\, \boldsymbol{\nu}\, dA\right\rangle\right] \tag{2.4.6}$$

where ΔV is an infinitesimal volume surrounding the point of coordinate $\mathbf{x}$, ΔA is the element of area of the void-solid interface within ΔV and $\boldsymbol{\nu}$ is a unit vector normal to ΔA, pointing from the void to the solid (Fig. 1.2d). Eq. (2.4.6) has been obtained by Saffman (1971). If w=0 on the void-solid interface, (2.4.6) becomes

$$\langle (\nabla w)\, h \rangle = \nabla (n\, w_f) \tag{2.4.7}$$

i.e. the phase average of the gradient is equal to the gradient of the average, whereas the last term in (2.4.6), which represents the average jump of w on A, is zero.

If we use a volume averaging procedure we arrive, instead of (2.4.6), at

$$\frac{1}{V} \int_V (\nabla w)\, h\, dV' = \nabla \Big[\frac{1}{V} \int_V w\, h\, dV' \Big] + \frac{1}{V} \int_A w\, \boldsymbol{\nu}\, dA' \tag{2.4.8}$$

where V is a finite volume whose centroid is at $\mathbf{x}$ and the gradient is with respect to $\mathbf{x}$. The relationship has been derived with the aid of Green's theorem by Slattery (1972), for instance. In the light of the discussion of Chap. 2.2 it is emphasized again that unlike (2.4.6), (2.4.8) is valid only for a slowly varying w.

In the case in which the function w is defined in the solid phase, we have in a similar fashion

$$\langle \nabla w\, (1-h) \rangle = \nabla \langle (1-n)\, w_s \rangle - \lim_{\Delta V \to 0} \Big[\frac{1}{\Delta V} \Big\langle \int_{\Delta A} w\, \boldsymbol{\nu}\, dA \Big\rangle \Big] \tag{2.4.9}$$

Finally, for a continuous function, defined everywhere in the porous medium, we have

$$\langle \nabla w \rangle = \nabla w \tag{2.4.10}$$

In the case of vectors of components w_i or tensors w_{ij} (i,j=1,2,3), the same relationships (2.4.7)-(2.4.10) apply to each component. In particular for a vector $\boldsymbol{w}$ and after summation of $\partial w_i/\partial x_i$ in (2.4.7) we get for the divergence operator

$$\langle (\nabla . \boldsymbol{w})\, h \rangle = \nabla . (n\, \mathbf{w}_f) + \lim_{\Delta V \to 0} \Big[\frac{1}{\Delta V} \Big\langle \int_{\Delta A} \boldsymbol{w} . \boldsymbol{\nu}\, dA \Big\rangle \Big] \tag{2.4.11}$$

Similar relationships can be derived for the average of time derivatives. Thus, for a function defined in the fluid domain we may write

$$\Big\langle \frac{\partial w}{\partial t}\, h \Big\rangle = \Big\langle \frac{\partial}{\partial t} (w\, h) \Big\rangle - \Big\langle w\, \frac{\partial h}{\partial t} \Big\rangle \tag{2.4.12}$$

The derivative $\partial h/\partial t$ is zero everywhere, except for the fluid-solid interface A. We may represent it, however by a limit process (see Chap. 1.5) as follows

$$\frac{\partial h}{\partial t} = \lim_{\Delta V \to 0} \left[\frac{1}{\Delta V} \int_{\Delta V} \frac{\partial h}{\partial t} dV\right] = \lim_{\Delta V \to 0} \frac{1}{\Delta V} \left\{ \lim_{\Delta t \to 0} \frac{1}{\Delta t} \int_{\Delta V} [h(t+\Delta t) - h(t)]\, dV \right\} =$$

$$= \lim_{\Delta V \to 0} \left[\frac{1}{\Delta V} \langle \int_{\Delta A} \boldsymbol{u}_A . \boldsymbol{\nu}\, dA \rangle\right] \tag{2.4.13}$$

where $\boldsymbol{u}_A$ is the velocity of the interface A, defined by its displacement along its normal. It results from h(t+t)-h(t) which is different from zero in the strip spanned by the interface during Δt. Hence, we have in (2.4.12)

$$\langle \frac{\partial w}{\partial t} h \rangle = \frac{\partial}{\partial t} (n\, w_f) + \lim_{\Delta V \to 0} \left[\frac{1}{\Delta V} \langle \int_{\Delta A} w\, (\boldsymbol{u}_A . \boldsymbol{\nu})\, dA \rangle\right] \tag{2.4.14}$$

Eq. (2.4.14) is the counterpart of (2.4.7) and it is seen that for a fixed solid matrix, i.e. $\boldsymbol{u}_A = 0$, the averaging and time derivative operators are commutative even for functions defined in one of the phases solely. In a similar manner we have for functions defined in the solid phase or everywhere

$$\langle \frac{\partial w}{\partial t} (1-h) \rangle = \langle \frac{\partial}{\partial t} [(1-n) w_s] \rangle - \lim_{\Delta V \to 0} \left[\frac{1}{\Delta V} \langle \int_{\Delta A} w\, \boldsymbol{u}_A . \boldsymbol{\nu}\, dA \rangle\right]$$

$$\langle \frac{\partial w}{\partial t} \rangle = \frac{\partial w}{\partial t} \tag{2.4.15}$$

respectively.

For purpose of illustration let us take $w=1$ in (2.4.14). Then, $\partial w / \partial t = 0$ and $w_f = 1$. Hence,

$$\frac{\partial n}{\partial t} = - \lim_{\Delta V \to 0} \left[\frac{1}{\Delta V} \langle \int_{\Delta A} \boldsymbol{u}_A . \boldsymbol{\nu}\, dA \rangle\right] \tag{2.4.16}$$

a relationship which expresses the rate of change of pore volume due to the motion of the solid-void interface A.

The various relationships derived in this section will be employed extensively in the sequel.

Exercises

2.4.1 Derive the medium average of the derivative of a function defined everywhere, continuous in each phase, but undergoing a finite jump on the void-solid interface A, in terms of its phase averages. What happens if the medium is statistically homogeneous and the jump is constant. Illustrate for the density of the porous medium for fluid and solid of constant densities.

2.4.2 Derive the equation of the space derivative of a random variable defined in the fluid phase by areal averaging rather than volume averaging in Eq. (2.4.8).

2.4.3 Derive the relationship between the rate of change of porosity and average grain surface velocities for a medium made up from disjoint spheres which (i) undergo pure compression and (ii) translate as rigid bodies with same velocity.

2.5 MACROSCOPIC VARIABLES AND MACROSCOPIC EQUATIONS OF MASS AND ENERGY CONSERVATION

2.5.1 Definition of macroscopic variables

The microscopic equations of Chap. 2.3 were written in terms of various physical variables. As a first step towards averaging these equations we have to define the average, i.e. macroscopic, variables. As a rule variables attached to the fluid or to the solid are defined as phase averages, according to Eqs. (2.4.1, 2.4.2, 2.4.5), whereas medium averages (Eq. 2.4.4) are adopted only for quantities which exist in the two phases. A notable exception is the specific discharge $\mathbf{q}$ which is a fluid variable averaged over the medium (see Eq. 2.5.5 below). Another convenient rule is to average physical variables of an extensive nature, i.e. those which are proportional to the mass, and which are additive. This has the advantage of preserving the structure of the conservation equations when passing from the microscopic to the macroscopic level. In all following definitions the macroscopic flow quantities are generally functions of space coordinate $\mathbf{x}$ or time t, either because of the dependence of the microscopic variables on them, or because of the dependence of the characteristic function h upon $\mathbf{x}$ and t.

Thus, the average fluid, solid and medium specific masses are defined as follows

$$\rho = \frac{\langle \underline{\rho}\,h \rangle}{n} \quad ; \quad \rho_s = \frac{\langle \underline{\rho}_s (1-h) \rangle}{1-n} \quad ; \quad \rho_t = \langle \underline{\rho}\,h + \underline{\rho}_s (1-h) \rangle = n\rho + (1-n)\rho_s \tag{2.5.1}$$

It is reminded again that the space average definitions of these variables are given in Eqs. (2.4.5), e.g.

$$\rho(\mathbf{x},t) = \frac{1}{V_v} \int_{V_v} \underline{\rho}(\mathbf{x}',t)\, d\mathbf{x}' \tag{2.5.2}$$

where $\mathbf{x}$ is the centroid of the averaging volume V. The two definitions are equivalent under the ergodicity conditions discussed in Sect. 2.2.2, which are assumed to prevail.

In a similar manner, the average solute concentration, and the fluid, solid and medium temperatures, are defined by the following equations

$$C = \frac{\langle C\, h\rangle}{n} \quad ; \quad T_f = \frac{\langle \underline{\rho}\, c_f\, T_f\, h\rangle}{\rho c_f n}$$

$$T = \frac{\langle \underline{\rho}\, c_f T_f h\rangle + \langle \underline{\rho}_s c_s T_s (1-h)\rangle}{\rho c_f n + \rho c_s (1-n)} = \frac{\rho c_f T_f + \rho_s c_s (1-n) T_s}{\rho_t c_t} \tag{2.5.3}$$

The average fluid velocity, known also in the literature as pore or filtration velocity, is defined by

$$\mathbf{u} = \frac{\langle \underline{\rho}\, \boldsymbol{u}\, h\rangle}{\rho n} \tag{2.5.4}$$

It is customary to use also, as an alternative, the specific discharge $\mathbf{q}$, which is defined as a medium average

$$\mathbf{q} = \frac{\langle \underline{\rho}\, \boldsymbol{u}\, h\rangle}{\rho} = n\, \mathbf{u} \tag{2.5.5}$$

$\mathbf{q}$ appears in the experimental context in a natural manner when measuring the outflow from a porous laboratory column (see Chap. 2.9) and dividing, subsequently the total mass flux by the average specific mass and by the total area. The corresponding space average definition of $\mathbf{q}$ is, therefore, given by

$$\bar{\mathbf{q}} = \frac{1}{\rho A} \int_{A_v} \underline{\rho}\, \boldsymbol{u}\, dA \tag{2.5.6}$$

as an areal average. As we have mentioned already in Sect. 2.2.2, it is one of the advantages of the ensemble averaging that it does not distinguish between the various types of space averages of Eqs. (2.2.21, 2.2.22).

The macroscopic molecular heat and mass fluxes are defined in a similar manner, i.e.

$$\mathbf{q}_{hf} = \frac{\langle q_{hf}\,h\rangle}{n} \quad ; \quad \mathbf{q}_{hs} = \frac{\langle q_{hs}\,(1-h)\rangle}{1-n}$$

$$\mathbf{q}_{h} = \langle q_{hf}\,h + q_{hs}\,(1-h)\rangle = n\,\mathbf{q}_{hf} + (1-n)\,\mathbf{q}_{hs} \quad ; \quad \mathbf{q}_{m} = \frac{\langle q_{m}\,h\rangle}{n} \tag{2.5.7}$$

while the natural definition by space averaging is the areal one of Eq. (2.5.6).

Finally, we consider the stresses $\underline{\sigma}_{f,ij}$ in the fluid, i.e. the pressure p and the viscous stress $\underline{\tau}_{ij}$. The usual, mechanical, definition of the stress is a ratio between force and area and we can, therefore, ensemble average both p and $\underline{\tau}_{ij}$ as follows

$$\sigma_{f,ij} = \frac{\langle \underline{\sigma}_{f,ij}\,h\rangle}{n} \quad ; \quad p = \frac{\langle ph\rangle}{n} \quad ; \quad \tau_{ij} = \frac{\langle \underline{\tau}_{ij}\,h\rangle}{n} \qquad (i,j=1,2,3) \tag{2.5.8}$$

It is seen that the macroscopic force exerted by the fluid on a planar areal element ΔA, of unit normal $\boldsymbol{\nu}$, is given by

$$\Delta F_i = \langle \Delta F_i\rangle = \langle \sum_{j=1}^{3} [(-p\,\delta_{ij} + \underline{\tau}_{ij})\nu_j\,h]\rangle \Delta A = [\sum_{j=1}^{3} (-p\,\delta_{ij} + \tau_{ij})\nu_j]\,n\,\Delta A \qquad (i = 1,2,3) \tag{2.5.9}$$

In this book we are generally concerned with transport through porous media with fixed, rigid, solid matrices. The case of a deformable matrix is nevertheless of considerable interest in relation with elastic storage (Chap. 2.6). Then, we shall make use of the average of the solid phase velocity, which is defined precisely like **u** in (2.5.4), i.e.

$$\mathbf{u}_s = \frac{\langle \underline{\rho}_s\,u_s\,(1-h)\rangle}{\rho_s\,(1-n)} \tag{2.5.10}$$

The definition of the macroscopic stress in the solid matrix is a more complex matter. In a granular matrix the stress is transmitted as an elastic stress throughout the grains and as contact, intergranular, forces at the small areas of contact between grains. The deformation of the matrix stems mainly from the sliding and rearrangement of the solid particles, rather than from their elastic deformation. It is the intergranular stress, therefore, that plays the dominant role in the deformation of the solid matrix, whereas the elastic deformation of the solid material is generally negligible. This concept can be traced back to Terzaghi (1925) who studied the phenomenon of consolidation, of wide interest in soil mechanics. A detailed analysis of the complex relationships between intergranular forces and the matrix deformation is beyond the scope of the present book. Unlike the case of fluid flow and of transport, in which the macroscopic equations are derived from well established microscopic relationships, we shall adopt phenomenological, macroscopic equations, from the outset, for the deformation of the solid matrix. At present we define the "effective stress", a concept established by Terzaghi (1925), by the following equation

$$\sum_{j=1}^{3} \frac{\partial \sigma'_{ij}}{\partial x_j} = \lim_{\Delta V \to 0} \frac{1}{\Delta V} \int_{\Delta A} \langle \underline{\sigma}_{ij} \nu_j \rangle \, dA \qquad (i=1,2,3) \tag{2.5.11}$$

where $\boldsymbol{\nu}$ is a unit vector normal to the intergranular contact area ΔA. In Eq. (2.5.11) the tensor $\underline{\sigma}_{ij}$ is the intergranular force per unit area acting on the surface of contact A between neighboring solid particles (Fig. 2.6.2) and σ'_{ij} is Terzaghi's effective stress. Eq. (2.5.11) defines it as the macroscopic stress equivalent to the average intergranular force. It is easy to visualize the vector $f_i = \Sigma \underline{\sigma}_{ij} \nu_j$ as the contact force between spherical particles in touch, but the concept loses this simple interpretation in the case of porous media of more complex structures in which the intergranular areas are not well defined. Still, we shall rely on this concept when analyzing the deformation of the solid matrix of porous media.

2.5.2 The macroscopic equations of state and of mass conservation

The macroscopic equation of state for the fluid is obtained from its microscopic version (2.3.1) by multiplication of the latter by h and subsequent ensemble averaging. With the definitions of Eqs. (2.5.1)-(2.5.3) we have immediately

$$\rho(C, T_f) = \rho_0 + (C - C_0) \frac{\partial \rho_0}{\partial C_0} + (T_f - T_{f0}) \frac{\partial \rho_0}{\partial T_{f0}} \quad ; \quad \rho_0 = \underline{\rho}(C_0, T_{f0}) \tag{2.5.12}$$

where $\partial \rho_0 / \partial c_0$ and $\partial \rho_0 / \partial T_{f0}$ are constant. It is seen that due to the linearity of (2.3.1), Eq. (2.5.12) is identical with (2.3.1). It is emphasized that even in the case of large temperature or concentration changes, which require using nonlinear equations of state, the result is essentially the same due to the small fluctuations of $\underline{\rho}$ within pores.

We shall derive now the macroscopic equation of mass conservation of the fluid by multiplying the microscopic equation (2.3.2) by h and ensemble averaging, i.e.

$$\langle h \frac{\partial \underline{\rho}}{\partial t} \rangle + \langle h \, \nabla.(\underline{\rho} \, \boldsymbol{u}) \rangle = 0 \qquad (2.5.13)$$

Since both time and space derivatives appear in (2.5.13), we have to make use of the averaging rules (2.4.14) and (2.4.11), respectively. By the boundary condition (2.3.24) the fluid, solid and interface velocities coincide at the void-solid interface. Consequently, after substituting $\boldsymbol{w} = \underline{\rho} \, \boldsymbol{u}$ in (2.4.11) and $w = \underline{\rho}$ in (2.4.14), it is seen that the contributions from the interfacial motion terms cancel out. Hence, with definition (2.5.4) of the macroscopic velocity, we obtain from (2.5.13)

$$\frac{\partial}{\partial t}(n \, \rho) + \nabla.(n \, \rho \, \mathbf{u}) = 0 \quad \text{i.e.} \quad \frac{\partial}{\partial t}(n \, \rho) + \nabla.(\rho \mathbf{q}) = 0 \qquad (2.5.14)$$

A few important particular cases can be obtained from the basic equation of conservation of fluid mass (2.5.14). Thus, if the matrix is not deformable, i.e. $\partial n/\partial t=0$, and homogeneous, i.e. $\nabla n=0$, we have

$$\frac{\partial}{\partial t}(n \, \rho) + \nabla.(\rho \mathbf{q}) = 0 \quad \text{and} \quad \frac{\partial \rho}{\partial t} + \nabla.(\rho \mathbf{u}) = 0 \qquad (2.5.15)$$

respectively.

Furthermore, if the fluid compressibility is negligible we arrive in (2.5.15) at the continuity equation

$$\nabla.(n \, \mathbf{u}) = 0 \quad \text{i.e.} \quad \nabla.\mathbf{q} = 0 \qquad (2.5.16)$$

By a similar reasoning one can derive the equation of conservation of mass of the solid matrix by the corresponding phase averaging (2.4.2). The result is

$$\frac{\partial}{\partial t}[(1-n)\rho_s] + \nabla.[(1-n)\rho_s \mathbf{u}_s] = 0 \qquad (2.5.17)$$

after making use of the definition of $\mathbf{u}_s$ (2.5.10). The compressibility of the solid phase is generally negligible, so that (2.5.17) reduces to the continuity equation of the solid volume

$$\frac{\partial n}{\partial t} - \nabla.[(1-n)\mathbf{u}_s] = 0 \qquad (2.5.18)$$

2.5.3 The macroscopic equation of conservation of energy

We proceed now with ensemble averaging of the microscopic equations (2.3.4) and (2.3.5) of conservation of energy in the fluid and solid phases, respectively. To simplify matters we shall assume that ρ and ρ_s do not depend on time whereas c_f and c_s are constant. We shall also neglect the influence of the matrix deformation upon heat transport, i.e. the fluid-solid interfaces are fixed. After multiplying Eqs. (2.3.4) by h and by ensemble averaging, with the definitions (2.5.3) and (2.5.7) taken into account, we obtain

$$\rho c_f n \frac{\partial T_f}{\partial t} = - \nabla.(n\, \mathbf{q}_{hft}) + \lim_{\Delta V \to 0} \frac{1}{\Delta V} \int_{\Delta A} \langle q_{hf} . \nu \rangle \, dA \tag{2.5.19}$$

with the total macroscopic heat flux in the fluid given by

$$\mathbf{q}_{fht} = \mathbf{q}_{hf} + \frac{1}{n} \langle \underline{\rho}\; c_f\; \boldsymbol{u}\; T_f\, h \rangle \tag{2.5.20}$$

In deriving (2.5.19) we have used the averaging rules of derivatives (2.4.11) and (2.4.14), taking into account, however, that the fluid velocity as well as the interface velocity are equal to zero, so that the only interfacial transfer term stems from the molecular heat conductive flux q_{hf}. The three terms of the macroscopic equation (2.5.19) represent heat storage, conduction-convection by the fluid motion and transfer by conduction to the solid phase, respectively. In the right-hand side of (2.5.20) the first term represents conduction and the last one convection by the fluid.

Multiplying the heat conduction equation in the solid (2.3.5) by 1-h and subsequent averaging, with (2.5.3), (2.5.7), (2.4.11) and (2.4.14) taken into account, yield

$$(1-n)\, \rho_s c_s \frac{\partial T_s}{\partial t} = - \nabla.[(1-n)\, \mathbf{q}_{hs}] - \lim_{\Delta V \to 0} \frac{1}{\Delta V} \int_{\Delta A} \langle q_{hs} . \nu \rangle \, dA \tag{2.5.21}$$

It is reminded that by boundary condition (2.3.21) the interfacial heat conduction in the two phases are equal and the same is true, therefore, for the last terms of Eqs. (2.5.19) and (2.5.21). An useful macroscopic equation is obtained, therefore, by summing up (2.5.19) and (2.5.21) and by using the definitions (2.5.3) and (2.5.7) of the macroscopic medium temperature T and total conductive flux $\mathbf{q}_h$, respectively. The result is

$$[\rho c_f n + \rho_s c_s (1-n)] \frac{\partial T}{\partial t} = - \nabla.\mathbf{q}_h - c_f \nabla.\langle \underline{\rho}\; T_f\; \boldsymbol{u}\; h \rangle \tag{2.5.22}$$

Eq. (2.5.22) is particularly useful in the case of a fluid in rest, when it becomes

$$\rho_t c_t \frac{\partial T}{\partial t} = - \nabla . \mathbf{q}_h$$
$$\rho_t = \rho n + \rho_s (1-n) \quad ; \quad \rho_t c_t = \rho c_f n + \rho_s c_s (1-n) \tag{2.5.23}$$

which is formally identical to the equation of heat conduction in a homogeneous material, provided that the heat capacity is replaced by the medium heat capacity $\rho_t c_t$. The relationship between the macroscopic conductive heat flux $\mathbf{q}_h$ and the temperature gradient ∇T will be discussed in Chap. 2.7.

2.5.4 The macroscopic equation of solute mass conservation

Following a similar procedure, we multiply Eqs. (2.3.6) by h and ensemble average, with the definitions of the macroscopic variables (2.5.3) and (2.5.7) taken into account. The interfacial transfer is by molecular diffusion through the fluid-solid interface. Hence,

$$\frac{\partial (nC)}{\partial t} = - \nabla . (n \, \mathbf{q}_m) - \nabla . \langle C \, \boldsymbol{u} \, h \rangle + \lim_{\Delta V \to 0} \frac{1}{\Delta V} \int_{\Delta A} \langle q_m . \boldsymbol{\nu} \rangle \, dA \tag{2.5.24}$$

where the four terms of (2.5.24) represent solute accumulation, molecular transfer, convective transport and adsorption, respectively. For the latter, we may use the equation of conservation of absorbed solute to replace the last term of (2.5.24) by

$$\lim_{\Delta V \to 0} \frac{1}{\Delta V} \int_{\Delta A} \langle q_m . \boldsymbol{\nu} \rangle \, dA = - (1-n) \frac{\partial C_s}{\partial t} \tag{2.5.25}$$

where C_s is the average adsorbed mass of solute, defined as mass of solute per unit volume of solid.

By using the equation of solid matrix continuity (2.5.14) and (2.5.25), the relationship (2.5.24) becomes

$$n \frac{\partial C}{\partial t} + \mathbf{u} . \nabla C = - \nabla . (n \, \mathbf{q}_m) - (1-n) \frac{\partial C_s}{\partial t} \tag{2.5.26}$$

Again, the simplest version of (2.5.26) is the one in absence of fluid motion, of distributed sources and of adsorption, i.e.

$$\frac{\partial C}{\partial t} = - \nabla.\mathbf{q}_m \tag{2.5.27}$$

Formally, Eq. (2.5.27) resembles the diffusion equation in a homogeneous material, but the relationship between the macroscopic molecular diffusive flux $\mathbf{q}_m$ and the gradient of the macroscopic concentration ∇C has to be established first, to warrant the analogy, and this is the object of Chap.2.8.

Exercises

2.5.1 Rewrite Eqs. (2.5.14) and (2.5.17) of conservation of fluid and solid mass, respectively, in terms of macroscopic material derivatives, when both fluid and solid matrix are in motion. The material (hydrodynamic) derivatives are defined as derivatives with respect to time in a frame moving with the average phase velocity.

Define the macroscopic velocity $\mathbf{u}_t$ of the fluid-solid mixture by writing down the equation of conservation of mass of the medium.

2.5.2 Consider a porous medium in which the solid phase does not have a bulk motion, but the solid material undergoes a phase transition by melting. Derive mass conservation equations for the two phases in this case. Assume that the rate of melting per unit area of solid-fluid interface is constant and given.

2.5.3 If one defines the macroscopic velocity by averaging the fluid microscopic velocity rather than the mass flux in (2.5.4), an additional term appears in the equation of mass conservation (2.5.14). Derive this term and interpret it.

2.5.4. Derive the equations of mass conservation for a flow of two immiscible fluids, e.g. water and oil, through a porous medium of rigid solid matrix.

2.6 THE MACROSCOPIC EQUATIONS OF CONSERVATION OF MOMENTUM

We proceed now with the ensemble averaging of the equations of momentum conservation in the fluid (2.3.7). In line with the assumption of small Reynolds number discussed in Sect. 2.3, we neglect the inertial terms making up the left-hand side of Eq. (2.3.7). As a matter of fact (2.3.7) becomes an equation of equilibrium between the total fluid stress $\underline{\sigma}_{f,ij}$ and the gravity. Phase averaging of the stress derivatives (see Eq. 2.4.7) in the fluid and employing the definition (2.5.8) of the macroscopic stress yields in (2.3.7)

$$\sum_{j=1}^{3}\left[\frac{\partial}{\partial x_j}(\sigma_{f,ij}\; n) - \lim_{\Delta V\to 0}\frac{1}{\Delta V}\int_{\Delta A}\langle\underline{\sigma}_{f,ij}\;\nu_j\rangle\, dA\right] = \rho\, g\, n\,\frac{\partial z}{\partial x_i} \qquad (i=1,2,3) \qquad (2.6.1)$$

where z is a vertical coordinate pointing upwards and x_i (i=1,2,3) are Cartesian coordinates. Eq. (2.6.1) can be rewritten in terms of the average pressure p and the average viscous stress τ_{ij} as follows

$$-\nabla(n\, p) - \rho\, g\, n\, \nabla z + \nabla.(n\, \boldsymbol{\tau}) = \mathbf{F} \qquad (2.6.2)$$

where the interfacial term of the integral in (2.6.1), representing the average force exerted by the fluid on the solid surface, per unit volume of medium, has been denoted by **F**. In the absence of motion, the viscous stress vanishes and the pressure is hydrostatic, i.e.

$$\nabla p + \rho\, g\, \nabla z = 0 \qquad (2.6.3)$$

Consequently, in Eq. (2.6.2) the vector

$$\mathbf{D} = \mathbf{F} + p\nabla n = \lim_{\Delta V\to 0}\frac{1}{\Delta V}\int_{\Delta A}\sum_{j=1}^{3}\langle\underline{\sigma}_{f,ij}\;\nu_j\rangle\, dA + p\nabla n \qquad (2.6.4)$$

vanishes for rest. Hence, **D** is the average drag per unit volume of medium. It is emphasized that **F** in (2.6.1, 2.6.2) results from both contributions of the viscous stress and of the isotropic pressure on the void-solid interface ΔA. The term $-p\nabla n$ in (2.6.2) is a static contribution which has to be subtracted from **F** in order to obtain the drag **D**. Obviously, for constant n, **F** and **D** coincide.

Eq. (2.6.1) can be obtained by space averaging as well by writing down the forces acting upon the fluid in an elementary macroscopic cube (Fig. 2.6.1): the derivatives of the average stress result from the balance of stresses acting on the pore area on the sides of the cube, the drag originates from the solid-fluid interface within the cube and the remaining term results from the weight of the fluid in the void domain.

Eq. (2.6.2) can be written in various useful particular forms. First, as we shall show in Sect. 2.9, the average viscous stress τ_{ij} is negligible for slowly varying velocity field **u** in space. Thus, this term can be dropped from the balance equation. Furthermore, if the fluid is homogeneous and compressible, but with the density a function of pressure solely, we have

$$-\rho\, n\, \nabla\left(\int\frac{dp}{\rho} + gz\right) = \mathbf{D} \qquad (2.6.5)$$

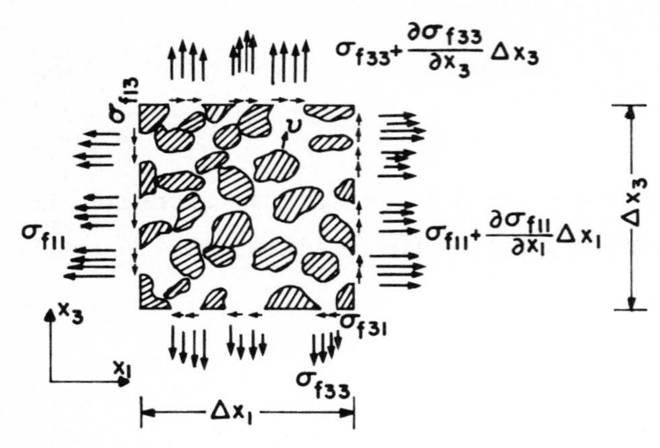

Fig. 2.6.1 An elementary macroscopic cube of dimensions Δx_1, Δx_2 and Δx_3 and the stresses acting on the fluid phase on its sides. The interior forces acting on the fluid are the interfacial stresses and the weight.

Finally, the simplest and most common form of (2.6.5) is for a homogeneous and incompressible fluid, i.e. ρ=const,

$$\nabla\phi = -\frac{\mathbf{D}}{\rho g n} \qquad \text{with } \phi = \frac{p}{\rho g} + z \tag{2.6.6}$$

where ϕ is the macroscopic fluid head (known also as piezometric or static head) and the right-hand side of (2.6.6) represents the drag per unit weight of fluid.

We proceed now with the derivation of the equation of motion of the solid phase. After neglecting the inertial terms, which is justified for the small and slow deformations of the solid matrix considered here, we arrive at a microscopic equation similar to (2.3.7). Phase averaging of such an equation is not useful as it introduces new variables, the average stress in the solid phase. This stress can be related to the deformation of the solid matrix if, for instance, it is assumed that the latter is elastic. We have pointed out, however, in Sect. 2.5.1 that for a granular material the deformation of the solid particles is negligible in comparison with the volumic deformation resulting from the relative motion of the particles. Thus, the solid particles can be considered to be rigid and the stress within the solid is immaterial. In contrast, the deformation of the matrix is dependent on Terzaghi's effective stress σ'_{ij}, defined by Eq. (2.5.11). To arrive at a macroscopic relationship between the effective stress and the other forces acting on the solid, we consider an elementary macroscopic volume whose envelope crosses the solid matrix through the small contact areas between particles (Fig. 2.6.2). Then, the stress acting on the envelope is $\underline{\sigma}_{f,ij}$ on the entire area (with neglect of the contact area) and σ'_{ij}, resulting from the intergranular stress. The equation of equilibrium of the medium can be written in terms of macroscopic stresses as follows

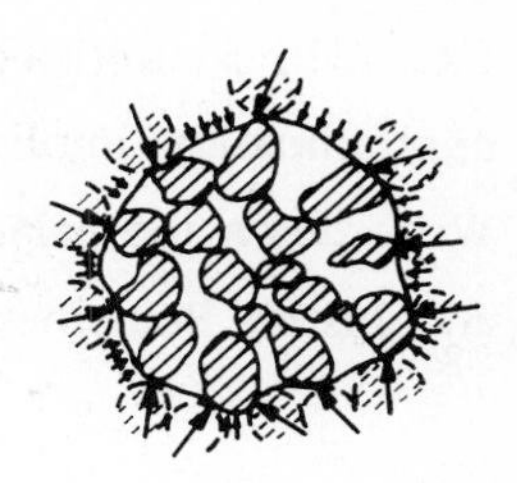

Fig. 2.6.2 An elementary macroscopic volume whose envelope crosses the solid matrix at the contact between grains. The stresses acting on the envelope are $\boldsymbol{\sigma}_f$ and the intergranular stresses. The interior forces are the weight of the medium.

$$\sum_{j=1}^{3} \left(\frac{\partial \sigma_{f,ij}}{\partial x_j} + \frac{\partial \sigma'_{ij}}{\partial x_j} \right) = [\rho\, n + \rho_s (1-n)]\, g\, \frac{\partial z}{\partial x_i} \tag{2.6.7}$$

Subtracting the equation of stress change in the fluid (2.6.1) from (2.6.7), we obtain the following equations of balance of forces acting on the solid phase

$$\sum_{j=1}^{3} [(1-n)\, \frac{\partial \sigma_{f,ij}}{\partial x_j} + \frac{\partial \sigma'_{ij}}{\partial x_j} - \sigma_{f,ij}\, \frac{\partial n}{\partial x_j}] = \rho_s (1-n)\, g\, \frac{\partial z}{\partial x_i} - F_i \quad (i=1,2,3) \tag{2.6.8}$$

which form the basis of the theory of consolidation. For a homogeneous medium, i.e. $\nabla n=0$, and with neglect of the gradient of τ_{ij}, we arrive at the usual form of the equation of equilibrium of the solid skeleton

$$\nabla \boldsymbol{\sigma}' - (1-n)\, \nabla p - \rho_s\, g\, (1-n)\, \nabla z + \mathbf{D} = 0 \tag{2.6.9}$$

This section concludes the derivation of the macroscopic equations of conservation. The following sections deal with the more complex problem of averaging of constitutive equations.

Exercises.

2.6.1 Generalize Eqs. (2.6.2) and (2.6.9) for the case in which the local term of the accelerations, i.e. $\partial \boldsymbol{u}/\partial t$ and $\partial \boldsymbol{u}_s/\partial t$, for the fluid and solid, respectively, are no more negligible. This equations have applications to problems of vibrations of porous media. Take into account the motion of the solid phase represented by $\mathbf{u}_s$.

2.6.2 Establish the relationships between the average stress in the solid phase and the effective stress σ'_{ij}

2.6.3 In a stratified fluid in which ρ changes with z, but not with the pressure p, Boussinesq approximation implies neglecting the variation of $\underline{\rho}$, except for the gravitational term in (2.6.2). Write down the balance equation (2.6.2) in this case with the aid of an appropriate generalized head ϕ_g. Justify Boussinesq approximation for a slow variation of $\underline{\rho}$ with z.

2.7 THE CONSTITUTIVE EQUATION OF HEAT TRANSFER (EFFECTIVE HEAT CONDUCTIVITY)

2.7.1 Definitions and experimental evidence

The constitutive microscopic equations of heat transfer (2.3.8), (2.3.9) relate the molecular heat fluxes q_{hf} and q_{hs} to the temperature gradients. By analogy, one may expect that a similar relationship exists between the average flux and temperature gradient, i.e.

$$\mathbf{q}_h = -K_h \nabla T \tag{2.7.1}$$

where $\mathbf{q}_h$ (2.5.7) is the average heat flux in the medium, T (2.5.3) is the average temperature of the medium and K_h is the effective heat conductivity of the saturated porous medium. For anisotropic media K_h is generally a tensor $\mathbf{K}_h$, whereas in the isotropic case it is a scalar.

The laboratory determination of K_h is usually carried out by applying a temperature difference on the boundary of a sample and measuring the heat discharge. Generally, the setup is such that for a homogeneous medium the heat flow would have been one-dimensional, e.g. by using a cylindrical sample insulated on its side and with uniform temperatures T_1 and T_2 applied on the two planar ends, respectively. With A the cross-sectional area, L the length and Q_h the heat discharge, the experimental definition of K_h is as follows

$$K_h = -\frac{q_h}{dT/dx} \quad ; \quad q_h = \frac{Q_h}{A} \quad ; \quad \frac{dT}{dx} = \frac{T_2-T_1}{L} \tag{2.7.2}$$

The basic assumption is that K_h thus measured is a property of the saturated medium solely and is equal to the statistical definition of Eq. (2.7.1). This assumption can be justified on dimensional grounds as follows. By the linearity of the microscopic constitutive equations (2.3.8) and (2.3.9), the flux q_h is a linear function of the gradient dT/dx. Consequently, one can write

$$q_h = -K_h(K_{hf}, K_{hs}, n, d, R, L)\frac{dT}{dx} \tag{2.7.3}$$

where R is the radius of the cylinder, K_{hf} and K_{hs} are the heat conductivities of the fluid and solid phases, respectively (Sect. 2.3), n is the porosity and d is the pore-scale (say, the diameter of uniform spherical particles). For the sake of simplicity we consider an isotropic medium characterized completely by d and n, but the argument will be the same for a medium whose structure is characterized by a few length scales related to the medium structure. Now, by dimensional analysis we can write

$$K_h = K_{hf} \, f(\frac{K_{hs}}{K_{hf}}, n, \frac{d}{R}, \frac{d}{L}) \tag{2.7.4}$$

where f is an unknown function. Next, the assumption is that d/L and d/R are very small compared to unity and K_h is insensitive to their magnitude, which can be taken equal to zero in (2.7.4). Then, for the given medium or media of same statistical structures, f depends only on K_{hs}/K_{hf} and n, i.e. it is a medium property solely. Furthermore, it is generally assumed that the same K_h relates the macroscopic flux and gradient for a nonuniform macroscopic flow, provided that T is slowly varying in the space (a detailed discussion of these assumptions is presented in section 2.9.3 for a related problem).

A large body of experimental data pertaining to granular materials, similar to unconsolidated natural media, have been collected from the literature by Crane and Vachon (1977). A few of these data, for materials for which $K_{hs}/K_{hf} < 7$, are reproduced in Table 2.7.1.

Data for natural formations are less systematic. As an example, Molz et al (1978) have measured the effective heat conductivity of samples from an aquifer which consists of medium to fine sand with interspersed clay and silt comprising about 15% by weight, and of porosity n=0.25. The effective conductivity was found to be $K_h=22.9\text{x}10^{-3}$ joule/cm.sec.degC, i.e. $K_h/K_w = 3.88$, where K_w is the water heat conductivity. The corresponding value for clay layers confining the formation was not too different, namely $K_h=25.6\text{x}10^{-3}$ j/cm secdegC, i.e. $K_h/K_w = 4.34$.

2.7.2 Theoretical derivation of the constitutive equation and of bounds of the effective conductivity

The problem of effective conductivity of a heterogeneous medium is encountered in various branches of physics and engineering and has been the object of numerous studies. A systematic and detailed presentation may be found in Beran (1968), whereas a brief review for two-phase materials is given by Batchelor (1974). It is beyond the scope of this book to carry out an extended survey of the subject and only a few aspects, considered to be mostly relevant, are discussed here. The difficulty of the problem may be understood by inspecting Fig. 2.2.1 which illustrates the complex structure of a porous medium.

TABLE 2.7.1 Experimental and theoretical (last column, Eq. 2.7.26) values of effective heat conductivity of saturated granular materials. The experimental data are reproduced from Crane & Vachon (1977). K_h is in units of (joule/sec.cm.degC) x10^3.

Fluid	K_{hf}	Solid	K_{hs}	K_{hs}/K_{hf}	n	K_h	K_h/K_{hm}
Hydrogen	1.73	$(C_6H_5)_2NH$	2.20	1.30	0.513	1.92	0.975
Water	6.33	Glass	8.50	1.72	0.408	8.50	0.965
Water	6.03	Glass	8.33	1.81	0.422	8.33	0.980
Water	6.00	Glass	8.30	1.81	0.420	8.30	0.972
Glycerin	5.38	Glass	8.52	2.03	0.428	8.52	1.05
Glycerol	5.28	Glass	8.30	2.06	0.420	8.30	1.02
Hydrogen	1.93	Coal	2.94	2.20	0.437	2.94	0.994
Et 0H	3.44	Glass	6.20	3.16	0.420	6.20	0.918
Et 0H	3.44	Glass	6.40	3.16	0.420	6.40	0.947
Et 0H	3.41	Glass	6.44	3.20	0.423	6.44	0.960
Oil	1.79	Glass	3.46	3.97	0.580	3.47	1.08
Water	5.91	Calcite	36.04	6.10	0.447	13.72	0.892

The theoretical problem of determining the effective conductivity of a saturated porous medium, or of any two-phase material, can be stated as follows. Consider a porous body and a temperature distribution $\mathrm{T} = -\mathbf{J}.\mathbf{x} + \text{const}$ applied on its boundary, with $\mathbf{J}$ a constant vector. In the case of a homogeneous material the ensuing temperature field is T everywhere, $-\mathbf{J}$ being a constant and uniform gradient. The actual porous material is statistically homogeneous and is completely characterized by the basic function $h(\mathbf{x})$ (Chap. 2.2) or by its various moments, and in particular by its expected value n, the porosity. With the microscopic temperature T satisfying the steady heat conduction equations in each phase and the interfacial boundary conditions (2.3.21, 2.3.22), the problem is to determine the effective heat conductivity defined by Eq. (2.7.1). It is reminded that $\mathbf{q}_{hm}$ and T are the ensemble averages of the heat flux and temperature, respectively, as defined in Chap. 2.5. The dimension of the body is much larger than the correlation scale d which characterizes the medium, such that the conditions under which the ergodic hypothesis holds (Chap. 2.2) are fulfilled. Then, ensemble and space averages can be interchanged and the macroscopic temperature gradient and heat flux can be defined by both averages as follows

$$\nabla \mathrm{T} = \langle \nabla T \rangle = \frac{1}{V} \int \nabla T \, dV \quad ; \quad \mathbf{q}_h = \langle q_h \rangle = \frac{1}{V} \int q_h \, dV \qquad (2.7.5)$$

where V is the volume of the body. For brevity, in (2.7.5) T and q_h stand for temperature and flux in the solid or fluid phase either. By applying Green's formula to the continuous function T in (2.7.5), it is immediately seen that

$$\nabla T = \frac{1}{V}\int T\,\boldsymbol{\nu}\,dA = -\frac{1}{V}\int(\mathbf{J}.\mathbf{x})\boldsymbol{\nu}\,dA = -\mathbf{J} \qquad (2.7.6)$$

where A is the boundary of the body, $\boldsymbol{\nu}$ is a normal unit vector pointing outwards and $\mathbf{x}$ is the coordinate vector. Hence, the average gradient is constant and equal to the applied gradient on the boundary.

The microscopic heat flux q_h is of a continuous normal component at the fluid-solid interface and in steady state it satisfies the continuity equation (2.3.4, 2.3.5)

$$\nabla.q_h = 0 \qquad (2.7.7)$$

We consider now, for the sake of simplicity, the configuration usually employed in the laboratory, namely a cylinder of cross-section A and of length L. The cylinder is insulated on its lateral surface and its axis is parallel to the applied gradient $\mathbf{J}$. By (2.7.7) the heat flux through any cross-section is constant, i.e. $\int q_h .(\mathrm{J}/\mathrm{J})\, dA = Q_h = \text{const}$. Hence, by integration over the entire volume we get

$$\mathbf{q}_h = \frac{1}{V}\int_0^L ds\int_A q_h .(\mathrm{J}/\mathrm{J})\,dA = Q_h/A \qquad (2.7.8)$$

Hence, the space average and the usual experimental definition of the flux are equivalent. The next step is to let V expand to infinity, to ensure that $Q_h=\langle Q_h\rangle$, i.e Q_h is ergodic. Under these conditions, the temperature and heat flux are decomposed as follows

$$T = -\mathbf{J}.\mathbf{x} + T' \quad ; \quad q_h = \mathbf{q}_h + q_h' \qquad (2.7.9)$$

with $\mathbf{J}$ and $\mathbf{q}_h$ constant and with T' and q_h' stationary random functions of $\mathbf{x}$. The problem of determining $\mathbf{q}_h$ for given moments of h($\mathbf{x}$) and for given K_{hf}, K_{hs} and $\mathbf{J}$, at the heart of the theory of heterogeneous materials, has not yet been solved. Furthermore, this problem is not even practical, since generally we do not possess the complete information about h. The most powerful, and relatively simple, results have been obtained for bounds of K_h, when the only available information is the porosity n. Towards the derivation of these bounds we shall show that K_h can be also defined with the aid of the "energy" flux defined by

$$E_h = -\langle q_h . \nabla T \rangle \tag{2.7.10}$$

Substituting now (2.7.9) into (2.7.10) and using the continuity equation (2.7.7) we arrive at

$$E_h = \mathbf{q}_h . \mathbf{J} - \nabla \langle q_h' T' \rangle \tag{2.7.11}$$

However, since the fluctuation fields are statistically homogeneous, the correlation $\langle q_h ' T' \rangle$ is constant. Hence, (2.7.11) yields

$$E_h = \mathbf{q}_h . \mathbf{J} = \mathbf{J} . \mathbf{K}_h . \mathbf{J}^T \tag{2.7.12}$$

where $\mathbf{J}^T$ is the transposed of $\mathbf{J}$. The last equality in (2.7.12) expresses in a general way the linear dependence of $\mathbf{q}_h$ upon $\mathbf{J}$ and may serve as a convenient definition of the effective conductivity tensor $\mathbf{K}_h$ in terms of E_h. Furthermore, by substituting the microscopic relationships between heat flux and temperature gradient in (2.7.10) we obtain

$$E_h = \langle \nabla T . \mathbf{K}_h . \nabla T^T \rangle = K_{hf} \langle \nabla T_f . \nabla T_f \ h \rangle + K_{hs} \langle \nabla T_s . \nabla T_s \ (1-h) \rangle \tag{2.7.13}$$

where we have written explicitly the contribution of the energy flux in the two phases, the fluid and the solid. It is seen that E_h is positive definite and the matrix $\mathbf{K}_h$ is, therefore, symmetrical and positive definite. Hence, $\mathbf{K}_h$ can be reduced to principal orthogonal axes and its principal values K_{hI}, K_{hII} and K_{hIII} are positive. Furthermore, $\mathbf{K}_h$ is invertible and we may also write

$$\mathbf{J} = \mathbf{K}_h^{-1} . \mathbf{q}_h \quad ; \quad E_h = \mathbf{q}_h . \mathbf{K}_h^{-1} . \mathbf{q}_h \tag{2.7.14}$$

where $\mathbf{K}_h^{-1}$ can be defined as the effective heat resistivity tensor.

After these preparatory steps, we shall derive the classical bounds for the effective conductivity by following the presentation of Batchelor (1974). Towards this aim we shall rewrite the energy flux

$$E_h = \mathbf{J} . \mathbf{q}_h = \mathbf{J} . \langle K_h \mathbf{J} - K_h \nabla T' \rangle = \langle K_h \rangle \mathbf{J} . \mathbf{J} - \mathbf{J} . \langle K_h \nabla T' \rangle \tag{2.7.15}$$

where, again, K_h stands for the conductivity of the fluid or solid. By the flux-gradient relationship we have $q_h = -K_h \mathbf{J} + K_h \nabla T'$, which leads after substitution in the last term of (2.7.15), to

$$\langle K_h \mathbf{J} . \nabla T' \rangle = \langle q_h \nabla T' \rangle + \langle K_h \nabla T' . \nabla T' \rangle \tag{2.7.16}$$

However, by using the continuity equation (2.7.7) we can also write that $\langle q_h . \nabla T' \rangle = \nabla . \langle q_h \; T' \rangle = 0$, since for a statistically homogeneous field the correlation $\langle q_h \, T' \rangle$ is constant. Hence, the energy flux (2.7.15) becomes

$$E_h = \langle K_h \rangle \, \mathbf{J}.\mathbf{J} - \langle K_h \; \nabla T'.\nabla T' \rangle \qquad (2.7.17)$$

The last term in this fundamental equality is positive and the following inequality, therefore, holds

$$E_h \leq \langle K_h \rangle \, \mathbf{J}.\mathbf{J} \qquad (2.7.18)$$

and comparison of (2.7.18) with (2.7.12) leads, for $\mathbf{J}$ in the principal directions of $\mathbf{K}_h$, to the inequalities

$$K_{hI} \; , K_{hII} \; , K_{hIII} \; \leq \; \langle K_h \rangle = K_A \qquad (2.7.19)$$

This result can be stated in words as follows: the arithmetic mean of the heat conductivity is an upper bound to the effective conductivity. This inequality has been obtained here in a general manner and it is valid for a multiphase medium or even for a continuous variation of K_h in space. In the case of a saturated porous medium we have for K_A

$$K_A = \langle K_{hf} \; h + K_{hs} (1-h) \rangle = K_{hf} \; n + K_{hs} \; (1-n) \qquad (2.7.20)$$

and indeed the upper bound is expressed in terms of porosity and conductivities of the fluid and of the solid, respectively.

To derive the lower bound we rewrite the energy flux (2.7.12) in terms of heat flux in the following two forms

$$E_h = \langle \frac{q_h . q_h}{K_h} \rangle = \langle \frac{\mathbf{q}_h . \mathbf{q}_h}{K_h} \rangle + \langle \frac{q'_h . q'_h}{K_h} \rangle + 2 \, \mathbf{q}_h . \langle \frac{q'_h}{K_h} \rangle$$

$$E_h = \mathbf{q}_h . \mathbf{J} = \mathbf{q}_h . \langle (q_h)/K_h \rangle = \langle 1/K_h \rangle \, \mathbf{q}_h . \mathbf{q}_h + \mathbf{q}_h . \langle q'_h / K_h \rangle \qquad (2.7.21)$$

Elimination of $\mathbf{q}_h . \langle q_h / K_h \rangle$ between these two equations yields

$$E_h = \langle 1/K_h \rangle \, (\mathbf{q}_h . \mathbf{q}_h) - \langle q'_h \cdot q'_h / K_h \rangle \qquad (2.7.22)$$

The last term in (2.7.22) is positive definite and, therefore, the following inequality holds

$$E_h \leq \frac{1}{K_H} q_h \cdot q_h \tag{2.7.23}$$

where $K_H = 1/(\langle 1/K_h \rangle)$ is the conductivity harmonic mean. Comparison with (2.7.14), applied to the principal directions of K_h, for which the resistivity tensor $\boldsymbol{K}_h^{-1}$ has the components $1/K_{hI}$, $1/K_{hII}$ and $1/K_{hIII}$, yields now

$$K_{hI}, K_{hII}, K_{hIII} \geq K_H \tag{2.7.24}$$

which stated in words reads: the conductivity harmonic mean is a lower bound to the effective conductivity. Again, for the particular case of a two-phase saturated porous medium we have

$$K_H = \frac{1}{\left[\langle \frac{h}{K_{hf}} \rangle + \langle \frac{1-h}{K_{hs}} \rangle\right]} = \frac{1}{\left[\frac{n}{K_{hf}} + \frac{1-n}{K_{hs}}\right]} \tag{2.7.25}$$

which is expressed again in terms of porosity and of the conductivity of the two phases. The first derivation of these bounds is attributed to Wiener (1912).

The next question is whether these bounds are "best bounds", in the sense that an anisotropic porous medium which has effective conductivity equal to K_A (2.7.19) and K_H (2.7.25) in two orthogonal directions, can be effectively constructed. We shall show in the next section that this is indeed the case.

How can the interval between the two bounds (2.7.18) and (2.7.24) be narrowed down ? An obvious means is to incorporate in the calculations more information about the structure than just the porosity, but this avenue is both difficult and impractical (see, e.g. Beran, 1968). Another possibility is to take advantage of the isotropy of the medium, when it exists. This is precisely the line of attack followed by Hashin and Shtrikman (1962) who have derived by variational methods conductivity bounds for multiphase heterogeneous materials. Their results for a two phase medium are reproduced here, while a heuristic derivation is given in the next section,

$$K_{hmin} = -2K_{hf} + \frac{1}{\frac{1-n}{2K_{hf}+K_{hs}} + \frac{n}{3K_{hf}}}$$

$$K_{hmax} = -2K_{hs} + \frac{1}{\frac{n}{2K_{hs}+K_{hf}} + \frac{1-n}{3K_{hs}}} \tag{2.7.26}$$

$$K_{hmin} \leq K_h \leq K_{hmax}$$

where it was assumed that $K_{hf} < K_{hs}$, as it is usually the case. Hashin and Shtrikman (1962) have proved that these are best bounds and in the next section we shall describe their models of porous medium which have effective conductivities equal to these bounds.

In Table 2.7.1 we have compared the lower bound K_{hmin} (2.7.26) and the measured values K_h for various granular materials. The striking result is that the actual conductivity is very close to the lower bound, the standard deviation being 5%. An explanation of this finding is offered in next section. The fact that many measurements are slightly lower that the bound can be attributed to anisotropy or experimental errors.

In Fig. 2.7.1 we have represented the ratio between the effective conductivity bounds and between the fluid conductivity K_h/K_{hf} for a given ratio K_{hs}/K_{hf}=4, which is common for water and some typical rocks, as function of the porosity n. The narrowing down of the bounds achieved by accounting for isotropy is clearly shown in the figure.

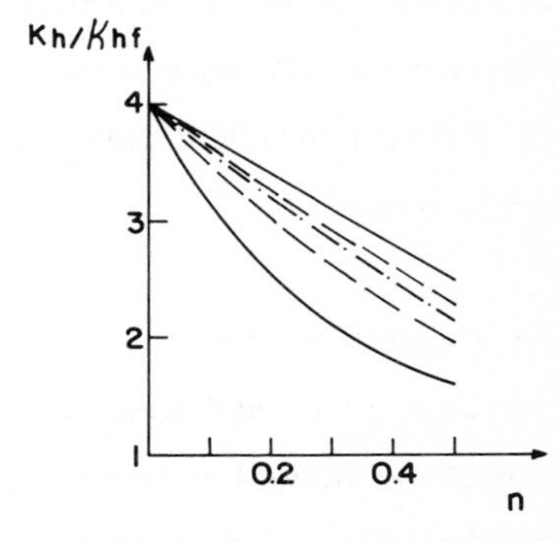

Fig. 2.7.1 The dependence of the ratio between the effective heat conductivity K_h and the fluid conductivity K_{hf} upon the porosity of the medium n, for K_{hs}/K_{hf}=4; ——— the bounds (2.7.19) and (2.7.25), - - - - the bounds (2.7.26), -·-·- the self-consistent approximation (2.7.35).

2.7.3 *Evaluation of the effective heat conductivity with the aid of models of porous media*

Because of the the lack of an exact solution of the problem of determining the effective conductivity on the basis of the medium structure, many schematical models of porous media, which lend themselves to simple formulae for K_h, have been developed in the past (a review of such models can be found in Van Brakel, 1975). Such models may be useful for explaining transport phenomena in physical terms and are, therefore, valuable pedagogical tools. At best they may produce results of a quantitative nature, which have to be validated by comparison with experiments or with exact solutions. A few simple models, considered to be particularly relevant, will be described in the sequel.

(i) Parallel layers

One of the simplest conceivable models of porous media consists from alternating layers of solid and fluid of constant thickness set in parallel (Fig. 2.7.2). If a temperature difference is applied to this body, such that the heat flow is normal to the layers (similar to a set of electrical resistances in "series"), the heat transfer is one-dimensional and the continuity equation (2.7.7) yields immediately q_h =const. By using the flux-gradient relationship in each layer it can be easily found that the conductivity of the body is equal to the harmonic mean K_H (2.7.25). Since this result is independent of the layers setting, it is a space and ensemble average as well. In a similar fashion, by applying the temperature gradient parallel to the layers (setting in parallel) it is found that the temperature gradient is constant, i.e. $d\mathcal{T}/dx = dT/dx = -J$, and the conductivity is equal to the arithmetic mean K_A (2.7.19).

It is seen that the present model represents an anisotropic axisymmetric porous medium whose effective conductivity has principal axes normal and parallel to the layers planes. This proves that the two bounds derived in the preceding section are indeed best bounds. Furthermore, any alteration of the structure will result in an increase and decrease of the minimal and maximal conductivities, respectively.

(ii) Composite spheres.

This is an ingenious model suggested by Hashin and Shtrikman (1962), in conjunction with the bounds (2.7.26). Let consider a sphere of radius R_s and of thermal conductivity K_{hs}, surrounded by a spherical shell defined by $R_s < r < R_f$ and of conductivity K_{hf} (Fig. 2.7.3). This composite sphere is submerged into a matrix of conductivity K_h extending to infinity. An uniform temperature gradient $-J$ is applied at infinity and the salient question is whether for given K_{hf}, K_{hs}, R_s and R_f, there exists an ambient conductivity K_h for which the temperature field in the entire surrounding matrix is of constant gradient, i.e. the composite sphere does not disturb the temperature field which prevails in a homogeneous medium. The temperature satisfies Laplace equation (i.e. Eqs. 2.3.13 and 2.3.14 for a fluid in rest and steady heat flow) at any point as well as the temperature (2.3.21) and normal flux (2.3.22) continuity conditions at the interfaces $r=R_f$ and $r=R_s$. With r,θ spherical coordinates (Fig. 2.7.3), the temperature fields satisfying Laplace equation are given by

$$\mathcal{T} = -J\,r\cos\theta \qquad (r > R_f\ \text{, uniform heat flow parallel to } \theta=0)$$

$$\mathcal{T}_f = \left(C_1 r + \frac{C_2}{r^2}\right)\cos\theta \qquad (R_s < r < R_f) \tag{2.7.27}$$

$$\mathcal{T}_s = C_3 r\cos\theta \qquad (r < R_s)$$

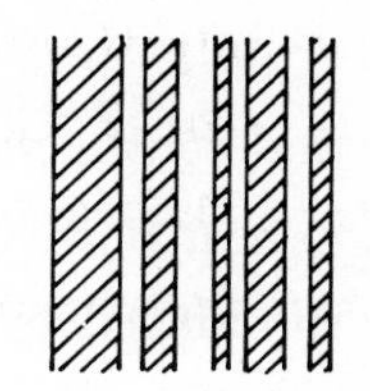

Fig. 2.7.2 Sketch of a layered two-phase material.

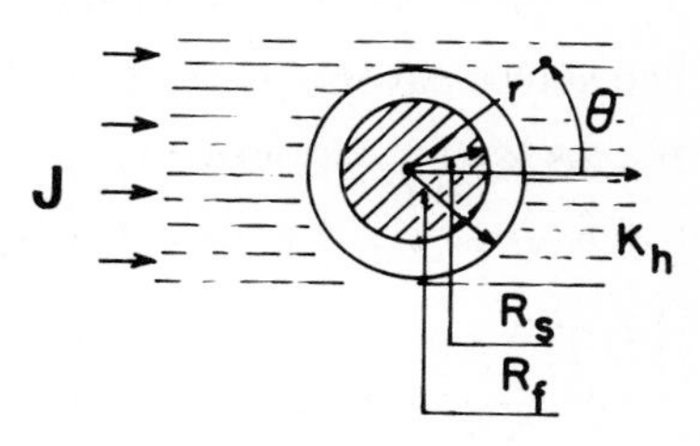

Fig. 2.7.3 The definition sketch of a composite sphere embedded into a matrix of conductivity K_h.

The constants C_1, C_2 and C_3 have to be determined with the aid of the following boundary conditions

$$T = T_f \; ; \; K_h \frac{\partial T}{\partial r} = K_{hf} \frac{\partial T_f}{\partial r} \qquad (r = R_f)$$

$$T_f = T_s \; ; \; K_{hf} \frac{\partial T_f}{\partial r} = K_s \frac{\partial T_s}{\partial r} \qquad (r = R_s) \tag{2.7.28}$$

Substituting (2.7.27) into (2.7.28) results into a linear system of four equations with the four unknowns C_1, C_2, C_3 and K_h. The solution for K_h is found to be

$$K_h = K_{hf} + \frac{(R_s/R_f)^3}{\dfrac{1}{K_{hs} - K_{hf}} + \dfrac{1-(R_s/R_f)^3}{3K_{hf}}} \tag{2.7.29}$$

Since the composite sphere does not disturb the uniform heat flow in the surrounding matrix, the latter can be filled with various composite spheres of same ratio R_f/R_s between the radii, which does not influence each other. The average energy flux per unit volume in a composite sphere is equal to $K_h J^2$ and, therefore, the effective conductivity of the swarm of composite spheres is precisely K_h. Furthermore, the porosity n is equal to the volume ratio $1-(R_s/R_f)^3$. Substituting n into (2.7.29) and after a few algebraic manipulations we arrive at

$$K_h = -2K_{hf} + \frac{1}{\dfrac{n}{3K_{hf}} + \dfrac{1-n}{2K_{hf}+K_{hs}}}. \tag{2.7.30}$$

If $K_{hf} < K_{hs}$, K_h (2.7.30) is precisely the lower bound K_{hmin} (2.7.26) and since the medium thus constructed is isotropic, it is found that this model represents the setup of maximum resistance to heat flow. By the same token, a medium in which spherical inclusions filled with fluid in rest are surrounded by solid shells, leads to an effective conductivity equal to K_{hmax} (2.7.26). Thus the bounds (2.7.26) are best bounds and these models provide a physical picture about the structure which achieves them.

A granular porous medium is very similar to the model of spherical particles surrounded by fluid concentric layers. This may explain the closeness of the measured values of Table 2.7.1 to the lower bound K_{hmin} (2.7.26).

(iii) The self-consistent approximation.

In some media, like unconsolidated formations, the solid particles cannot be regarded as completely surrounded by fluid films and they are in contact over part of their envelopes. Obviously, such a medium has an effective conductivity larger than K_{hmin} (2.7.26). A simple conceptual model for such a medium can be devised by regarding it as made up from an union of spheres of various radii of conductivity K_{hs} and K_{hf}. Now, according to this model, for each sphere the surrounding medium is replaced by a fictitious homogeneous matrix of conductivity K_0. For a temperature gradient $-\mathbf{J}$ applied at infinity, the temperature field for an isolated sphere (Fig. 2.7.4) is given by

$$T_s = -J \frac{3K_0}{2K_0+K_{hs}} r \cos\theta \qquad (r < R_s)$$

$$T_0 = -J \left[r + \frac{R_s^3}{r^2} \frac{K_0 - K_{hs}}{2K_0+K_{hs}} \right] \cos\theta \qquad (r > R_s) \tag{2.7.31}$$

where $\theta=0$ in the direction of $\mathbf{J}$. Eq. (2.7.31) can be easily derived along the lines of the preceding paragraph. By using the definitions (2.7.6) of the average temperature gradient and heat flux we may calculate them for the entire space for the field (2.7.31), the result being

$$\nabla T = \lim_{V\to\infty} \frac{1}{V} \int \nabla T \, dV = -\mathbf{J} - \frac{K_0 - K_{hs}}{2K_0+K_{hs}} \frac{R_s^3}{V} \mathbf{J}$$

$$q_h = \lim_{V\to\infty} \frac{1}{V} \int q_h \, dV = K_0 \mathbf{J} - 2K_0 \frac{K_0 - K_{hs}}{2K_0+K_{hs}} \frac{R_s^3}{V} \mathbf{J} \tag{2.7.32}$$

Fig. 2.7.4 The definition sketch of heat flow past a sphere embedded into a matrix of conductivity K_0.

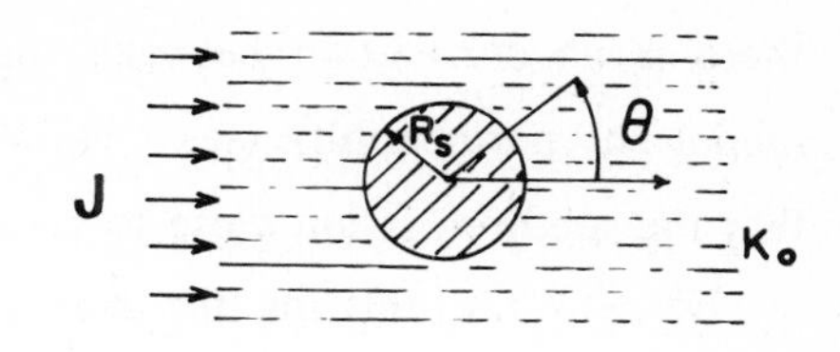

The last terms in each expression of (2.7.32) represent the disturbance of the otherwise uniform heat flow caused by the spherical inclusion (for a detailed derivation see Dagan, 1979b). An expression similar to (2.7.32) can be derived for a fluid spherical inclusion, with R_f and K_{hf} replacing R_s and K_{hs}, respectively.

Now, the medium is regarded as an union of such spheres and the average gradient and flux results from the summation of the disturbances created by all the spheres. Summing up first and dividing the resulting $\mathbf{q}_h$ by $\mathbf{J}$ yields the following expression for the effective conductivity

$$K_h = -2K_0 + \frac{1}{\dfrac{n}{2K_0+K_{hf}} + \dfrac{1-n}{2K_0+K_{hs}}} \tag{2.7.33}$$

where n is the ratio between the total volume of spheres of conductivity K_{hf} and the total spheres volume.

In the self-consistent approximation (also termed the renormalization approximation, see Sect. 3.4.3) it is assumed that the conductivity K_0 of the matrix surrounding each inclusion is equal to the effective conductivity K_h. Then, we obtain from (2.7.33)

$$K_h = \frac{1}{3}\,\frac{1}{\dfrac{n}{2K_h+K_{hf}} + \dfrac{1-n}{2K_h+K_{hs}}} \tag{2.7.34}$$

which can be solved explicitly to yield

$$K_h = \frac{-b+(b^2+8K_{hf}K_{hs})^{1/2}}{4} \quad ; \quad b = -K_{hf}(3n-1) - K_{hs}(2-3n) \tag{2.7.35}$$

A survey of the past various derivations of the self-consistent approximation may be found in Beran (1968) and Landauer (1978). The latter coined the term effective medium approximation for the same approach. It seems that (2.7.34) has been suggested by Böttcher (1952), who followed an earlier derivation by Bruggeman (1935). In a study of water flow through heterogeneous formations, similar to the heat conduction problem, Dagan (1981) arrived at the conclusion that the self-consistent approximation is appropriate for a "completely random" medium, i.e. a material for which there is no correlation between the conductivity of neighboring spheres. In a study of a network model for permeability (see Chap. 2.9) Koplik (1982) has arrived at a similar conclusion, provided that the medium is not close to zero conductivity.

Eq. (2.7.33), relating the effective conductivity K_h to that of the embedding matrix K_0 has a few interesting properties. First, if K_0 is assumed to be equal to the smallest conductivity of the various phases, i.e to K_{hf} in the present case, the lower bound (2.7.26) is obtained. By the same token the upper bound is achieved for $K_0 = K_{hs}$. These results are intuitively appealing, since submerging each inclusion in a matrix of the lowest conductivity would result in a medium of lower effective conductivity than the actual medium and vice-versa. Furthermore, it is easy to ascertain that K_h (2.7.33) is an increasing function of K_0. It follows that the self-consistent approximation (2.7.34) lies between the two bounds. This point is illustrated by Fig. 2.7.1, where the self-consistent approximation is also represented.

Exercises

2.7.1 The directional heat conductivity for an anisotropic medium is defined by $\mathbf{q}_h . \mathbf{J}/J^2$, where $\mathbf{J}$ may have any direction in space. Regarding K_h thus defined as a vector in the $\mathbf{J}$ direction, prove that it spans an ellipsoid whose principal axes are those of the tensor $\mathbf{K}_h$.

Similarly, show that if the temperature gradient $\mathbf{J}$ and the heat flux $\mathbf{q}_h$ are parallel, this is a principal direction.

Prove, along these lines, that a medium built from cubes of different conductivities in a bricklayering pattern, is isotropic.

2.7.2 Derive the best two-dimensional bounds of the effective conductivity of an isotropic material by using Hashin and Shtrikman (1962) model, i.e. by regarding the medium as made up from a bundle of parallel cylinders with heat flow taking place normal to their axes.

Derive the two-dimensional self-consistent approximation by a similar reasoning.

2.7.3 Derive the effective conductivity of an unsaturated porous medium, i.e. a three-phase material made up from solid, water and air by the self-consistent approximation. Represent graphically the dependence of K_h upon the degree of saturation with water for usual values of the phase conductivities and for n=0.4.

2.8 THE CONSTITUTIVE EQUATION OF MASS TRANSFER (EFFECTIVE DIFFUSION COEFFICIENT)

The definitions of the macroscopic equation of mass transfer and of effective diffusion coefficient are similar to those of heat transfer of Sect. 2.7 and the arguments leading to them will not be repeated here. The macroscopic equation is, similarly to (2.7.1),

$$\mathbf{q}_m = - \mathrm{D}_m \nabla C \tag{2.8.1}$$

where, according to the definitions of Sect. 2.5.1 $\mathbf{q}_m$ is the phase ensemble average of the solute flux. In terms of space averages or laboratory experiments of one-dimensional transfer, $\mathbf{q}_m$ is the total solute mass discharge divided by the void area. The essential difference between heat and mass transfer in a porous medium stems from the fact that diffusion occurs only through the fluid phase. From considerations of dimensional analysis which led to eq. (2.7.4) it is easy to see that for media of similar structure, e.g. a mixture of spheres of uniform diameters, the effective diffusion coefficient can be written as follows

$$\mathrm{D}_m = D_m \, \mathrm{f}(n) \tag{2.8.2}$$

where D_m is the molecular diffusion coefficient in the fluid. The experimental determination of the ratio D_m/D_m under steady state conditions is not a simple task because of the smallness of D_m and because of the difficulty in measuring the solute flux. Thus, for weak concentrations of aqueous solutions of NaCl D_m is equal to 1.1×10^{-5} cm²/sec and the establishing of a steady one-dimensional flow is very slow. Therefore, most laboratory experiments are conducted under unsteady conditions by bringing a porous column saturated with pure water in contact with a solution of constant concentration. The effective diffusivity D_m/D_m is determined indirectly from the measurement of the breakthrough curve, which is defined as the concentration at the outlet as function of time. This procedure will be discussed in some detail in Sect. 2.10.

It is easier to determine experimentally D_m by taking advantage of the analogy between mass and electricity transfer. Indeed, if the solid matrix is of zero or very low electrical conductivity and if the fluid is an ionic solution, a complete analogy exists between the two processes. It is common to define the formation factor F (for a detailed discussion, see Dullien, 1979) as the ratio between the conductivity of the solution and that of the porous medium, i.e.

$$\mathrm{F} = D_m / (n \, \mathrm{D}_m) \tag{2.8.3}$$

Thus, measurements of F can be easily converted into those of D_m. Laboratory determinations of D_m by diffusion experiments are discussed in Sect. 2.10. The data collected by Pfannkuch (1963) for granular, unconsolidated media, suggest the value $\mathrm{D}_m/D_m = 2/3$ as close to measured ones. In contrast, smaller values have been found by formation factor measurements in consolidated media. Thus, Table 4.3 in Dullien (1979) for various sandstones presents values of $X = D_m/\mathrm{D}_m$ in excess of 1.5. From a theoretical standpoint the bounds for the effective heat conductivity can be applied to the effective diffusion coefficient by setting the conductivity of the solid equal to zero in the various relevant formulae. It is also necessary to divide the effective heat conductivity by n in the same formulae, because the average heat flux $\mathbf{q}_h$ (2.7.1) is defined as a medium average, whereas $\mathbf{q}_m$ (2.8.1) is a phase average. Thus, the bounds (2.7.19) and (2.7.24) yield the trivial result

$$0 \le \mathrm{D}_m \le D_m \tag{2.8.4}$$

which is easy to grasp with the aid of the models which led to these bounds. Indeed, for diffusion normal to parallel, alternating, layers of solid and fluid (Fig. 2.7.2), the mass transfer is prevented by the solid. In contrast, for diffusion parallel to the layers, the process in each layer is identical with the one prevailing in a free solution.

The bounds for isotropic media can be obtained from (2.7.26) by substituting $K_{hs} = 0$, $K_{hf} = D_m$, $K_{hmin}/n = \mathrm{D}_{mmax}$ and $K_{hmax}/n = \mathrm{D}_{mmin}$. The result is

$$0 \le \mathrm{D}_m \le \frac{2}{3-n} D_m \tag{2.8.5}$$

The lower bound is easily understood with the aid of the composite spheres models of Sect. 2.7.3. Indeed, the lower bound corresponds to spherical cavities filled with fluid and surrounded by solid shells and it is obvious that such a medium prevents diffusion. The upper bound can be traced back to Maxwell (1881) and it is seen that for small n it is close to measured values $\mathrm{D}_m/D_m = 2/3$ for granular materials mentioned before. The self-consistent approximation (2.7.34) yields, by a limit process for $K_{hs} \to 0$ and after dividing by n, to

$$\mathrm{D}_m = \frac{3n-1}{2n} D_m \quad (n > \tfrac{1}{3}) \quad ; \quad \mathrm{D}_m = 0 \quad (n < \tfrac{1}{3}) \tag{2.8.6}$$

It is seen that the self-consistent approximation underestimates the effective diffusion coefficient for low n. This is understandable in view of the observation of Sect. 2.7.3 about its limitations at low conductivity threshold. In fact, we shall show in Chap. 3.4, dealing with heterogeneous formations, that the self-consistent approximation is bound to be accurate for broad distributions of conductivities of various phases. Still, the result is indicative of the drop of the value of the effective diffusivity of consolidated materials as compared to unconsolidated ones. Finally, many models of porous media for which flow occurs only in the fluid phase have been suggested in the literature (see, e.g., reviews in Van Brakel, 1975 and Dullien, 1979) mainly for deriving Darcy's law and the permeability. Some of these models will be discussed in Sect. 2.9. In particular, Dullien (1979) presents an elaborate model for diffusion which accounts for the pore-size distribution and which is shown to predict quite accurately the formation factor of sandstone.

We shall depict here a very simple model which yields the experimental value mentioned before for unconsolidated media. We consider a fissured porous medium created by planar elements, of openings much smaller than their areal dimension, which are interconnected. Such a model, although a simplification of actual porous media, has the distinctive feature that it resembles a class of media encountered in practice. The use of such models for deriving Darcy's law can be apparently traced back to Irmay (1958). In contrast, the popular model of a bundle of capillary tubes is of a lesser similarity to actual porous media. In Fig. 2.8.1 a regular, cubic, arrangement of blocks and associated fissures is depicted. For a concentration gradient parallel to one of the planes of symmetry, and with neglect of the contribution of the junctions compared to that of the fissures to mass transfer, it is easy to ascertain that $\mathrm{D}_m/D_m = 2/3$. Indeed, in steady state $\mathbf{q}_m$ in the fissures which are parallel to the gradient is constant and equal to $-\mathrm{D}_m \nabla C$. Furthermore, these fissures contribute 2/3 of the void space in the volume average of $\mathbf{q}_m$. It is easy to ascertain that this medium is isotropic (see Exercise 2.7.1). In Sect. 2.9, dealing with Darcy's law, we shall prove that an identical result is obtained if the fissures have random orientations.

As we shall also show in Sect. 2.9, similar results for macroscopic coefficients can be obtained with completely different models, which should serve as a warning against the belief that agreement with experiments for such coefficients is indicative of the degree of realism of the model.

Exercises

2.8.1 Prove that for a model of porous medium consisting of capillary tubes arranged in a cubic grid the ratio D_m/D_m is equal to 1/3.

2.8.2 Derive the expression of the effective diffusion coefficient for a fissured medium similar to that of Fig. 2.8.1, but with blocks in shape of parallelipipeds rather than cubes. Arrive at similar results for the cubic blocks of Fig. 2.8.1, but with fissures of various apertures, characterized by probability distribution functions.

2.8.3 Derive the effective diffusion coefficient for a medium in which there is "mobile" and "immobile" water. The mobile phase is represented by ordinary voids, whereas the immobile part is modeled as a porous material of secondary porosity. Employ the isotropic bounds (2.7.26) and the self-consistent approximation (2.7.34).

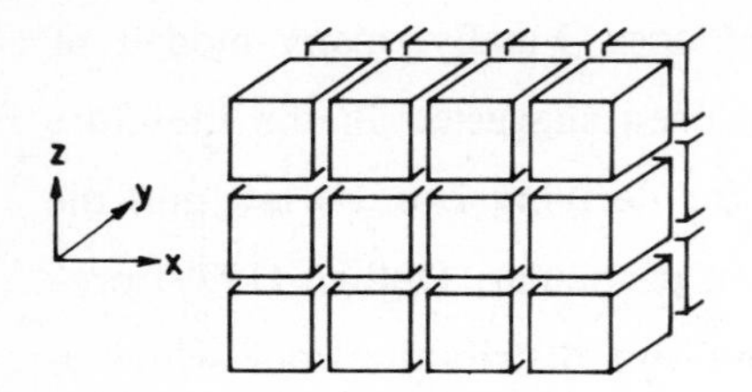

Fig. 2.8.1 Schematical representation of a periodic fissured porous medium made up from cubical blocks.

2.9 DARCY'S LAW.

2.9.1 Definitions and experimental evidence

The macroscopic equations of conservation of momentum (2.6.2) or (2.6.4) express the pressure or head gradient in terms of the average drag **D**. The drag is a function of the microscopic velocity field $\boldsymbol{u}$ and its dependence upon the statistical moments of h and $\boldsymbol{u}$ has to be established in an independent manner. Towards this aim we consider first the case of a uniform average flow, such that n=const and the average viscous stress $\tau_{ij}=0$. The last equality stems from the constitutive equation (2.3.12), which has the same form after its averaging (see Sect. 2.9.4). Hence, the macroscopic momentum equation is given by (2.6.6), which relates the drag **D** to the head gradient.

Darcy's law, established by Darcy (1856) on empirical grounds, stipulates that **D** is a linear function of the specific discharge **q** or equivalently, of $\mathbf{u}=\mathbf{q}/n$. The experimental setup, which served for deriving this law, consists of a porous column of cross-sectional area A subjected to a head difference $\phi_1-\phi_2$ at its ends, as depicted in Fig. 2.9.1. If the head is maintained constant with the aid of two constant level reservoirs, then $\phi_1=z_1$ and $\phi_2=z_2$, where z is the elevation (the head losses in the free fluid are negligible compared to those occurring in the porous column). By maintaining a one-dimensional steady flow of discharge Q and by adopting space averages to define macroscopic variables, the following relationships prevail

$$\nabla\phi = \frac{\phi_2 - \phi_1}{L}\,\mathbf{s} \quad ; \quad \mathbf{q} = \frac{Q}{A}\,\mathbf{s} \quad ; \quad \mathbf{u} = \frac{\mathbf{q}}{n} \tag{2.9.1}$$

where **s** is a unit vector in the direction of the mean flow (Fig. 2.9.1).

Darcy's law stipulates that $\nabla\phi$ is proportional to **q**, i.e.

$$\mathbf{q} = -\,K\,\nabla\phi \qquad \text{i.e.}\quad \frac{Q}{A} = -\,K\,\frac{\phi_1-\phi_2}{L} \tag{2.9.2}$$

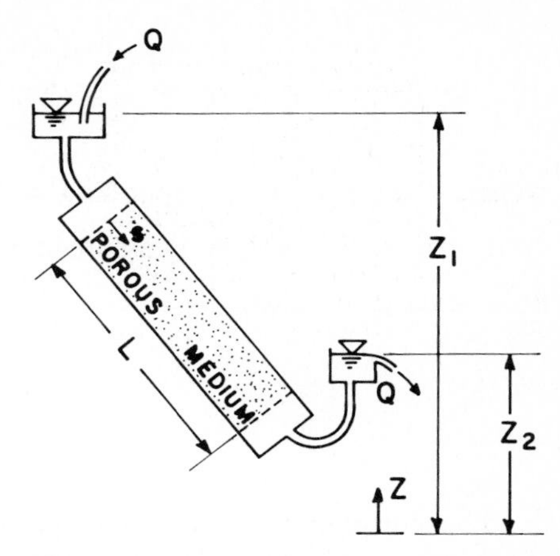

Fig. 2.9.1 Sketch of experimental setup (laboratory column) for deriving Darcy's law and for measuring permeability.

The constant of proportionality K, called hydraulic conductivity in the case in which the fluid is water at standard temperature, is assumed to depend on the medium structure, i.e. to be the same for columns of different cross-sections and lengths and to be independent of the head drop. These assumptions have been found to hold for considerable ranges of values of the parameters of interest and Darcy's law (2.9.2) forms the basis of the theory of flow through porous media. Some circumstances in which Darcy's law needs generalizations will be discussed in Sect. 2.9.4.

The nature of the constant of proportionality in Darcy's law can be analyzed in a general manner with the aid of dimensional analysis. Thus, for a statistically homogeneous medium and for average uniform flow we may write for the average drag per unit volume

$$\mathbf{D} = f(\mathbf{u}, \nu, \rho, n, d_1, d_2, R, \ldots) \tag{2.9.3}$$

where R is some length scale of the porous body (e.g. the column radius), $d_1, d_2, \ldots$ are length-scales associated with the statistical moments of $h(\mathbf{x})$ (see for a similar procedure Sect. 2.7.1), ν is the coefficient of kinematic viscosity of the fluid and ρ is its density. Thus, in the simplest case of a porous medium consisting of uniform spheres, their diameter $d=d_1$ is the only length scale. From considerations of dimensional analysis, (2.9.3) can be rewritten as follows

$$\frac{\mathbf{D}}{\rho u^2 d_1} = f\left(\frac{u\rho}{d_1\mu}, n, \frac{d_1}{R}, \frac{d_2}{d_1}, \ldots\right) \tag{2.9.4}$$

However, by the linearity of the microscopic equation of viscous flow (2.3.18), $\mathbf{D}$ is a linear function of $\mathbf{u}$. Assuming for the time being that the medium is isotropic, this linearity implies in (2.9.4)

$$\mathbf{D} = \frac{\rho \nu n^2}{k}\,\mathbf{u} \quad ; \quad k = d_1^2\, f\left(n\,,\, \frac{d_1}{R}\,,\, \frac{d_2}{d_1}\,, ...\right) \tag{2.9.5}$$

where k is a constant depending only on the porous structure. By the same argument as in Sect. 2.7.1, for $d_1/R << 1$, k is assumed to be independent of R. Thus, for the simplest case of a unique pore scale d we have

$$k = d^2\, f(n) \tag{2.9.6}$$

The constant k is known as permeability or intrinsic conductivity and unlike K, it depends on the geometry of the medium solely. Darcy's law may be now rewritten by using the macroscopic momentum equation (2.6.6) and (2.9.5) as follows

$$\mathbf{q} = -\frac{k}{\mu}\,\nabla(p + \rho g z) = -\frac{kg}{\nu}\,\nabla\phi = -K\,\nabla\phi \tag{2.9.7}$$

where the relationship between k and K through the fluid kinematic viscosity ν has also been written down. Since in this book we are concerned with flow of water, we shall use mostly the hydraulic conductivity K as the medium property, keeping in mind however, that k is the basic property of the medium.

The unit of k is cm^2, but in reservoir engineering it is customary to use the darcy, with 1 darcy= 0.987×10^{-8} cm^2. The hydraulic conductivity has dimension of velocity. For water at 20 ^{0}C, $\nu = 10^{-2}$ cm^2/sec, such that a medium of k=1cm^2 has K=9.81×10^4 cm/sec and K=9.613×10^{-4} cm/sec corresponds to k=1darcy.

The laboratory measurement of permeability with the aid of cores is a standard procedure and many data, as well as empirical formulae, are available in the literature, e.g. Dullien (1979). Table 9.1, reproduced from Bear (1972), gives a classification of permeability of soils and rocks used by the Bureau of Reclamation (1 md = 10^{-3} darcy).

The extension of Darcy's law to anisotropic media can be carried out by the same considerations of dimensional analysis as those employed in (2.9.5). Indeed, the most general linear dependence of the vector **D** on the vector **u** is expressed with the aid of a tensor. Thus, in a Cartesian system we may replace (2.9.7) by

$$q_i = -\sum_{j=1}^{3} \frac{k_{ij}\, g}{\nu}\,\frac{\partial\phi}{\partial x_j} = -\sum_{j=1}^{3} K_{ij}\,\frac{\partial\phi}{\partial x_j} \qquad (i=1,2,3) \tag{2.9.8}$$

where k_{ij} and the associated $K_{ij} = k_{ij}\, g/\nu$ are the permeability and hydraulic conductivity tensors, respectively. In the next section we shall prove that k_{ij} is a symmetrical, positive definite, tensor

TABLE 9.1 Typical values of hydraulic conductivity and permeability.

<table>
<tr><td>$-\log_{10} \cdot K$(cm/s)</td><td>−2 to −1</td><td>−1 to 0</td><td>0 to 1</td><td>1 to 2</td><td>2 to 3</td><td>3 to 4</td><td>4 to 5</td><td>5 to 6</td><td>6 to 7</td><td>7 to 8</td><td>8 to 9</td><td>9 to 10</td><td>10 to 11</td></tr>
<tr><td>Permeability</td><td colspan="4">Pervious</td><td colspan="4">Semipervious</td><td colspan="5">Impervious</td></tr>
<tr><td>Aquifer</td><td colspan="5">Good</td><td colspan="4">Poor</td><td colspan="4">None</td></tr>
<tr><td rowspan="2">Soils</td><td colspan="2">Clean gravel</td><td colspan="3">Clean sand or sand and gravel</td><td colspan="4">Very fine sand, silt, loess, loam, solonetz</td><td colspan="4"></td></tr>
<tr><td colspan="4"></td><td colspan="2">Peat</td><td colspan="3">Stratified clay</td><td colspan="4">Unweathered clay</td></tr>
<tr><td>Rocks</td><td colspan="4"></td><td colspan="3">Oil rocks</td><td colspan="2">Sandstone</td><td colspan="2">Good limestone, dolomite</td><td colspan="2">Breccia, granite</td></tr>
<tr><td>$-\log_{10} \cdot k$(cm²)</td><td>3 to 4</td><td>4 to 5</td><td>5 to 6</td><td>6 to 7</td><td>7 to 8</td><td>8 to 9</td><td>9 to 10</td><td>10 to 11</td><td>11 to 12</td><td>12 to 13</td><td>13 to 14</td><td>14 to 15</td><td>15 to 16</td></tr>
<tr><td>$\log_{10} k$(md)</td><td>8 to 7</td><td>7 to 6</td><td>6 to 5</td><td>5 to 4</td><td>4 to 3</td><td>3 to 2</td><td>2 to 1</td><td>1 to 0</td><td>0 to −1</td><td>−1 to −2</td><td>−2 to −3</td><td>−3 to −4</td><td>−4 to −5</td></tr>
</table>

which can be reduced, therefore, to three orthogonal principal axes and to three corresponding principal values k_I, k_{II} and k_{III}. In the case of natural porous formations, anisotropy, at laboratory scale, is primarily a result of microstratification.

2.9.2 *Theoretical derivation of Darcy's law*

Unlike the case of effective thermal conductivity (Chap. 2.7), no theoretical bounds have been derived so far for permeability and in spite of the efforts invested in the subject, the connection between k and the structure of the porous medium has been investigated mainly with the aid of specific models (a few such models are described in the next section). The main purpose of the present section is to establish a general framework for the derivation of Darcy's law.

Following the reasoning of Chap. 2.7, we consider a porous body of dimension which is large compared to the pore scale and which is in contact with the free fluid. The head ϕ on the boundary is assumed to be given by

$$\phi = - \mathbf{J}.\mathbf{x} + \text{const} \tag{2.9.9}$$

where $\mathbf{J}$ is the constant head gradient (e.g., in Fig. 2.9.1 $\mathbf{J} = \mathbf{s}\ (\phi_1-\phi_2)/L$). Strictly speaking, specification of the head as a boundary condition is not sufficient to determine uniquely the viscous flow within the void space. However, outside a thin boundary layer near the area of contact between the porous body and the free fluid, it can be assumed that the flow outside the body is uniform and the shear-stresses are zero, due to the contrast between the resistance to flow of the two media. Thus, as an additional boundary condition we may assume that $\underline{\tau}_{ij}=0$ in the free fluid.

It is convenient to employ the microscopic energy equation (2.3.20) as a vehicle towards deriving the expression of k, rather than the momentum equation (2.3.16). Towards this aim we rewrite (2.3.20) in terms of $\underline{\phi}$ for an incompressible flow in vectorial notation as follows

$$\nabla(\underline{\phi}\, \boldsymbol{u}) = \frac{\nu}{g} [\, \nabla.(\boldsymbol{u}.\nabla \boldsymbol{u}) - \nabla \boldsymbol{u}.\nabla \boldsymbol{u}\,] \tag{2.9.10}$$

Integration of (2.9.10) over the void space by using Green's theorem, Eq.(2.9.9) and the vanishing of $\boldsymbol{u}$ on the void-solid interface, yields

$$- \mathbf{J}.\int_S \mathbf{x}\, (\boldsymbol{u}.\boldsymbol{\eta})\, dS = \frac{\nu}{g} \int_S \boldsymbol{\eta}(\boldsymbol{u}.\nabla \boldsymbol{u})\, dS - \frac{\nu}{g} \int_V (\nabla \boldsymbol{u}.\nabla \boldsymbol{u})\, h\, dV \tag{2.9.11}$$

where S is here the boundary of the porous body, V is its volume and $\boldsymbol{\eta}$ is a unit vector normal to S. Since we have assumed that in the free fluid adjacent to the body the flow is uniform and of velocity $\mathbf{q}$, the first term of the r.h.s. of (2.9.11) vanishes, whereas in the l.h.s. we get by Green's theorem

$$\int_S \mathbf{x}\, (\mathbf{q}.\boldsymbol{\eta})\, dS = \mathbf{q}\, V \tag{2.9.12}$$

Thus we may write finally

$$\mathbf{q}.\mathbf{J} = \mathrm{E} \quad ; \quad \mathrm{E} = \frac{\nu}{g} \frac{1}{V} \left(\int_V (\nabla \boldsymbol{u}.\nabla \boldsymbol{u})\, h\, dV \right) \tag{2.9.13}$$

where E is the energy per weight of fluid and per unit volume dissipated by viscous stresses (see Chap. 2.6). Substituting Darcy's law (2.9.7) into (2.9.13) we arrive at the following definition of K or k

$$\mathrm{K} = \frac{\mathrm{E}}{\mathrm{J}^2} \quad ; \quad \mathrm{k} = \frac{\nu}{g} \frac{\mathrm{E}}{\mathrm{J}^2} \tag{2.9.14}$$

By the same token we may define k by ensemble average of E. Towards this aim let us separate the velocity and head fields into their average and fluctuations and assume stationarity, i.e. the various covariances of fluctuations and of their derivatives are constant, the same being true for the average velocity. Thus, we may write

$$\boldsymbol{u} = \mathbf{u} + \mathbf{u}' \quad ; \quad \underline{\phi} = \phi + \phi' \quad ; \quad \langle \boldsymbol{u}'\mathrm{h}\rangle = \langle \phi'\mathrm{h}\rangle = 0$$

$$\langle \boldsymbol{u}'\phi'\mathrm{h}\rangle = \mathrm{const} \quad ; \quad \langle \boldsymbol{u}'.\nabla\boldsymbol{u}'\mathrm{h}\rangle = \mathrm{const} \tag{2.9.15}$$

Multiplying (2.9.10) by h, ensemble averaging and taking into account the fact that $\boldsymbol{u}$=0 on the fluid-solid interface (Eq. 2.3.30), yields with the aid of (2.9.15)

$$n\ \mathbf{u}.\nabla\phi = -\ \mathrm{E} \quad ; \quad \mathrm{E} = \frac{\nu}{g}\ \langle \nabla\boldsymbol{u}.\nabla\boldsymbol{u}\ \mathrm{h}\rangle \tag{2.9.16}$$

Eq. (2.9.16) is the statistical equivalent of (2.9.13) and the two definitions coincide, of course, under ergodic conditions. There are advantages in using the energy E rather than **D** to derive the permeability: first, as we shall show immediately, it permits us to prove the symmetry of the permeability tensor and secondly, it defines the permeability even for models in which the solid phase is not represented by a collection of disjunct particles, for which **D** is clearly defined.

Substituting Darcy's law (2.9.8) into (2.9.16), yields the following definition of K_{ij}

$$\sum_{i=1}^{3}\sum_{j=1}^{3} K_{ij}\ \frac{\partial\phi}{\partial x_i}\ \frac{\partial\phi}{\partial x_j} = \mathrm{E} \tag{2.9.17}$$

Since E (2.9.16), an average of a sum of squared quantities, is positive for any **J**, it is seen that the K_{ij} matrix is positive definite. Now, if we ensemble average the microscopic equation (2.3.19), which expresses the rate of work of a flow by the pressure gradient of a different flow, under the same conditions of stationarity (2.9.15), we arrive immediately at

$$\mathbf{q}^{I}.\nabla\phi^{II} = \mathbf{q}^{II}.\nabla\phi^{I} = -\ \frac{\nu}{g}\ \langle \nabla\boldsymbol{u}^{I}.\nabla\boldsymbol{u}^{II}\rangle \tag{2.9.18}$$

which may be stated in words as follows: the rate of work of flow I by the head gradient of flow II is equal to the rate of work of flow II by the head gradient I, both being equal to the last term of (2.9.18). Substituting again Darcy's law (2.9.8) into (2.9.18), leads to the conclusion that K_{ij}=K_{ji}, i.e. the hydraulic conductivity tensor is symmetric and positive definite. It can be reduced, therefore, to three positive principal values in orthogonal principal directions K_I, K_{II} and K_{III}. The head gradient **J** and average velocity **u** vectors, are parallel in, and only in, these principal directions.

2.9.3 Derivation of permeability with the aid of models

In view of the lack of a general theory which relates permeability to the geometry of the medium, considerable effort has been invested in developing schematical models of porous media which lend themselves to the calculation of the permeability. Succinct reviews of such models can be found for instance in Van Brakel (1975) or Dullien (1979). It should be emphasized that such models neither attempt nor are they able to reflect the complexity of the structure of natural porous media, like the one illustrated in Fig. 2.2.1.

A few requirements are considered by this writer as essential for rendering models useful : they should capture the main mechanisms involved in the phenomenon of interest, they should be simple, they should resemble even though schematically actual media and they should lead to the right order of magnitude of the coefficients of interest.

Models of porous media can be classified in different manners. Thus, a first division is in models of periodic structure or of random media. Periodic media, in which a basic unit cell is reproduced in space, lend themselves to general and interesting mathematical properties and some results of this nature can be found, for instance, in Sanchez-Palencia (1980) and in Brenner and Adler (1985). We shall present in the sequel an illustration of such a model, which has served as a basis for a numerical solution of the equations of viscous flow. However, this book is concerned with natural porous formations which display the random structure illustrated vividly by Fig. 2.2.1.

Another classification distinguishes between models of networks of channels, representing the void space, and of arrays of simple bodies (e.g. spheres) standing for the solid phase.

It is beyond the scope of this book to review the various models which can be found in the literature. Instead, a simple model of fissured media, which is believed to satisfy the basic requirements enumerated above, will be developed in detail. Two additional recent models, of considerable theoretical interest, are also reviewed.

A type of model which enjoys some popularity in the literature and which will not be considered here consists of networks of tubes of circular cross-sections. First, these models do not satisfy the requirement of resemblance with actual media. Furthermore, as we have already indicated in Chap. 2.8, they underestimate significantly the value of effective diffusivity of granular materials and the same is true for permeability. To remedy this deficiency, the concept of tortuosity, which implies the replacement of straight tubes by winding ones of increased resistance, has been introduced in the past. It is suggested here that the tortuosity ratio is mainly a "fudge" factor which lacks a sound physical foundation. This point has been emphasized by Koplik (1982), whose work will be discussed in the sequel, and who found that networks of higher "tortuosity" are of lesser resistance to flow, contrary to the usual belief.

(i) Fissured media.

We have presented already in Chap. 2.8 a model of an ordered set of fissures, which led to the value of effective diffusivity consistent with experiments. As we already emphasized, such a model, although highly schematical, at least resembles a class of existing natural media. Its use to derive the value of the permeability, can be traced back to Irmay (1958).

We shall present here in detail a model of a random three-dimensional setting of planar fissures (Fig. 2.9.2). This model is underlain by the following simplifying assumptions: (i) the fissures are planar and of constant aperture b. Thus, an individual fissure is characterized by its orientation through an unit vector normal to it $\boldsymbol{\eta}$, by its width b and by its volume v_f, which is equal to b time the area; (ii) the viscous flow in each fissure is assumed to be identical with the two-dimensional flow prevailing between two rigid plates, of parabolic velocity distribution; (iii) the dissipation occurring in the junctions is neglected; (iv) the fissures are oriented at random, their distribution being described by a probability distribution function to be discussed below and (v) an average uniform flow under an average head gradient $-\mathbf{J}$ prevails. Furthermore, the local head gradient in a fissure is the projection of the average gradient over the plane of the fissure. This latter assumption holds exactly in the case of a regular network and is bound to hold approximately in an irregular network as well (Koplik, 1982, 1983).

We can proceed now with the computation of the permeability of such a medium. According to assumption (v) above, the local head gradient J in a fissure is given by the projection of $\mathbf{J}$ on the fissure plane, i.e.

$$J = \mathbf{J} - (\mathbf{J}.\boldsymbol{\eta})\boldsymbol{\eta} \tag{2.9.19}$$

whereas the local average velocity, resulting from the parabolical profile for flow between parallel plates (e.g. Batchelor, 1967), and the energy dissipated per unit weight of fluid, are given by

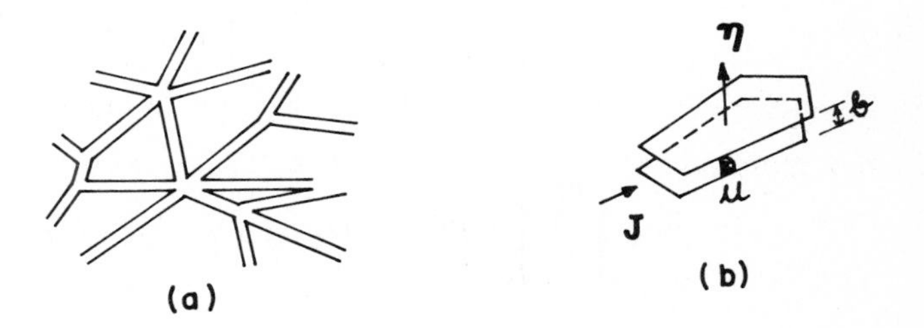

Fig. 2.9.2 Sketch of a model of fissured porous medium: (a) the fissures network and (b) the geometry of a fissure.

$$\bar{\boldsymbol{u}} = \frac{b^2 g}{12\nu} \boldsymbol{J} \quad ; \quad E = \bar{\boldsymbol{u}} . \boldsymbol{J} = \frac{b^2}{12} \frac{g}{\nu} J^2 \tag{2.9.20}$$

The orientation of the fissures in space is described with the aid of the elementary solid angle $d\omega$, defined as the elementary area on a sphere of unit radius, surrounding $\boldsymbol{\eta}$. In a general manner $f(b, v_f, \boldsymbol{\eta})$ is the joint probability distribution function of channels of aperture b and volume v_f, whose normal lie within the solid angle $d\omega$ centered at $\boldsymbol{\eta}$. Hence, the average energy density in a statistically homogeneous medium is given by

$$E = \langle E\, h \rangle = n \iiint E(\mathbf{b}, \boldsymbol{\eta})\, f(\mathbf{b}, v_f, \boldsymbol{\eta})\, db\, dv_f\, d\omega \tag{2.9.21}$$

To simplify matters we shall restrict the treatment to the case in which b, v_f and $\boldsymbol{\eta}$ are uncorrelated. Furthermore, the aperture b is taken as constant, the same for all channels. Hence,

$$f(b, v_f, \boldsymbol{\eta}) = f(b)\, f(v_f)\, f(\boldsymbol{\eta}) \quad ; \quad f(b) = \delta(b - b) \tag{2.9.22}$$

Substituting (2.9.22) in (2.9.21) we obtain for the energy

$$E = \frac{nb^2}{12} \frac{g}{\nu} \int J^2 f(\boldsymbol{\eta})\, d\omega \tag{2.9.23}$$

We shall refer $\boldsymbol{\eta}$ and $d\omega$ to a spherical system of coordinates, as depicted in Fig. 2.9.3. Thus,

$$d\omega = \sin\theta\, d\theta\, d\phi \quad ; \quad \boldsymbol{\eta}(\cos\theta\, ,\, \sin\theta \cos\phi\, ,\, \sin\theta \sin\phi) \tag{2.9.24}$$

This representation allows to write the integral in (2.9.23) explicitly as follows

$$\int J^2 f(\omega)\, d\omega = \int [J^2 - (\mathbf{J}.\boldsymbol{\eta})^2]\, f(\omega)\, d\omega =$$

$$= \int_0^{\pi/2}\int_0^{2\pi} \left[J^2 - (J_x \cos\theta + J_y \sin\theta \cos\phi + J_z \sin\theta \sin\phi)^2\right] f(\theta,\phi) \sin\theta\, d\theta\, d\phi \tag{2.9.25}$$

where $f(\theta,\phi)$ is the probability density function of $\boldsymbol{\eta}$. Next, we consider a completely random, isotro-

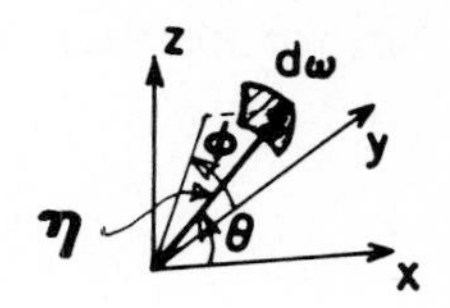

Fig. 2.9.3 Definition sketch of spherical coordinates

pic, medium such that there is equal probability for η to lie in any $d\omega$. Furthermore, we shall select the x axis parallel to **J**. Under these conditions, we get for E (2.9.23), (2.9.25)

$$f = \frac{1}{2\pi} \quad ; \quad J_x = J \, , \, J_y = J_z = 0 \quad ; \quad \int J^2 f(\omega)\, d\omega =$$

$$= \frac{J^2}{2\pi} \int_0^{2\pi} \int_0^{\pi/2} (1-\cos^2\theta) \sin\theta \, d\theta \, d\phi = \frac{2}{3} J^2$$

$$E = \frac{n\, g\, b^2}{18\, \nu} J^2 \tag{2.9.26}$$

Finally, from the definition of the permeability (2.9.14) we have

$$k = \frac{\nu}{g} \frac{E}{J^2} = \frac{n\, b^2}{18} \tag{2.9.27}$$

rendering k in terms of b. To compare this result with existing empirical formulae and with other models, we shall express k in terms of the hydraulic diameter D_H, which is equal to four times the ratio between the void volume and the void-solid contact area. This quantity is used in hydraulics as an unifying geometrical parameter for resistance to viscous flow in channels of various cross-sections. The relationship between D_H and the porosity n, the specific area σ and the solid specific area σ_s, defined as area of solid-void contact divided by the solid volume, is given below. For a fissure of large areal extent compared to b, D_H is simply related to b as follows

$$D_H = \frac{4V_v}{A} = \frac{4n}{\sigma} = \frac{4n}{\sigma_s(1-n)}$$

$$D_H = \frac{4ab}{2a} = 2b \quad ; \quad b = \frac{D_H}{2} = \frac{2n}{\sigma_s(1-n)} \tag{2.9.28}$$

Substitution of the last relationship (2.9.28) into (2.9.27), renders k in terms of these quantities as follows

$$k = \frac{nD_h^2}{72} = \frac{2}{9} \frac{n^3}{\sigma^2} = \frac{2}{9} \frac{n^3}{(1-n)^2} \frac{1}{\sigma_s^2} \tag{2.9.29}$$

and, for instance, in the case of a medium made up from uniform spheres of diameter d

$$\sigma_s = \frac{6}{d} \quad ; \quad k = \frac{1}{162} \frac{n^3}{(1-n)^2} d^2 \tag{2.9.30}$$

These results can be compared with existing semi-empirical formulae. Thus, the Carman-Kozeny (see Dullien, 1979) and Blake-Kozeny (see Bird et al, 1960) formulae are

$$k_{CK} = \frac{nD_h^2}{180} \quad ; \quad k_{BK} = \frac{3nD_h^2}{200} \qquad \text{i.e.} \quad k = 1.11\, k_{CK} = 0.93\, k_{BK} \tag{2.9.31}$$

and the value of k (2.9.29) derived here is very close to, and falls between the two, as shown in the last equation of (2.9.31). It is emphasized that the Carman-Kozeny formula is underlain by a model of capillary tubes and is brought to agreement with experimental values for granular materials with the aid of an empirical tortuosity factor. It is encouraging to find out that the simple model adopted here, which yielded realistic values for the effective diffusivity (Chap. 2.8) works as well for permeability, at least for unconsolidated media.

The present model can be generalized in a few ways. Thus, if the fissures form a regular cubic network, which is a particular case of f (2.9.26), precisely like in Fig. 2.8.1, the result for k is the same. Various anisotropic media can be modeled by selecting particular forms of the p.d.f. f (2.9.26) and such an example is suggested in Exercise 2.9.3.

(ii) Networks of spherical cavities and tubes (Koplik, 1982, 1983).

One of the weaknesses of models of porous media of tubes or channels is their inability to reproduce the complex structure of actual media, which are characterized by large variations of the cross-sections of the streamtubes. A model which attempts to account more realistically for such flow patterns has been developed recently by Koplik (1982,1983). The void space is represented as a regular network of spherical cavities connected by capillary tubes. Such a spherical cavity, with three openings, is shown in Fig. 2.9.4.

After solving in detail the equation of slow viscous flow (2.3.16) for a two-dimensional (cylindrical) cavity, Koplik arrived at the conclusion that for sufficiently small "throat" apertures compared to the radius of the cavity, most of the pressure drop and dissipation take place in the neighborhood of the throat-cavity junction. Furthermore, the pressure drop or dissipation can be easily calculated by approximating each junction as one between a straight tube and a half-space (Fig. 2.9.4b). In such a case the pressure drop along any streamline ensuing (or leaving) the tube into a cavity is given by

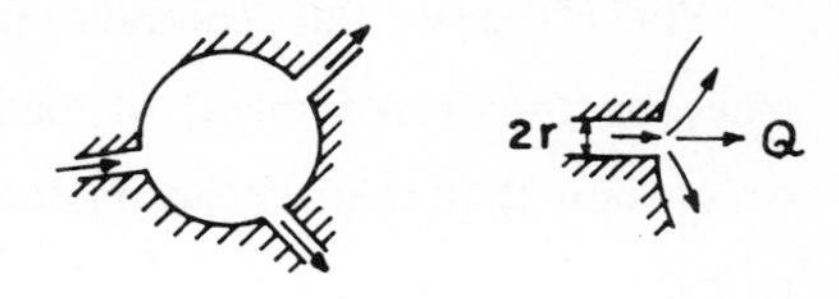

Fig. 2.9.4 An element of Koplik's (1982, 1983) network: (a) a spherical cavity and three tubes and (b) detail of flow at a junction.

$$\Delta p = \frac{8\,\mu Q}{\pi r^3} \tag{2.9.32}$$

where Q is the discharge in the tube.

The model of porous media suggested by Koplik consists of a regular network of tubes connecting spherical cavities. The total pressure (head) drop results from the summing up of the elementary losses occurring in each tube and cavity, the latter being the sum of losses of the different junctions adjacent to the cavity. In addition, conservation of mass is obeyed by equating the sum of discharges Q entering and leaving each cavity to zero.

The bulk of Koplik (1982) work deals with the computation of the effective conductivity of regular networks of elements which have given relationships between the discharge and the conductivity. The networks are random in the sense that the conductivity of various elements is a random variable, with no correlation between adjacent links. By using concepts of the theory of electrical networks, Koplik (1982) evaluated the effective conductivity of the network by a method similar to the self-consistent approximation of Chap. 2.7. The final result is that the actual network of random distributed conductivities k can be replaced by a similar network of elements of equal, effective, conductivity k_{ef}. The latter is given by

$$\left\langle \frac{k_{ef} - k}{k + (1/2m-1)k_{ef}} \right\rangle = 0 \tag{2.9.33}$$

where m is the number of links adjacent to each node, e.g. m=4 for a planar, square network, and m=6 for a cubic one.

The validity of (2.9.33) has been tested by carrying Monte-Carlo simulations of flow through a finite network subjected to a given pressure drop on the boundary. The p.d.f. of the elementary conductivities was rectangular, of a broad distribution, the coefficient of variation being equal to 0.58. The effective conductivity based on the Monte Carlo simulations and on Eq. (2.9.33) were in good agreement. Koplik (1982) mentions that previous works have shown that the self-consistent type of approach may fail if the p.d.f. of conductivities is strongly weighted near k=0, when other approaches, like percolation theory, may be appropriate.

The effective link conductivity k_{ef} increases with m, the coordination number, and the same is true for the "tortuosness" of paths. As we have already mentioned before, Koplik arrives at the conclusion that there is no analogy between tortuosity and the change of network effective conductivity.

Although the network model presented by Koplik (1982) reflects in a more realistic manner than many other models the complexity of the porous structure, it is difficult to compare it with experimental data since the relationships between the geometrical elements of the model and those of porous media are not obvious. Nevertheless, the model offers a theoretical basis for deriving properties of porous media and provides a promising tool for the study of other transport phenomena.

An additional conclusion derived by Koplik (1982) will be discussed in the next Section.

(iii) Periodic array of spheres, numerical solution (Zick and Homsy, 1982).

This is an example of a periodic arrangement of spheres in three possible packings: simple-cubic, body-centered cubic and face-centered cubic. In each type of packing the only independent parameter is the porosity, which can be varied by increasing the distance between the centers of the spheres. For tightest packings, the minimal porosities are 0.4764, 0.3198 and 0.2595, respectively.

Zik and Homsy (1982) have developed an integral equation method for solving numerically the problem of viscous flow of average uniform velocity within a cell. This solution is far from being simple, but the great interest of the results stems from the fact that the solution is exact, within numerical errors, and may serve as a basis of comparison with simple models.

To illustrate this point we have reproduced in Table 2.9.2 the numerical results for permeability (arrays of tightest packings).

TABLE 2.9.2 Permeability values derived by Zick & Homsy (1982) and its comparison with other models.

Arrangement	Porosity n	$\frac{k(1-n)^2}{n^3 d^2}$			
		Numerical solution	Carman-Kozeny	Blake-Kozeny	Eq. (2.9.29)
Simple-cubic	0.4764	$\frac{1}{156.5}$	$\frac{1}{180}$	$\frac{1}{150}$	$\frac{1}{162}$
Body-centered	0.3198	$\frac{1}{141.1}$	$\frac{1}{180}$	$\frac{1}{150}$	$\frac{1}{162}$
Face-centered	0.2595	$\frac{1}{184.8}$	$\frac{1}{180}$	$\frac{1}{150}$	$\frac{1}{162}$

The dimensionless permeability is compared in the same Table 2.9.2 with the figures based on the Carman-Kozeny, Blake-Kozeny and fissured medium formulae, (2.9.30) and (2.9.29), respectively.

In spite of the far-reaching approximations involved in the simple model of fissures, the agreement with numerical results is quite good and the differences are of the order of the ones existing for same porosity but different packings.

2.9.4. Generalizations of Darcy's law

Darcy's law has been enunciated under a few limiting conditions. In the present section we are going to examine briefly a few circumstances in which there is a need to generalize Darcy's law. Only those generalizations which are considered of some interest for problems of water flow in saturated formations are recalled here.

(i) Nonlinear inertial effects.

As we have already mentioned in Chap. 2.3, Darcy's law is a consequence of the linearity of the equations of slow viscous flow (2.3.16), which are obtained from the Navier-Stokes equations (Batchelor, 1967) by neglecting the inertial terms. As it is well-known from similar problems encountered in Fluid Mechanics, like flow past bodies or through conduits of rapidly varying cross-section, the neglect of inertial terms is justified only for low Reynolds numbers. The value at which the slow viscous flow equations cease to be valid depends on the particular problem at hand. It is emphasized that the influence of inertial terms is felt much earlier than the inception of turbulence, a notable exception being the case of one-directional flow through tubes of constant cross-section. It is worthwhile to stress here that a porous medium resembles a series of rapid expansions and constrictions, rather than a collection of straight tubes. We can expect, therefore, that nonlinear effects start to manifest at values of Reynolds characterizing flow past bodies.

Experimental data collected by Rose (1945) indicate that Darcy's law holds for $Re=ud/\nu < 1$, similarly to the range of linear dependence of the drag of a sphere upon Re. However, Darcy's law is in good agreement with experimental data even for $Re<10$. In contrast, the head gradient becomes proportional to the velocity squared for $Re>600$. In the intermediate range some semi-empirical formulae have been suggested in the past, but this topic is of limited interest in applications considered in the present book.

(ii) Nonuniform average velocity field.

One of the prerequisites to deriving Darcy's law, both experimentally and theoretically, is the prevalence of an average uniform flow. In a general manner we have already mentioned in Chap. 2.2 that the statistical nonstationarity associated with the spatial variation of the medium properties or of the average flow can be neglected, if the spatial scale of these variations is large compared to the pore scale. It is of interest to find out, in quantitative terms, what is going to be the deviation from Darcy's law if the average velocity varies rapidly in space. Such cases may occur at the boundary between a porous medium and another type of medium. As a matter of fact, the interest in this topic has been caused by the study of the problem of shear flow in a channel past a porous bottom. In such flows a thin layer, of the order of the pore scale, in which the velocity drops from its value at the channel bottom to zero, is formed in the porous body. The Darcy's law cannot explain the formation of such a layer in which the pressure gradient is zero.

An immediate generalization of the macroscopic momentum equation to account for the variation of the average velocity is achieved by keeping in it (see Eq. 2.6.2) the gradient of the average viscous stress. Thus, the additional term in (2.6.6) is given by

$$\langle \underline{\tau}_{ij}\, h\rangle = \mu <(\frac{\partial u_i}{\partial x_j} + \frac{\partial u_j}{\partial x_i})\, h> = \mu \left[\frac{\partial}{\partial x_j}\langle u_i h\rangle + \frac{\partial}{\partial x_i}\langle u_j h\rangle\right] = \mu\, n \left(\frac{\partial u_i}{\partial x_j} + \frac{\partial u_j}{\partial x_i}\right) \tag{2.9.34}$$

The generalized macroscopic flow equation (2.6.6) thus obtained, in which the drag **D** is still related to the average velocity by Darcy's law, is known as Brinkman equation.

This generalization is, however, inadequate since the drag (or the dissipation) depends also on the gradient of the average velocity. Furthermore, the additional terms which result from the modification of Darcy's law are generally more important than (2.9.34), which is negligible. This point has been raised by Saffman (1971) and elaborated by Dagan (1979a). The latter has suggested the following generalization of Darcy's law (2.9.5) for an isotropic medium

$$\mathbf{D} = \mu \left(- \frac{n}{k}\, \mathbf{q} + r^{(2)}\, \nabla^2 \mathbf{q}\right) \tag{2.9.35}$$

which reduces to Brinkman equation for $r^{(2)} = 1$. The magnitude of the additional medium coefficient $r^{(2)}$ has been evaluated with the aid of a model of porous media suggested by Happel and Brenner (1965). It has been found indeed that under most circumstances $r^{(2)} >> 1$, i.e. the modification of the drag is more important than Brinkman correction.

An important conclusion of application of the generalized Darcy's law (2.9.35) to a few flow problems by Dagan (1979), is that the effect of nonuniformity of the average velocity manifests in very thin boundary layers or small zones, of the order of the pore-scale, near boundaries, and such effects can be generally neglected. A similar conclusion about the robustness of Darcy's law has been reached by Koplik (1982), who has applied a discontinuity of the pressure on the boundary of a network and has solved the flow problem by Monte Carlo simulations (see preceding paragraph).

(iii) Moving solid matrix.

Darcy's law has been written down for a fixed solid matrix. If the solid matrix is also moving, the immediate, and usual, generalization of Darcy's law stems from replacing the fluid velocity $\mathbf{u}$ in the expression of the drag (2.9.5) by the relative velocity of the fluid, i.e.

$$\mathbf{D} = \frac{\rho \nu n^2}{k} (\mathbf{u}_f - \mathbf{u}_s) \quad ; \quad \mathbf{u} = \mathbf{u}_f - \mathbf{u}_s = - \frac{kg}{\nu n} \nabla \phi \tag{2.9.36}$$

where $\mathbf{u}_f$ and $\mathbf{u}_s$ are the macroscopic fluid and solid velocities, respectively, with respect to a fixed frame.

Eq. (2.9.36) is rigorous if the solid skeleton undergoes a translational motion. It is adopted as a general relationship in view of the presumably negligible dissipation associated with rotational velocities of the solid grains.

Exercises

2.9.1 In the case of unsteady flow at small Reynolds number, the local term $\partial \boldsymbol{u}/\partial t$ of the acceleration $\mathbf{D}\boldsymbol{u}/\mathbf{D}t$ has to be kept in the equation of motion. Exercise 2.6.1 required the derivation of the macroscopic momentum equations in this case.

Derive an approximate criterion of validity of Darcy's law for unsteady flow. As a guideline, the solution of the problem of an impulsive application of a pressure gradient on a viscous fluid in a pipe, can be used. This solution can be found in Batchelor (1967), Chap. 4.3. The relevant result is that for the dimensionless time $\nu t/d^2 > 0.2$, the velocity profile is practically the parabolic steady one (ν is the kinematic viscosity and d is the pipe diameter).

2.9.2 In an experiment similar to the classical one of Fig. 2.9.1, leading to Darcy's law, the head drop applied to the porous column starts to change linearly with time after being maintained fixed. Derive the expression of the velocity in the porous medium assuming one-dimensional flow. Under what circumstances do the velocity and discharge differ significantly from the result based on neglect of unsteadiness in the momentum equation? Use a few values of hydraulic conductivity of Table 2.9.1 for applications.

2.9.3 A fissured porous medium consists of a system of planar fissures normal to three distinct directions in space. Assuming that there is equal probability to encounter fissures in each of these directions, derive first the probability density function of $f(\theta,\phi)$ (see Eq. 2.9.24). Subsequently, derive the expression of the permeability tensor under the assumptions of Sect. 2.9.3. Finally, determine the principal directions and the principal values of the tensor if two of the vectors are orthogonal and the third one makes angles of $\cos\alpha = 1/\sqrt{3}$ with the first two.

2.9.4 Derive the permeability of a fissured medium under the assumptions of Sect. 2.9.3 for the case of completely random orientations, but with a rectangular probability distribution function of the apertures. The fissures area can be assumed to be linearly correlated to their aperture.

2.10 CONVECTIVE-DIFFUSIVE TRANSPORT (HYDRODYNAMIC DISPERSION)

2.10.1 Definitions and experimental evidence

In this chapter we are going to discuss the phenomenon of transport of a solute through a saturated porous medium, by both mechanisms of molecular diffusion and of convection by the fluid motion. We shall limit the discussion at present to the simplest case of an inert solute, i.e. a tracer which does not undergo decay or any other change, which does not interact with the solid matrix and which is at sufficiently low concentration, such that the fluid properties are not influenced by the presence of the solute. The equation satisfied by the concentration (2.5.26), in absence of adsorption and for constant porosity, is rewritten here as follows

$$\frac{\partial C}{\partial t} + \mathbf{u}.\nabla C = - \nabla.(\mathbf{q}_m + \langle \boldsymbol{u}' C \; h\rangle) \qquad (2.10.1)$$

In (2.10.1) $\mathbf{u}$ is the macroscopic pore-velocity, which may vary with $\mathbf{x}$ or t, but satisfies the continuity equation (2.5.16), while $\boldsymbol{u}'=\boldsymbol{u}-\mathbf{u}$ stands for the fluid velocity fluctuation. The molecular macroscopic flux $\mathbf{q}_m$ can be expressed with the aid of the effective diffusion coefficient D_m (2.8.1), whereas the presence of the fluid motion manifests in the macroscopic convection via the term $\mathbf{u}.\nabla C$ of (2.10.1) and in transport by the fluctuating velocity field $\boldsymbol{u}'$ in the last term of (2.10.1). The analysis of this latter term is the main objective of this chapter.

Like in other branches of physics, it is customary, under certain conditions, to represent the macroscopic effect of convection by the fluctuating field, as a gradient type, Fickian, process similar to molecular diffusion. In other words it is assumed that

$$\langle \boldsymbol{u}' C \; h\rangle = -\mathbf{D} \; \nabla C \qquad (2.10.2)$$

where $\mathbf{D}$ is the tensor of effective diffusion coefficients. To distinguish the latter from the effective macroscopic coefficient of molecular diffusion (see Sect. 2.8), $\mathbf{D}$ has been termed the hydrodynamic dispersion tensor in the literature.

A qualitative analysis of the process of dispersion can be best carried out along the general theory of diffusion by continuous motions (see Part 4). Towards this aim let us depict a uniform

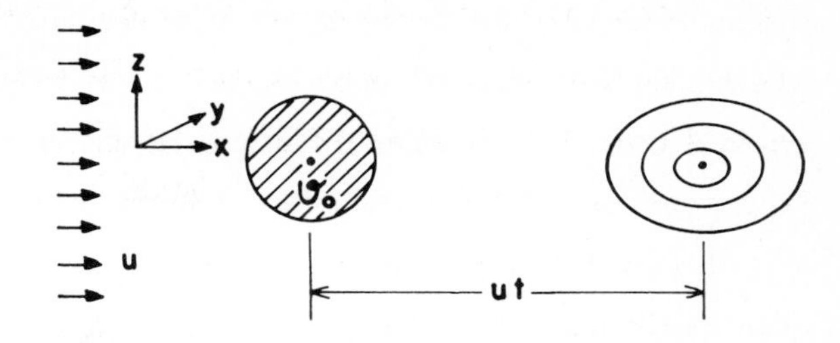

Fig. 2.10.1 Definition sketch for the motion of a solute body in average uniform flow of velocity **u**(u,0,0).

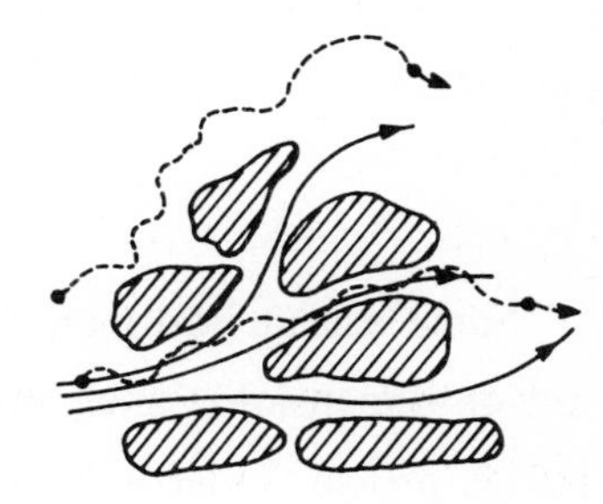

Fig. 2.10.2 Schematical representation of microscopic trajectories of solute particles. Solid lines : hypothetical paths by convection only; broken lines : actual trajectories resulting from convection and molecular diffusion.

macroscopic flow taking place in a homogeneous and isotropic medium and a solute body of constant concentration C_0 inserted at t=0 in a volume V_0, which for simplicity is a sphere of radius R_0 (Fig. 2.10.1). Let consider now a hypothetical solute particle, viewed as indivisible, and follow its motion through the intricate porous structure (Fig. 2.10.2). This motion is the superposition of convection by the fluid microscopic velocity $\boldsymbol{u}$ along streamlines, and a diffusive, brownian type, of motion representing the effect of molecular diffusion. The overall diffusive nature of the motion can be understood by considering simultaneously an additional particle belonging to the same cloud. If such a particle lies at a distance larger than the pore-scale from the first one (Fig. 2.10.2), it will follow a sinous path which is uncorrelated with the first one, due to the random nature of the porous structure. As we show in Part 4, such a motion results at the macroscopic level in a diffusive process, provided that the particles have travelled for a time which is sufficiently large compared to the Lagrangian macroscale. The latter is of the order d/u, where d is the pore-scale, so that for $\mathrm{tu}/d \gg 1$ the process would obey a Fickian type of transport, with an effective diffusivity coefficient of order $O(\mathrm{u}d)$. An interesting question is whether the transport is irreversible, and this can be understood if we follow a particle starting its motion at t=0. Assume now that at time t the flow is reversed. Since the microscopic velocity field is steady, the particle would follow exactly the same path along the streamline passing through the point of departure, in absence of molecular diffusion. However, due to the effect of molecular diffusion, the particle travels across streamlines and its displacements become uncorrelated if t is larger than the typical time d^2/D_m, required to move across a pore by molecular diffusion. Hence, in an experiment in which $td^2/D_m \ll 1$, reversing the flow would return the particles and the cloud to their initial position, whereas an irreversible process of a diffusive nature takes place if $td^2/D_m \gg 1$. We shall return to these points in a more detailed quantitative manner in the next section. Assuming at present that the aforementioned three requirements are met, i.e. $R_0/d \gg 1$, $tu/d \gg 1$ and $td^2/D_m \gg 1$, where t is the travel time and $L \simeq ut$ is the associated travel distance, the macroscopic process can be assumed to be Fickian. Then C satisfies the dispersion equation resulting from (2.10.1), (2.10.2) and (2.8.1)

$$\frac{\partial C}{\partial t} + \mathbf{u}.\nabla C = \nabla.[(D_m + \mathbf{D}).\nabla C] \qquad (2.10.3)$$

Some useful information about the structure of the dispersion tensor **D** can be derived with the aid of dimensional analysis and of requirement of tensorial invariance, along the lines of Poreh's (1965) article. Indeed, let us assume that for similar, isotropic, media defined by the pore scale d solely, $\boldsymbol{D}$ is a function of **u**, d and D_m, i.e.

$$D_{ij} = D_{ij}(\mathbf{u}, d, D_m, n) \qquad (i,j = 1,2,3) \qquad (2.10.4)$$

To investigate the nature of the tensor **D**, Poreh (1965) follows a method which is used extensively in the theory of turbulence (Robertson, 1940), by which D_{ij} is first contracted by two arbitrary vectors **R** and **S**. The inner product $D_{ij}R_iS_j$ is an invariant under a transformation of the Cartesian system of axes and depends, therefore, only on the invariants defined by the three vectors **R**, **S** and **u**, i.e. **R.S** , **R.u** , **S.u** , $|\mathbf{R}|$, $|\mathbf{S}|$ and $|\mathbf{u}|$. Furthermore, $\Sigma\,\Sigma D_{ij}S_iR_j$ is a bilinear function of R_j and S_i, which leads for two arbitrary vectors to the most general dependence

$$D_{ij} = A_1\delta_{ij} + A_2u_iu_j \qquad (i,j = 1,2,3) \qquad (2.10.5)$$

where A_1 and A_2 are scalars depending only on $u = |\mathbf{u}|$, d , D_m and n, while δ_{ij} is the Kronecker delta. Eq. (2.10.5) can be rewritten as follows

$$D_{ij} = D_T\,\delta_{ij} + (D_L - D_T)\,\frac{u_iu_j}{u^2} \qquad (i,j = 1,2,3) \qquad (2.10.6)$$

with D_L and D_T, the only two independent coefficients, being the longitudinal and transverse dispersion coefficients. Indeed, if for simplicity the mean flow is taken in the x direction, i.e.

u(u,0,0), it is seen that in (2.10.6) we get $D_{11}=D_L$, $D_{22}=D_{33}=D_T$, whereas $D_{12}=D_{13}=D_{23}=0$. Thus, in this coordinate system and for constant **u**, the dispersion equation (2.10.3) becomes

$$\frac{\partial C}{\partial t} + u \frac{\partial C}{\partial x} = D_m \nabla^2 C + D_L \frac{\partial^2 C}{\partial x^2} + D_T \left(\frac{\partial^2 C}{\partial y^2} + \frac{\partial^2 C}{\partial z^2} \right) \tag{2.10.7}$$

Thus, D_L , D_T and D_T are the principal values of the dispersion tensor, whereas the direction of the mean flow and those normal to it are the directions of its principal axes. Finally, by considerations of dimensional analysis we have

$$\frac{D_L}{D_m} = f_L\left(\frac{ud}{D_m}, n\right) \quad ; \quad \frac{D_T}{D_m} = f_T\left(\frac{ud}{D_m}, n\right) \tag{2.10.8}$$

where f_L and f_T are dimensionless functions of the Peclet number ud/D_m and of n.

Simple, asymptotic, expressions can be derived for f_L and f_T. If these functions are continuous and analytical in Pe, we can write for Pe<<1 by a power expansion

$$\frac{D_L}{D_m} = \lambda_L \, Pe^2 \quad ; \quad \frac{D_T}{D_m} = \lambda_T \, Pe^2 \qquad (Pe \ll 1) \tag{2.10.9}$$

where λ_L and λ_T are constant.

On the other hand, if Pe>>1, it is reasonable to assume that the dispersion process is dominated by the microscopic convective mechanism and $\boldsymbol{D}$ is independent of the molecular diffusion. Then, if D_m has to drop out from (2.10.8), the functions f_L and f_T must be linear in Pe, and we necessarily have

$$\frac{D_L}{D_m} = \beta_L \, Pe \quad ; \quad \frac{D_T}{D_m} = \beta_T \, Pe \qquad (Pe \gg 1) \tag{2.10.10}$$

where β_L and β_T are dimensionless and constant.

If the regime of (2.10.10) prevails, D_L and D_T can be rewritten in terms of new coefficients

$$\alpha_L = \beta_L \, d \ ; \ \alpha_T = \beta_T \, d \qquad \text{i.e.} \qquad D_L = \alpha_L \, d \ ; \ D_T = \alpha_T \, d \tag{2.10.11}$$

α_L and α_T, which have dimension of length, are called longitudinal and transverse dispersivities, respectively, and they depend on the geometry of the structure solely.

Before analyzing some experimental data, it is worthwhile to mention that Poreh (1965) employs the technique of tensorial invariance to derive the general expression of **D** for anisotropic media (see Exercise 2.10.1). Another point of interest is related to the magnitude of the Reynolds number. Indeed, (2.10.4) is valid only for sufficiently small Re, for which Darcy's law is obeyed (see Sect. 2.9). This requirement may be conflicting with Pe>>1, which underlies (2.10.10). At any rate, viscosity, and subsequently Re, have to be added as independent variables in (2.10.4) and (2.10.8), respectively, if inertial effects become significant. We are not going to deal here with any of these two cases which require generalization of (2.10.8).

The measurement of **D** in the laboratory has been the object of numerous studies. In principle, an experiment like the one depicted in Fig. 2.10.1 can serve for identifying D_L and D_T indirectly. Indeed, it is easy to solve Eq. (2.10.7) analytically, for the initial conditions of Fig. 2.10.1, in order to determine C(x,y,z,t). Subsequently, matching of the measured distribution of C with the analytical solution may provide the values of D_L and D_T by best fit. This is not easy, however, because of experimental difficulties in measuring the concentration in the porous medium.

The common experimental setup employed in order to determine the longitudinal dispersion coefficient is a column filled with porous material in which the fluid at concentration C=0 flows at constant pore-velocity u (Fig. 2.9.1). At t=0, fluid with constant concentration C_0 is introduced at the inlet, at x=0. Subsequently, the concentration of the effluent, i.e. C(L,t), is measured at the outlet x=L. The resulting graph of concentration as a function of time is known as the breakthrough curve. The solution of the dispersion equation with boundary and initial conditions appropriate to this case, i.e.

$$\frac{\partial C}{\partial t} + u\frac{\partial C}{\partial x} = D\frac{\partial^2 C}{\partial x^2} \qquad \text{where } D = D_m + D_L$$

$$C(0,t) = C_0 \quad ; \quad C(x,0) = 0 \tag{2.10.12}$$

is standard and can be obtained by the Laplace transformation method, the result being

$$\frac{C(L,t)}{C_0} = \frac{1}{2}\left[\mathrm{erfc}\left(\frac{L-ut}{2\sqrt{Dt}}\right) + \exp\left(\frac{uL}{D}\right)\mathrm{erfc}\left(\frac{L+ut}{2\sqrt{Dt}}\right)\right] \tag{2.10.13}$$

The second term in (2.10.13) is generally much smaller than the first, their ratio being of order $\sqrt{D/uL}$, and can be neglected if the column is sufficiently long compared to the pore-scale (this condition has to be satisfied anyway to ensure the validity of the dispersion equation). In a standard experiment D_L is determined by a best fit between the measured breakthrough curve and (2.10.13) (see Exercise 2.10.2).

It is easy to ascertain by straightforward integration that in the case in which C satisfies the one-dimensional Eq. (2.10.16), the second moment obeys the following equation

$$\frac{1}{2}\frac{dm_2}{dt} = (D_m + D_{TR})\, m_0 \qquad (2.10.19)$$

where m_0 is the solute total mass divided by A, the area of the cross-section.

One of the main results of Aris (1956), obtained by integration of (2.10.15) over the cross-section of the tube and along it, is that for $tD_m/a^2 >> 1$, m_2 satisfies

$$\frac{1}{2}\frac{dm_2}{dt} = (D_m + \alpha\,\frac{a\bar{u}^2}{D_m})m_0 \qquad (2.10.20)$$

where α is a dimensionless coefficient resulting from the auxiliary function $\alpha^*(r)$. The latter is a function of r which satisfies the following equation in a circle of unit radius

$$\frac{1}{r}\frac{d}{dr}(r\frac{d\alpha^*}{dr}) = -\frac{u'(r)}{\bar{u}} \quad ; \quad \frac{d\alpha^*(1)}{dr} = 0 \qquad (2.10.21)$$

which leads for a circular tube to Taylor's result

$$\alpha = \frac{1}{A}\int|\nabla\alpha^*|^2\, dA = 2\int_0^1 (\frac{d\alpha^*}{dr})^2\, r\, dr = \frac{1}{48} \qquad (2.10.22)$$

Besides its generality and rigor, Aris method permits one to calculate $m_2(t)$ for any t and it also allows to evaluate the time for which it reaches the asymptotic regime (2.10.20). As a matter of fact, the transients decay exponentially with the dimensionless time tD_m/a^2. Another important feature is the evaluation of the skewness moment m_3, as well as of higher odd moments. These also have to decay to warrant the validity of Eq. (2.10.16), and it is found that m_3 tends to zero much slower than the transients of m_2.

2.10.3 Saffman's (1960) model of dispersion in porous media

Unlike the case of dispersion in a tube, there is no general theory or model for dispersion in porous media. A few attempts have been made in the past to derive the structure and magnitude of the dispersion coefficients by some simplified models, similar to those described in the previous sections. It is beyond the scope of this book to review the various models (such reviews may be found in Fried & Combarnous, 1971, or Dullien, 1979), which are of limited success in predicting the main features of the experimental findings. Instead, we shall describe only one model, developed by Saffman (1960), which in spite of its limitations, is believed to capture the essence of the dispersion phenomenon.

Before describing the model, it is worthwhile to mention that it is generally believed that three mechanisms contribute to dispersion: molecular diffusion, convection by nonuniform velocity distribution across streamtubes and branching.

The role played by molecular diffusion has been discussed already in the previous sections. For low Pe numbers diffusion is the main transport mechanism, as shown in detail in Chap. 2.8. At high Pe its direct impact is negligible, but molecular diffusion plays an important indirect role in ensuring the lateral spread of the solute particles across streamtubes.

The nonuniform velocity distribution across streamtubes stems mainly from the nonslip condition at the solid boundary, precisely like in the case of Taylor dispersion. However, unlike a tube, a porous medium is characterized by large variations of the cross-sections of streamtubes, including presence of dead-end pockets. Hence, dispersion related to nonuniform velocity distribution is expected to be much larger in a porous medium than in a tube of constant cross-section.

The branching phenomenon is typical to the intricate three-dimensional structure of a porous medium. It manifests in the separation of neighboring streamtubes, as depicted schematically in Fig. 2.10.2. In its simplest form, this mechanism is captured by the representation of the porous medium as a network (see Chap. 2.9). It is believed that the branching phenomenon is the main factor in explaining lateral dispersion, whereas the nonuniformity of the velocity may explain in principle the longitudinal dispersion along the lines of Taylor dispersion or similar models.

Saffman (1960) models the porous medium as a network of capillary tubes of same length ℓ, of same radius a, and of random orientation (Fig. 2.10.8). The dispersion coefficient is calculated along the lines of the theory of diffusion by continuous movements (see Part 4). Thus, the dispersion coefficient is given by

$$D_m \delta_{ij} + D_{ij} = \int_0^\infty v_{ij}(t',0)\, dt' \quad ; \quad v_{ij}(t,0) = \langle v_i(t,\underset{\sim}{\xi})\, v_j(0,\underset{\sim}{\xi}) \rangle \tag{2.10.23}$$

where v_{ij} is the Lagrangian correlation, **V** is the velocity of the solute particle which at t=0 was at $\mathbf{x}=\underset{\sim}{\xi}$ and **v** is its fluctuation.

Saffman (1960) made the following assumptions on the flow and on the velocity of a particle (Fig. 2.10.8) : (i) in each tube slow viscous flow of parabolic velocity profile (2.10.14) takes place, (ii) the pressure gradient is the projection of the average gradient on the direction of the tube axis (see discussion in Chap. 2.9), (iii) while in a tube, a solute particle undergoes molecular diffusion

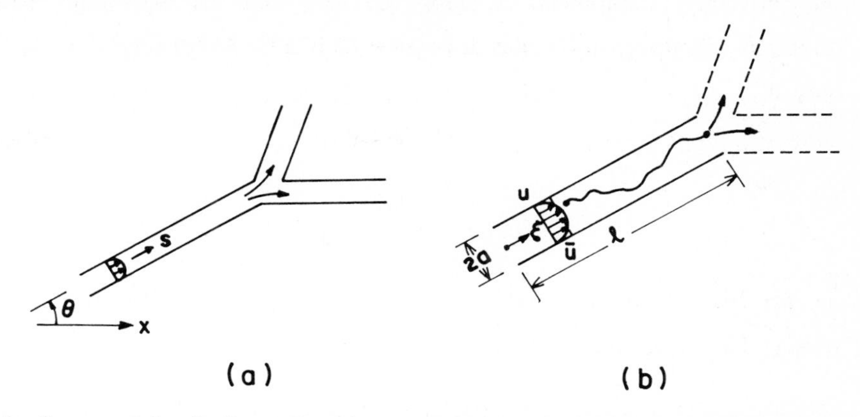

Fig. 2.10.8 Definition sketch for model of dispersion in a network of capillary tubes: (a) the branching phenomenon ; (b) nomenclature for Saffman (1960) model.

by brownian motion and convection by the fluid velocity. These two combine in axial Taylor dispersion which obeys (2.10.16) and (2.10.17), (iv) solute particles arriving at a junction mix uniformly into the fluid and have equal probability of entering any of the tubes which leave the junction and (v) the orientation of the tubes is random and isotropic, with no correlation between directions of neighboring capillaries.

It is emphasized that assumption (iv) has an important impact upon the transport model and it can be justified by the effect of molecular diffusion in junctions, for not too large Pe. This shows that the problem of transport by convective motion differs essentially from that of molecular transport and simple models, like that of Fig. 2.8.1, is not able to predict the observed features of dispersion.

The orientation of a tube in space is given by the unit vector **s**, which in a spherical coordinates system has the Cartesian components $\cos\theta$, $\sin\theta.\cos\phi$, $\sin\theta.\sin\phi$ (see Eq. 2.9.24). The average cross-sectional velocity in a tube is given by

$$\bar{\boldsymbol{u}} = 3\cos\theta\; \mathrm{u}\; \mathbf{s} \tag{2.10.24}$$

where u is the average, uniform, velocity in the x direction. This relationship follows from assumption (ii) above. An ensemble average for all orientations is carried out like in Sect. 2.9.3 by using the p.d.f. $f=1/2\pi$ and integrating over the solid angle $d\omega=\sin\theta\; d\theta\; d\phi$. Thus, it is seen that

$$\int \bar{u}_x\; f\; d\omega = \frac{3u}{2\pi}\int_0^{2\pi} \cos^2\theta\; \sin\theta\; d\theta\; d\phi = u \quad ; \quad \int \bar{u}_y\; f\; d\omega = \int \bar{u}_z\; f\; d\omega = 0 \tag{2.10.25}$$

Saffman regards the velocity of a solute particle as made up, according to assumption (iii), from the convective component $\bar{u}$ along the tube and the Brownian motion velocity u_m in the axial direction, resulting in Taylor dispersion (2.10.17). According to (2.10.17) this representation is valid only if

$$\frac{\ell}{u} >> \frac{a^2}{8D_m} \qquad \text{i.e.} \qquad \frac{u\ell}{D_m} << \frac{8\ell^2}{a^2} \tag{2.10.26}$$

i.e. for small to intermediate Pe numbers, depending on the ratio ℓ/a between the tube length and radius. The components of the velocity fluctuations are given, therefore, by

$$v_x = (\bar{u} + u_m)\cos\theta - u = u(3\cos^2\theta - 1) + u_m \cos\theta$$

$$v_y = (\bar{u} + u_m)\sin\theta\cos\phi = 3\,u\cos\theta\sin\theta\cos\phi + u_m\sin\theta\cos\phi \tag{2.10.27}$$

$$v_z = (\bar{u} + u_m)\sin\theta\sin\phi = 3\,u\cos\theta\sin\theta\sin\phi + u_m\sin\theta\sin\phi$$

The Lagrangian correlation v_{ij} is obtained in two stages. First, ensemble averaging is carried out for a fixed tube over the brownian motion velocity regarded as a random variable. Thus, we have in the axial direction

$$v_{11}(t|\theta,\phi,\xi) = \langle v_x(t,\xi)\, v_x(0,\xi)\rangle =$$

$$= u^2(3\cos^2\theta-1)^2\, P(\xi,\theta,t) + \langle u_m(t)\, u_m(0)\rangle \cos^2\theta \tag{2.10.28}$$

where $P(\xi,\theta,t)$ is defined as the probability that a solute particle released at time t=0 at a distance ξ from the entrance of the capillary has not reached one of the ends after time t. Hence, the first term in the r.h.s. of (2.10.28) expresses the fact that as long as two particles are in the same tube their convective velocity u' (see 2.10.24) is the same, and after that they are uncorrelated, as a consequence of assumptions (iv) and (v) above. Since there is no correlation between $\bar{u}$ and u_m, the brownian motion component in (2.10.27) results in the last term of (2.10.28). It is assumed now that $u_m(t)$ and $u_m(0)$ are correlated for such a short time, that $\langle u_m(t)\, u_m(0)\rangle$ is different from zero when the particles are anywhere in the same tube. We have for this term (see Part 4)

$$\langle u_m(t)\, u_m(0)\rangle = 2\,(D_m + D_{TR})\,\delta(t) \tag{2.10.29}$$

where is δ is the Dirac-delta and $D_{TR} = a^2\bar{u}^2/48D_m = 3a^2u^2/16D_m$ is the Taylor dispersion coefficient (2.10.17).

port, and the error incurred by the simplified relationships (2.10.40) and (2.10.41), is bound to be rather small.

The validity of (2.10.40, 2.10.41), expressing the analogy between dispersive transport of solute and heat, has been confirmed by the experimental study of Green (1963).

Similar generalizations for solutes which undergo decay or adsorption will be discussed in Sect. 2.12.

Exercises

2.10.1 Derive the general expression of the dispersion tensor for an anisotropic, axisymmetric, medium (see Poreh, 1965). Assume that the geometry of such a medium is characterized by a length scale d and a direction in space. Write down the components of the tensor for the particular case of a medium made up from parallel rods.

2.10.2 Let t_{C/C_0} be the time for which the concentration is equal to C/C_0 on the breakthrough curve (Eq. 2.10.13). Derive the relationship between $t_{0.5}$ and the fluid velocity on one hand, and between $t_{0.1}-t_{0.9}$ and the longitudinal dispersion coefficient on the other.

2.11 SUMMARY OF MACROSCOPIC EQUATIONS OF WATER FLOW

In the previous sections we have derived several macroscopic equations of state, of conservation and constitutive equations, which comprise various flow variables. In the present and next section we shall combine these relationships in sets of partial differential equations which permit one to solve, in principle, problems of flow or transport. Since the macroscopic equations result from the averaging of the equations of slow viscous flow, of heat transfer and of diffusion, it is expected that they should also result in closed sets in which the number of dependent variables does not exceed the number of equations. As a matter of methodology we shall start with simple cases and move gradually to more general conditions. The appropriate boundary conditions will be elaborated in Chap. 2.13.

2.11.1 Rigid solid matrix, incompressible and homogeneous fluid

This is the simplest, but nevertheless the most frequent, flow condition. It applies to steady flow of water, in absence of large temperature or concentration gradients (see next section).

The macroscopic flow variables are the specific discharge $\mathbf{q}$ or the velocity $\mathbf{u}=\mathbf{q}/n$ vector, on one hand, and the head ϕ or the pressure $p = \rho g(\phi - z)$, on the other. These are generally functions of $\mathbf{x}$ and t and they satisfy the equation of continuity (Sect. 2.5.2)

$$\nabla.\mathbf{q} = 0 \tag{2.5.16}$$

and Darcy's law (Chap. 2.9)

$$\mathbf{q} = - \mathbf{K}\ \nabla\phi \tag{2.9.8}$$

which have been reproduced here for convenience. Elimination of $\mathbf{q}$ from the above equations leads to the unique equation for the scalar function $\phi(\mathbf{x},t)$

$$\nabla.(\mathbf{K}\ \nabla\phi) = 0 \tag{2.11.1}$$

Eq. (2.11.1) has been written for the general case of an anisotropic and nonhomogeneous matrix, i.e. for the hydraulic conductivity tensor $\mathbf{K}$ depending on $\mathbf{x}$. If $\mathbf{K}$ is isotropic, (2.11.1) becomes

$$\nabla.(K\ \nabla\phi) = 0 \tag{2.11.2}$$

whereas in the case of a homogeneous matrix, for constant K, we obtain

$$\nabla^2\phi = 0 \tag{2.11.3}$$

Thus, in the simplest case of flow through a homogeneous and isotropic medium, the head or the pressure satisfy the Laplace equation, one of the classical partial differential equations of physics. Furthermore, by Darcy's law (2.9.8), $K\phi$ or $K\phi/n$ can be regarded as the potential for $\mathbf{q}$ or $\mathbf{u}$, respectively, i.e. the macroscopic flow is irrotational. The kinematical analogy between flow through porous media and inviscid flow of a fluid (see, for instance, Batchelor, 1967, for a derivation of the latter) is of a mathematical nature, the physical background being entirely different.

Indeed, in the case of inviscid flow, the viscous stresses and the dissipation are neglected, and pressure changes balance accelerations. In contrast, at the microscopic level, the flow through porous media is a slow viscous motion (Chap. 2.3), in which accelerations are neglected. The pressure gradient balances the viscous stresses and the pressure drops along streamlines, even in uniform flow. Vorticity, which is generally associated with presence of solid boundaries, is present at the microscopic level. It disappears at the macroscopic one because of the change of sign throughout the space and mutual cancellation under averaging. The change of the nature of the flow, from a rotational to an irrotational one, is also accompanied by a lowering of the order of the differential equation satisfied by the velocity. Indeed, while ***u*** in slow viscous flow of an incompressible fluid satisfies the biharmonic equation (e.g. Batchelor, 1967 and also Eq. 2.3.18), **u** obeys Laplace equation (2.11.3). This lowering of order has a profound effect upon the macroscopic boundary conditions and this topic will be discussed in Chap. 2.13.

2.11.2 Rigid matrix, incompressible but nonhomogeneous fluid

Here we address again the case in which the compressibilities, i.e. volumic deformations of the matrix or of the fluid, resulting from pressure changes, are negligible. Still, the specific mass of the fluid may vary due to temporal or spatial changes of the temperature or of the concentration of a solute. The change of ρ due to these variations are very small for liquids, and with ρ_0 the average fluid density, we may rewrite (Chap. 2.5)

$$\rho = \rho_0 + \rho' \quad ; \quad \rho'(C,t) = \frac{\partial \rho_0}{\partial C_0}(C - C_0) + (T_f - T_{f\,0})\frac{\partial \rho_0}{\partial T_{f\,0}} \tag{2.5.12}$$

and for the concentrations and temperature variations of interest here, we have $\rho'/\rho_0 \ll 1$. Substitution of ρ (2.5.12) into the equation of conservation of mass (2.5.14) yields

$$\frac{\partial(n\,\rho')}{\partial t} + \rho'(\nabla.\mathbf{q}) + \mathbf{q}.\nabla\rho' + \rho_0(\nabla.\mathbf{q}) = 0 \tag{2.11.4}$$

Let L be a length scale characterizing the gradients of the macroscopic variables **q** or ρ. Inspection of the various terms of (2.11.4) shows that the first three are of order $O(\rho' q/L)$, whereas the last term is $O(\rho_0 q/L)$. Hence, the first are negligible compared to the last, and we can adopt the continuity equation (2.5.17), given above, for an incompressible and homogeneous fluid in this case as well.

Turning now to Darcy's law (2.9.7), we rewrite it as follows by substituting (2.5.12) in it,

$$\mathbf{q} = -\frac{k}{\mu}(\nabla p + \rho g\,\nabla z) = -\frac{k\rho_0 g}{\mu}\nabla\left(\frac{p}{\rho_0 g} + z\right) - \frac{k\rho_0 g}{\mu}\frac{\rho'}{\rho_0}\nabla z \tag{2.11.5}$$

If we define in this case the water head and the hydraulic conductivity by

$$\phi = \frac{p}{\rho_0 g} + z \quad ; \quad K = -\frac{k\,\rho_0 g}{\mu} \tag{2.11.6}$$

Eq. (2.11.5) can be rewritten as follows

$$\mathbf{q} = - K \nabla\phi - K \frac{\rho'}{\rho_0} \nabla z \tag{2.11.7}$$

Eqs. (2.5.17) and (2.11.7) form a system of four equations with five unknowns $\mathbf{q}$, ϕ (or p) and ρ'. The latter depends on C and/or T_f through (2.5.12). To close the system we have to add the macroscopic equations of heat and solute transport (Chap. 2.12).

2.11.3 Deformable elastic matrix, homogeneous and elastic fluid

Although the compressibility of the solid matrix and of liquids is small, in some circumstances it may play an important role in flow through porous formations. This happens, for instance, in unsteady flow through confined formations of large thickness, caused by large and rapid drops of the pressure at the boundary.

The starting point consists now of the equations of conservation of mass of fluid and solid of Chap. 2.5, which are reproduced here for convenience

$$\frac{\partial(n\,\rho)}{\partial t} + \rho\, \nabla.(n\, \mathbf{u}) = 0 \tag{2.5.14}$$

$$\frac{\partial n}{\partial t} - \nabla.[(1-n)\, \mathbf{u}_s] = 0 \tag{2.5.18}$$

where $\mathbf{u}=\mathbf{u}_f$ is the fluid velocity and $\mathbf{u}_s$ stands for the solid velocity, both defined as macroscopic phase averages. It is reminded that ρ_s, the solid specific mass, is assumed to be constant and (2.5.18) reflects the change of porosity which results from the matrix deformation solely (see discussion in Chap. 2.6).

Since Darcy's law (2.9.36) expresses the drag in terms of the fluid-solid relative velocity $\mathbf{u}-\mathbf{u}_s$, Eqs. (2.5.14) and (2.5.18), are subtracted to yield, after multiplication by ρ,

$$n\,\frac{\partial\rho}{\partial t} + \rho(\nabla.\mathbf{u}_s) = -\,\rho\, \nabla.[n\,(\mathbf{u}-\mathbf{u}_s)] \tag{2.11.8}$$

Substitution of Darcy's law (2.9.36) in (2.11.8) and elimination of $\mathbf{u}$ leads to

$$n\,\frac{\partial\rho}{\partial t} + \rho(\nabla.\mathbf{u}_s) = \rho\, \nabla.(K\, \nabla\phi) \tag{2.11.9}$$

In absence of fluid compressibility, i.e. for $\partial\rho/\partial t=0$, and for a rigid matrix, i.e. for $\mathbf{u}_s=0$, Eq. (2.11.9) reduces to the above Eq. (2.11.2) for ϕ. To arrive at an equation for ϕ solely in the general case, we have to express the terms of the left-hand side of (2.11.9) with the aid of constitutive relationships between deformations and stresses for the fluid and for the matrix.

For water, as well as for other liquids, the behavior is elastic and Eq. (2.3.11) leads to

$$\frac{\partial \rho}{\partial t} = \beta \rho \frac{\partial p}{\partial t} \tag{2.11.10}$$

where β, the coefficient of compressibility, is equal to the inverse of the elasticity modulus. For water the appropriate value is $\beta = 4.9 \times 10^{-11}$ cm²/dyn.

As for the solid matrix, the term $\nabla . \mathbf{u}_s$ represents the rate of change of the volumic deformation of the solid matrix e_s, regarded as a continuum, i.e.

$$\nabla . \mathbf{u}_s = \partial e_s / \partial t \tag{2.11.11}$$

since the velocity $\mathbf{u}_s$ is precisely the displacement per unit time of the solid matrix.

The relationship between the deformation of the solid matrix and between the solid stress and the fluid pressure has been the object of many studies in both contexts of soil consolidation and of hydrology. A general approach, in which the matrix is regarded as an elastic medium obeying Hooke's law, has been developed by Biot (1941), and its implications for water flow have been analyzed by Verrujt (1969). The behaviour of the solid matrix of a porous medium in which deformations result from the rearrangement of the grains is very complex and its representation by Hooke's law should be regarded as an approximation at best. In the case of water flow through large porous formations, in which the thickness is much smaller than the horizontal extent, the fluid velocity is essentially horizontal, while the matrix deformation is mainly vertical. Such cases are usually treated by Terzaghi's (1925,1943) simplified theory and by its extension to hydrology by Jacob (1940, 1950), which rely on the concept of effective stress, defined in Chap. 2.6. Starting with the equation (2.6.9) of solid matrix equilibrium in the vertical direction, we have

$$\frac{\partial \sigma'_{zz}}{\partial z} - \frac{\partial p}{\partial z} - g\rho_s (1-n) = 0 \tag{2.11.12}$$

To arrive at (2.11.12) from (2.6.9) we have assumed that the drag $\mathbf{D}$ is essentially horizontal and does not contribute in the z-direction balance and that the solid deformation is vertical, so that the only component of the effective stress contributing to $\nabla \boldsymbol{\sigma}'$ is σ'_{zz}.

Next, we consider an initial state, prior to the deformation of the matrix, and denote the prevailing effective stress and pressure by $\sigma_{zz,o}$ and p_0, respectively. Thus, we may write

$$\sigma'_{zz} = \sigma'_{zz,o} + \sigma''_{zz} \quad ; \quad p = p_0 + p'' \tag{2.11.13}$$

Substitution of (2.11.13) in (2.11.12), with hydrostatic pressure distribution (2.6.3) in the fluid, yields for the initial state

$$\frac{\partial \sigma'_{zz,o}}{\partial z} - \frac{\partial p_o}{\partial z} - g\rho_s(1-n) = 0 \qquad \text{i.e.} \qquad \frac{\partial \sigma'_{zz,o}}{\partial z} = g\rho_t \tag{2.11.14}$$

where $\rho_t = \rho_s(1-n)+\rho n$ (2.5.1) is the medium specific mass.

Subtracting (2.11.14) from (2.11.12) yields for the incremental stress and pressure of (2.11.13)

$$\frac{\partial \sigma''_{zz}}{\partial z} - \frac{\partial p''}{\partial z} = 0 \qquad \text{i.e.} \qquad \sigma'_{zz} - p'' = w \tag{2.11.15}$$

where integration in the z direction leads to the second relationship in (2.11.15). The meaning of these equations is illustrated by consolidation experiments depicted schematically in Fig. 2.11.1. Thus, a saturated sample is confined in a vertical rigid cylinder and is loaded on the top with a weight W, while the hydrostatic pressure is set with the aid of an open vessel. With W and h_o given, the initial field corresponding to Fig. 2.11.1a is completely defined. The determination of $\sigma_{zz,o}$ as a function of z, is left as an exercise at the end of this section.

In Figs. 2.11.1b and c we have indicated separately two factors which may cause incremental stress and pressure: either an additional loading $w=W''/A$ or a change in the pressure by $\Delta h''$. The first case is typical for a consolidation test, whereas the second is representative of groundwater flow. Referring to the last case, i.e. in the absence of incremental loading, we may write from (2.11.15)

$$\frac{\partial \sigma'_{zz}}{\partial t} = \frac{\partial p}{\partial t} \tag{2.11.16}$$

which is the well known Terzaghi (1925) relationship. Next, the volumic deformation in this case stems from vertical deformation solely and for a supposedly elastic matrix we have

$$e_s = e_{zz} = \alpha\, \sigma'_{zz} \tag{2.11.17}$$

where the vertical compressibility coefficient α is related to the Lamme' coefficients μ and λ by

$$\alpha = \frac{1}{2\,\mu + \lambda} \tag{2.11.18}$$

We are now in a position to relate the matrix deformation to p. Indeed, from (2.11.11), (2.11.16) and (2.11.17) we have

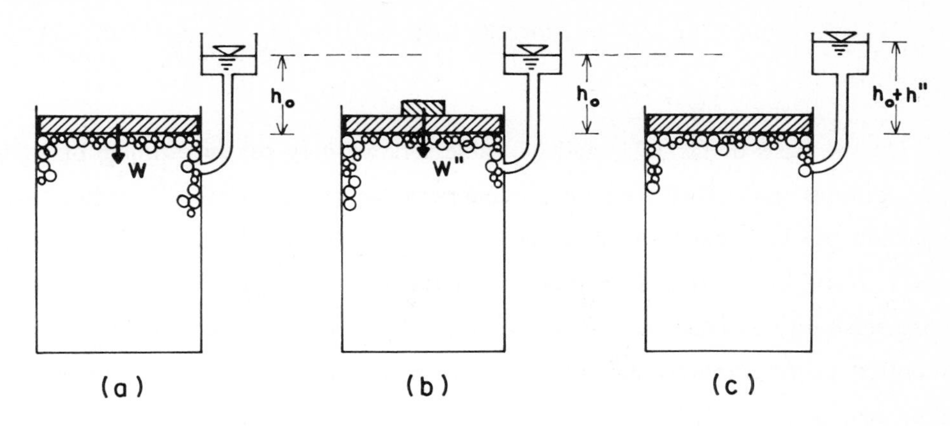

Fig. 2.11.1 Explanatory experiment of consolidation : (a) a confined saturated porous medium subjected to loading W and head h_0. In this state the prevailing effective stress and water pressure are $\sigma'_{zz,o}$ and p_0, respectively ; (b) the incremental effective stress σ''_{zz} results from the additional loading $w = W''/S$ and (c) σ'_{zz} results from the additional head h".

$$\nabla.\mathbf{u}_s = \frac{\partial e_s}{\partial t} = \alpha \frac{\partial p}{\partial t} \tag{2.11.19}$$

and substitution in (2.11.9) yields

$$(n\,\beta\,\rho + \rho\,\alpha)\,\frac{\partial p}{\partial t} = \rho\,\nabla.(K\;\nabla\phi) \tag{2.11.20}$$

Now, since $\partial p/\partial t$ differs from $\partial\phi/\partial t$ by a term of order $p\beta$, which for the value of β for water is exceedingly small for usually encountered pressures, we may rewrite (2.11.20) in the following final form

$$s\,\frac{\partial \phi}{\partial t} = \nabla.(K\;\nabla\phi) \tag{2.11.21}$$

$$s = \rho\,g\,(\alpha + n\,\beta) \tag{2.11.22}$$

The coefficient s, which has dimension of the inverse of a length, is known in the literature as specific storativity. It can be defined in words as the volume of water released by a unit volume of saturated porous medium for a unit drop of the water head. Measurements indicate that clayely soils are more compressible than sand and Freeze (1975) has proposed a linear correlation between the average of the logarithms of α and K. On the basis of Fig. 9 of Freeze (1975), the relationship is approximately as follows

$$\log_{10}\alpha = -\frac{\log_{10}K + 31}{3} \qquad (\alpha \text{ - cm}^2/\text{dyn, K - cm/sec}) \tag{2.11.23}$$

On the basis of (2.11.23) and of Table 9.1, we may get an estimate of s for a few soils. Thus, for pervious sand (K=0.1 cm/sec), semi-pervious sandy soil ($K=10^{-3}$ cm/sec) and poorly permeable soil ($K=10^{-5}$ cm/sec) we get $\rho g\alpha = 9.8\times10^{-8}$, 4.6×10^{-7}, 2.1×10^{-6} cm^{-1} while $\rho g\beta = 4.8\times10^{-8}$ cm-1, respectively. It is seen that except for very pervious formations, the contribution of water compressibility is small compared to that of the matrix. Values larger than the above ones are reported in the literature and they are attributed by some authors to air-bubbles entrapped by the solid matrix.

It is worthwhile to emphasize that Eqs. (2.11.10) and (2.11.20) are underlain by the assumption that the fluid velocity and the solid deformations are small. Otherwise, in the case of large deformations, we have to adopt a Lagrangian approach, i.e. to replace partial time derivatives by total derivatives, in these equations. Such an approach has been followed by Gambolati (1968) for one-dimensional flow, but this generalization may be required for the study of unusually large soil subsidence.

It is also seen that for constant K, the equation (2.11.21) satisfied by the head becomes the classical heat conduction equation. We can define in (2.11.22) a "diffusivity" K/s, which in the above three examples and for, say, n=0.3 has the values K/s= 8.9×10^{5}, 2.1×10^{3} and 4.7 cm²/sec, respectively.

The types of flow summarized in this section are valid for most conditions encountered in groundwater flow. Their application to flow in porous formations is the object of the remaining chapters of this book.

Exercises.

2.11.1 Write Eq. (2.11.1) for a homogeneous and anisotropic medium in a Cartesian system with axes in the principal directions of the hydraulic conductivity tensor. Show that in the case of a homogeneous and anisotropic medium, by a proper linear scaling of the coordinates, the ensuing equation can be reduced to Laplace equation (2.11.3).

2.11.2 Derive the equations satisfied by the water head for horizontal flow of a nonhomogeneous liquid in a rigid aquifer, by using (2.11.4) and (2.11.7) as the starting point.

2.11.3 Derive the distribution of the effective stress $\sigma_{zz,o}$ for the hydrostatic case depicted in Fig. 2.11.1.

2.11.4 The streamfunction $\psi(\mathbf{x})$ is defined in two-dimensional flow by the equation $\mathbf{q} = \text{rot}(\psi\mathbf{k})$, where **k** is a unit vector normal to the plane of the flow. Derive the equations satisfied by ψ in steady flow of a homogeneous fluid in an incompressible medium, for anisotropic and isotropic media.

2.12 SUMMARY OF MACROSCOPIC EQUATIONS OF SOLUTE AND HEAT TRANSPORT.

In this section we are going to combine the various macroscopic state, conservation and constitutive equations of the previous sections into unique equations for the solute concentration or the temperature. Along the lines of the preceding section we will start from simple cases and develop gradually the more complex ones.

2.12.1 Solute transport

The simplest case is the one of molecular diffusion in a fluid in rest. The two basic equations (2.5.27) and (2.8.1) are reproduced herein

$$\frac{\partial C}{\partial t} + \nabla.\mathbf{q}_m = 0 \qquad (2.5.27)$$

$$\mathbf{q}_m = - \mathrm{D}_m \ \nabla C \qquad (2.8.1)$$

It is reminded that C is defined as mass of solute per volume of fluid and the molecular flux $\mathbf{q}_m$ is a fluid phase average as well. Elimination of $\mathbf{q}_m$ from the above equations yields, for a homogeneous and isotropic medium,

$$\frac{\partial C}{\partial t} = \mathrm{D}_m \ \nabla^2 C \qquad (2.12.1)$$

where D_m, the effective diffusion coefficient, is discussed in Chap. 2.8. It is reminded that D_m is proportional to D_m, and for a granular uniform material $\mathrm{D}_m / D_m \simeq 2/3$ (see also Chap. 2.10).

Thus, in this simple case C satisfies the diffusion equation (2.12.1), one of the classical partial differential equations of physics.

We consider now the case in which the solute is not conservative, i.e. its mass undergoes changes due to physical, chemical or biological processes. There is a very large number of such process. They have been quantified by various models and relationships describing the transformation undergone by the solute. A comprehensive and recent review of the state-of-the-art may be found in Abriola (1987). It is beyond our scope to enumerate these processes and we shall refer to a few simple ones only, which are encountered in many applications. The first case is of reversible exchange between the solution and the matrix, e.g. by adsorption and desorption. By (2.5.24) and (2.5.25), the pertinent transport equation is now

$$\frac{\partial C}{\partial t} = D_m \nabla^2 C - \frac{1-n}{n} \frac{\partial C_s}{\partial t} \tag{2.12.2}$$

where it is reminded that C_s was defined as mass of adsorbed solute per unit volume of solid. Eq. (2.12.2) is quite general and may serve as starting point for other phenomena, like exchange between "mobile" and "immobile" water (see exercises at the end of this section).

To eliminate C_s from (2.12.2) we need an additional equation relating it to C. There are many studies of the exchange processes in the soil physics and groundwater quality literature (e.g. Bresler et al, 1982, Abriola, 1987). We shall adopt here the relatively simple, but common, relationship

$$\frac{\partial C_s}{\partial t} = b\,(K'_d\, C - C_s) \tag{2.12.3}$$

known as first-order kinetics. The constant K'_d is known as the slope of the linear isotherm. It represents the ratio C_s/C under equilibrium conditions, for constant temperature (the order of magnitude of the maximum capacity of adsorption of clay minerals is 10% by weight). The constant b (of dimension time^{-1}), characterizing linear chemical kinetics, is of order minutes^{-1} for adsorption by clay, but may be much smaller for organic compounds and interaction with various minerals (see for a few field values Chap. 4.9).

The linear equation (2.12.3) can be easily solved. Assuming, for simplicity that $C_s(\mathbf{x},0)=0$, we get

$$C_s(\mathbf{x},t) = b\,K'_d \int_0^t C(\mathbf{x},t-\tau)\, e^{-b\tau}\, d\tau \tag{2.12.4}$$

Substitution of C_s (2.12.4) into (2.12.3) yields a unique, linear, integro-differential equation for C. A considerable simplification is achieved if the time of interest t is much larger than b^{-1}. Then, we may expend C in (2.12.4) in a Taylor expansion in τ as follows

$$C(\mathbf{x},t-\tau) = C(\mathbf{x},t) - \tau\, \frac{\partial C(\mathbf{x},t)}{\partial t} + \ldots \tag{2.12.5}$$

leading to

$$C_s(\mathbf{x},t) \simeq K'_d\, C(\mathbf{x},t) + \frac{K'_d}{b} \frac{\partial C(\mathbf{x},t)}{\partial t} \qquad (2.12.6)$$

Substitution of the first term of (2.12.6) in (2.12.2) yields the final form

$$R \frac{\partial C}{\partial t} = D_m \nabla^2 C \quad ; \quad R = 1 + \frac{(1-n)\rho_s}{n} K_d \qquad (2.12.7)$$

where $K_d = K'_d/\rho_s$, whose dimension is volume per mass, is also known as sorption distribution coefficient, while the product $(1-n)\rho_s$ is the bulk density of the solid. K_d is the coefficient commonly used in the literature and it is less variable than K'_d.

The constant coefficient $R > 1$ is known as a retardation factor (see, for instance, Freeze and Cherry, 1979, for a detailed discussion), since a scaling of the time variable, i.e. t/R, transforms (2.12.7) into (2.12.1). Eq. (2.12.7) could be obtained from the outset by assuming an instantaneous sorption process. Various generalizations of (2.12.7) can be obtained if the adsorption isotherm is nonlinear, if the kinetics is retarded (see exercises) or if the adsorption and desorption processes follow different isotherms. Such generalizations are not considered here.

A similar extension applies to the case of a radioactive tracer, which decays exponentially with time, i.e. $\partial C/\partial t = -aC$. The constant "a" is related to the half-life , $a = \ell n2/(\text{half-life})$. We now have

$$\frac{\partial C}{\partial t} = D_m \nabla^2 C + S \quad ; \quad S = -\, aC \qquad (2.12.8)$$

replacing (2.12.2). By the simple transformation

$$C(\mathbf{x},t) = C_e(\mathbf{x},t)\, e^{-at} \qquad (2.12.9)$$

we can bring (2.12.8) to the standard form

$$\frac{\partial C_e}{\partial t} = D_m \nabla^2 C_e \qquad (2.12.10)$$

Thus, C is obtained by solving the usual diffusion equation (2.12.21), and multiplying subsequently the solution by exp(-at). The case of mass decay following a law different from the exponential one can be treated in a similar manner (see Chap. 4.9 for a linear rate).

We consider now the important case of transport by diffusion and convection. In the absence of interaction with the matrix and with the decomposition (2.10.1) and (2.10.2) of the convective term, C satisfies the dispersion equation

$$\frac{\partial C}{\partial t} + \mathbf{u}.\nabla C = \nabla.[(\mathbf{D}_m + \mathbf{D})\ \nabla C] \qquad (2.12.11)$$

where generally, for an anisotropic medium, the effective molecular diffusion coefficients form also a tensor $\mathbf{D}_m$. Following our discussion at the end of Chap. 10, $\mathbf{u}$ is assumed to be generally a function of $\mathbf{x}$ and t. If the effective porosity n is variable, the right-hand side of (2.12.11) should be written as $(1/n)\ \nabla.[n(\mathbf{D}_m+\mathbf{D})\nabla C]$. The n variability may be neglected, however, in most circumstances.

In the simple case of an isotropic medium and a uniform flow in the x direction, (2.12.11) becomes

$$\frac{\partial C}{\partial t} + u\ \frac{\partial C}{\partial x} = (D_m + D_L)\ \frac{\partial^2 C}{\partial x^2} + (D_m + D_T)\ (\frac{\partial^2 C}{\partial y^2} + \frac{\partial^2 C}{\partial z^2}) \qquad (2.12.12)$$

Furthermore, if the Peclet number is sufficiently large (see Sect. 2.10) D_m can be neglected as compared to $\mathbf{D}$, and the latter can be replaced, according to (2.10.11), by $D_L = \alpha_L u$ and $D_T = \alpha_T u$.

Finally, the most general case considered here is the one of convection in presence of adsorption and decay, the latter being present in both moving and absorbed states. Combining (2.12.7), (2.12.9) and (2.12.11) results in

$$R\ \frac{\partial C}{\partial t} + \mathbf{u}.\nabla C = \nabla.[(\mathbf{D}_m + \mathbf{D})\ \nabla C] - aRC \qquad (2.12.13)$$

Again, by the simple following transformation

$$t_e = t/R \quad ; \quad C = C_e \exp(-at) \qquad (2.12.14)$$

Eq. (2.12.13) is brought to the standard form (2.12.11), with C and t replaced by C_e and t_e, respectively.

2.12.2 Heat transport

Again, we begin with the simple case of heat conduction. The basic equations, given in Chaps. 2.5 and 2.7, are reproduced here for convenience

$$\rho_t c_t\ \frac{\partial T}{\partial t} + \nabla.\mathbf{q}_h = 0 \quad ; \quad \rho_t c_t = \rho\ c_f\ n + \rho_s c_s (1-n) \qquad (2.5.23)$$

$$\mathbf{q}_h = -\ K_h\ \nabla T \qquad (2.7.1)$$

It is reminded that T is the macroscopic medium temperature defined by Eq. (2.5.3) and K_h is the effective heat conductivity, discussed at length in Chap. 2.7. Elimination of the conductive heat flux $\mathbf{q}_h$ from the above equations, leads to the unique equation for T

$$\frac{\partial T}{\partial t} = D_{mh} \ \nabla^2 T \quad ; \quad D_{mh} = \frac{K_h}{\rho_t c_t} \tag{2.12.15}$$

where D_{hm} is the effective heat diffusivity of the saturated porous medium. Thus, the medium temperature satisfies the well-known heat conductivity equation in homogeneous materials.

We consider now the case in which the fluid is in motion. Eq. (2.5.22) can be rewritten, with the aid of (2.7.1) and (2.10.38) as follows

$$\rho_t c_t \frac{\partial T}{\partial t} + c_f \rho(\mathbf{u}.\nabla T_f) = K_h \ \nabla^2 T + n \rho c_f \ \nabla.(\mathbf{D}_h \nabla T_f) \tag{2.12.16}$$

where it is reminded that T_f is the fluid macroscopic temperature, $\mathbf{u}$ is the fluid velocity and $\mathbf{D}_h$ is heat dispersion tensor, analyzed at the end of Chap. 2.10. Since both T and T_f appear in (2.12.16), one has generally to consider separately the energy equation in the fluid and solid phase, (2.5.21) and (2.5.19), respectively. Furthermore, they have to be supplemented by an equation of transfer, similar to (2.12.3) for solute. However, the two phases reach thermal equilibrium quite quickly, at a time of the order d^2/K_{hs}, where d is the grain-size and K_{hs} is the heat diffusivity of the solid phase. Since the latter is of the order of 10^{-2} cm²/sec, it is seen that for usual porous media the macroscopic temperatures of the two phases equalize almost instantaneously, at the time scales characterizing typical heat transport phenomena, and we can assume that $T_f = T_s = T$. Then, Eq. (2.12.16) can be written in a compact form as follows

$$\frac{\partial T}{\partial t} + \mathbf{u}_h.\nabla T = \nabla.(\mathbf{D}_{eh} \ \nabla T_f) \tag{2.12.17}$$

where

$$\mathbf{u}_h = \frac{\rho c_f}{\rho_t c_t} \mathbf{u} \quad ; \quad \mathbf{D}_{eh} = \mathbf{D}_{mh} + \frac{n \rho c_f}{\rho_t c_t} \mathbf{D}_h \tag{2.12.18}$$

Eq. (2.12.17) resembles the equation of heat conduction and convection by a a free fluid. The velocity $\mathbf{u}_h$, which is smaller than the fluid velocity $\mathbf{u}$, can be regarded as the velocity of propagation of heat fronts. Indeed, in absence of heat diffusion, (2.12.17) shows that T is constant on surfaces moving with velocity $\mathbf{u}_h$. $\mathbf{D}_{eh}$ stands here for the effective heat diffusivity, with $\mathbf{D}_{mh}$, the effective heat conductivity, written for generality as a tensor.

In the case of an isotropic medium, for fluid flow in the x direction and for sufficiently large heat Peclet numbers, (2.12.17) becomes

$$\frac{\partial T}{\partial t} + u_h \frac{\partial T}{\partial x} = (D_{mh} + n\,\alpha_L u_h)\frac{\partial^2 T}{\partial x^2} + (D_{mh} + n\,\alpha_T u_h)\left(\frac{\partial^2 T}{\partial y^2} + \frac{\partial^2 T}{\partial z^2}\right) \tag{2.12.19}$$

The dispersion terms in (2.12.18) are based on the analogy between dispersion of solute and heat, discussed at the end of Chap. 2.10, and on the definition (2.12.18) of $\mathbf{u}_h$. It is reminded that unlike the case of solute transport, the two terms, the conductive and dispersive, may be of comparable magnitude and there is no a-priori reason to neglect one of them.

Exercises

2.12.1 Derive the relationships between the times $t_{0.5}$ and $t_{0.1} - t_{0.9}$ (defined in Exercise 2.10.2 for an inert solute) and the corresponding times for a solute undergoing instantaneous adsorption. Interpret the results in quantitative terms by using some usual values for the coefficients of interest.

2.12.2 Show that the expression of the convective term on the left-hand-side of (2.12.11) is valid even if the porosity n changes with $\mathbf{x}$ and t, provided that ρ depends linearly on C. Rewrite the transport equation in such a case, with proper care to the definition of the dispersion coefficient.

2.12.3 Derive an equation of heat transport by a moving fluid in the case in which the medium is made up from large solid blocks, such that the assumption of instantaneous thermal equilibrium between the two phases is not obeyed. Assume a linear constitutive equation for the heat transfer between phases, similar to (2.12.3).

Extend the approach to solute transport in a medium in which convection is due to flow of "mobile" water in part of the pore-volume, while the "immobile" water is stagnant.

2.13 FLOW AND TRANSPORT BOUNDARY CONDITIONS

In the preceding two sections we have summarized the macroscopic differential equations of flow and transport for water head (or pressure), solute concentration and temperature, respectively. These equations can be solved only if they are supplemented by appropriate initial and boundary conditions. The statement of the initial conditions, i.e. the distribution of ϕ (or p), C or T in space at t=0, is straightforward and does not need elaboration.

The matter is different with regard to boundary conditions, which may occur either at the boundary between the porous medium and a different medium (e.g. impervious solid or free fluid) or in the interior of the porous material (e.g. a free surface). Boundary conditions at the microscopic level are well defined and have been given in Chap. 2.3. The formulation of the corresponding macroscopic boundary conditions is a delicate matter, since statistical homogeneity ceases to prevail in the neighborhood of the boundary. The difficulty stems from the fact that the macroscopic equations of flow and transport have been derived under conditions of statistical homogeneity or slow spatial variation of the statistical moments of quantities of interest. These conditions underlie the derivation of the effective heat and solute diffusion coefficients, of permeability and of dispersion coefficients, in Chaps. 2.7-2.10. Hence, the formulation of the boundary conditions requires a careful examination of the equations of flow and transport, which may change their structure near a boundary. We shall examine in the sequel a few boundary conditions encountered frequently in applications.

2.13.1 Flow condition at an impervious boundary.

A planar boundary between a porous medium and an impervious solid is depicted schematically in Fig. 2.13.1. Whereas ensemble averages of flow variables are defined at any point, ergodicity may prevail only if space averaging is carried out over planes parallel to the boundary. Starting with porosity, let us assume that the medium is homogeneous up to the boundary, i.e. n is constant for $z>0$ and jumps to zero at z=0. As a matter of fact, even if the medium is homogeneous far from the boundary, n may vary in a thin layer adjacent to the boundary.

In the case of a viscous fluid, the boundary condition prevailing at z=0 is (2.3.25), i.e. u=0, and by ensemble or space averaging on the plane z=0, we arrive at an identical equation for the macroscopic velocity, i.e. $\mathbf{u}=0$. Referring first to the normal component and to Darcy's law (2.9.7), we have

$$u_z = 0 \quad ; \quad q_z = 0 \quad ; \quad \frac{\partial \phi}{\partial z} = 0 \tag{2.13.1}$$

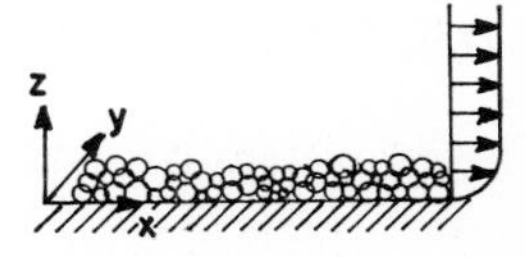

Fig. 2.13.1 Definition sketch for flow in a porous medium near an impervious planar boundary.

This basic condition does not depend, of course, on the nature of the fluid, i.e. on viscosity, and it follows from the very definition of the impervious boundary. Since the water head ϕ satisfies Laplace (2.11.3) or the parabolic (2.11.21) equations, the boundary condition (2.13.1) of Neuman type is sufficient for solving for ϕ. The additional requirement $u_x = u_y = 0$, i.e. ϕ=const, is superfluous and leads to an improperly posed problem. Furthermore, as it is well known from theory of flow of inviscid fluids, there is generally slip along the boundary if ϕ obeys (2.13.1), i.e. $\partial\phi/\partial x$ and $\partial\phi/\partial y$ are different from zero at z=0. This apparent contradiction is related to the lowering of the order of the equation of flow, which has been already discussed in Sect. 2.11.1. It is well known from similar problems of fluid mechanics and heat transfer that the transition from the boundary to the Darcian velocity is achieved through a boundary layer, as depicted in Fig. 2.13.1. In this layer Darcy's law is no more valid, and the drag **D** depends not only on **u**, but also on its derivatives. This point has been discussed in Sect. 2.9.4, where a generalization of Darcy's law (2.9.35), proposed by Dagan (1979a), has been forwarded. Based on this generalization and on a simplified model of porous medium, Dagan (1979a) has arrived at the following velocity distribution in the boundary layer

$$u_x(z) = u_x^D(0)\,[1 - \exp(-z/\sqrt{\lambda})] \tag{2.13.2}$$

where u_x is the macroscopic velocity component parallel to the wall, u_x^D is the Darcian velocity resulting from the solution of Laplace equation, and λ is a medium constant which has the dimension of a square of a length. It has been found for a simplified model of spherical particles that $\lambda \simeq d^2/80$, where d is the particles diameter. Thus, the boundary layer is extremely thin and can be neglected in most circumstances. The usual approach, in which Darcy's law is applied up to the wall and (2.13.1) is adopted as boundary condition, is seen to be very accurate and shall be adopted in the sequel. Still, the elaboration of this point is of fundamental interest and may have also meaningful applications in other circumstances (see Sect. 2.13.3).

2.13.2 Boundary between two homogeneous porous bodies.

This is the case of two homogeneous porous media which are in contact along a plane z=0 (Fig. 13.2). In reality, it is hard to figure out that the two media are separated by a sharp boundary and it is rather conceivable that they interpenetrate and the transition is gradual.

Assuming, nevertheless, that the boundary is sharp, two flow conditions have to be obeyed at z=0. First, mass conservation across the plane requires

$$q_z^{(1)} = q_z^{(2)} \quad \text{i.e.} \quad K^{(1)} \frac{\partial\phi^{(1)}}{\partial z} = K^{(2)} \frac{\partial\phi^{(2)}}{\partial z} \tag{2.13.3}$$

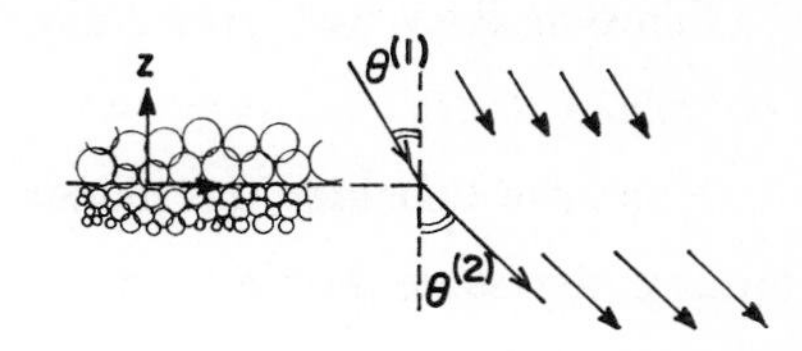

Fig. 2.13.2 Definition sketch for flow near a planar boundary between two different porous media.

where $K^{(1)}$ and $K^{(2)}$, are the hydraulic conductivities of the two media. Similarly, the pressure p, and consequently the head ϕ, are continuous in the fluid at both micro- and macroscopic levels. Thus, we have

$$\phi^{(1)} = \phi^{(2)} \quad \text{i.e.} \quad \frac{\partial\phi^{(1)}}{\partial s} = \frac{\partial\phi^{(2)}}{\partial s} \tag{2.13.4}$$

where s is the direction of the projection of the specific discharge **q** in either medium, on z=0.

Defining by θ the angle between **q** and the normal (Fig. 2.13.2), i.e.

$$\tan\theta^{(1)} = \frac{\partial\phi^{(1)}/\partial s}{\partial\phi^{(1)}/\partial z} \quad ; \quad \tan\theta^{(2)} = \frac{\partial\phi^{(2)}/\partial s}{\partial\phi^{(2)}/\partial z} \tag{2.13.5}$$

we get from (2.13.3) and (2.13.4) the relationship

$$\frac{\tan\theta^{(1)}}{\tan\theta^{(2)}} = \frac{K^{(1)}}{K^{(2)}}) \tag{2.13.6}$$

Hence, the vectors **q** and the macroscopic streamlines satisfy Snell laws of optical refraction, namely $\mathbf{q}^{(1)}$, $\mathbf{q}^{(2)}$ and the normal z are in the same plane while $\theta^{(1)}$ and $\theta^{(2)}$ are interrelated through (2.13.6). This completes the derivation of the boundary conditions necessary in order to match the water heads on the separation plane. Eq. (2.13.4) implies that the macroscopic velocity $\mathbf{u}=\mathbf{q}/n=-(K/n)\,\nabla\phi$ has a discontinuous tangential component at z=0. Like in the case of an impervious boundary, this result contradicts the requirement of continuous microscopic velocity obeyed by a viscous fluid. Again, this apparent paradox may be solved along the lines of the previous section, namely by generalizing Darcy's law in a thin boundary layer near z=0. It was shown in this case too (Dagan, 1979a) that the boundary layer is very thin, and although the streamlines bend continuously through it, (2.13.6) is an accurate approximation.

2.13.3 Boundary with free fluid

First, we consider the planar boundary between a saturated porous medium and a reservoir containing the same liquid, with flow normal to the boundary (Fig. 2.13.3). In fact this case can be considered as a limit of the previous one for $K^{(1)} \to \infty$, leading to the two boundary conditions

$$q_z = u_{z,ext} \quad ; \quad \phi = \frac{p_{ext}}{\gamma} + z \tag{2.13.7}$$

where $\mathbf{u}_{ext}$ and p_{ext} are the velocity and the pressure in the free fluid, respectively. It is emphasized that the viscous head loss in the free fluid is generally so small compared to that occurring in the porous medium, that we can neglect it and regard the fluid outside as inviscid. Furthermore, as far as the solution of the flow equation in the porous medium is concerned, only one of the two conditions of (2.13.7) can be imposed at z=0. The other condition can be used in order to determine the corresponding variable in the free fluid.

An interesting point is related to the discontinuity between the normal velocity $u_z = q_z/n = u_{z,ext}/n$ in the porous medium and between the outer normal velocity $u_{z,ext}$. In reality, there exists a thin layer around z=0 through which a continuous transition takes place, and as a matter of fact the boundary itself can be defined up to a distance of the order of the pore-scale along the normal.

Another case of theoretical interest in context of flow in channels is depicted in Fig. 2.13.4. Shear flow takes place in a fluid layer overlying a porous bed saturated with the same fluid. Unlike the previous case, there is no flow normal to the boundary. Furthermore, there is no pressure gradient along the boundary, the water head being therefore constant on z=0. It is easy to ascertain that the Laplace equation (2.11.3) admits in this case the unique solution ϕ=const within the porous medium, i.e. no flow takes place there. If this is the case, one would expect the velocity in the free fluid to be zero at the boundary, i.e. the porous body behaves like a solid boundary for the external flow. However, experiments have shown that is not the case and this problem has been investigated theoretically by Saffman (1971), who has suggested a general extension of Darcy's law. The same problem has been solved approximately by Dagan (1979a) with the aid of the generalized Darcy's law (2.9.35). The result is again a thin boundary layer within the porous medium in which the velocity drops from its value in the free fluid, at z=0, to zero. Thus, from a macroscopic point of view the presence of the porous bed manifests in a small slip velocity which depends on the properties of the medium. We shall not elaborate here upon the results, since this case is of limited interest in hydrological applications and we shall neglect the tangential motion induced by an external shear flow.

Finally, a case of importance in many applications, is the outflow from a porous medium saturated with a liquid in the free space, occupied by air at atmospheric pressure (Fig. 2.13.5). The plane z′=0 along which the water flows outwards is known in the literature as a seepage face. In

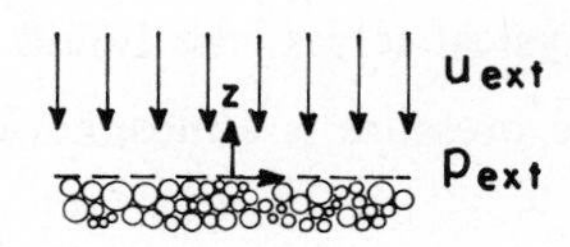

Fig. 2.13.3 Definition sketch for flow normal to a planar boundary between a free fluid and a saturated porous medium.

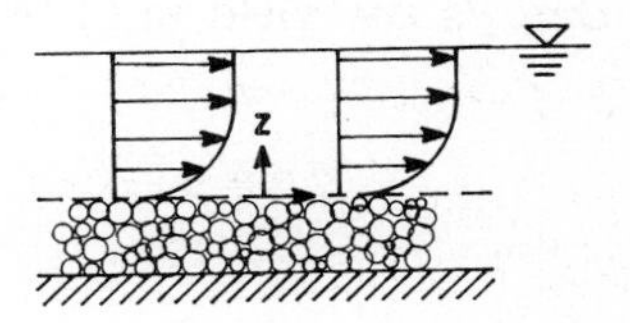

Fig. 2.13.4 Definition sketch for flow parallel to a planar boundary between a free fluid and a saturated porous medium (shear flow in a channel with a porous bottom).

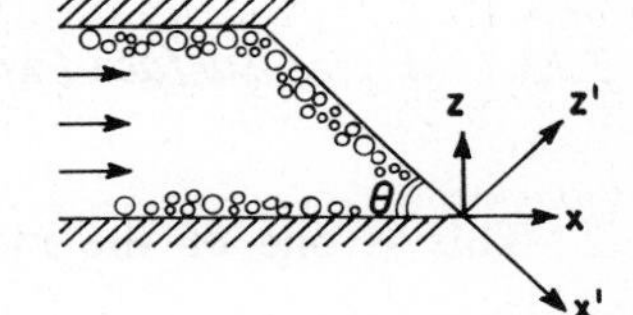

Fig. 2.13.5 Definition sketch for flow through a planar seepage face.

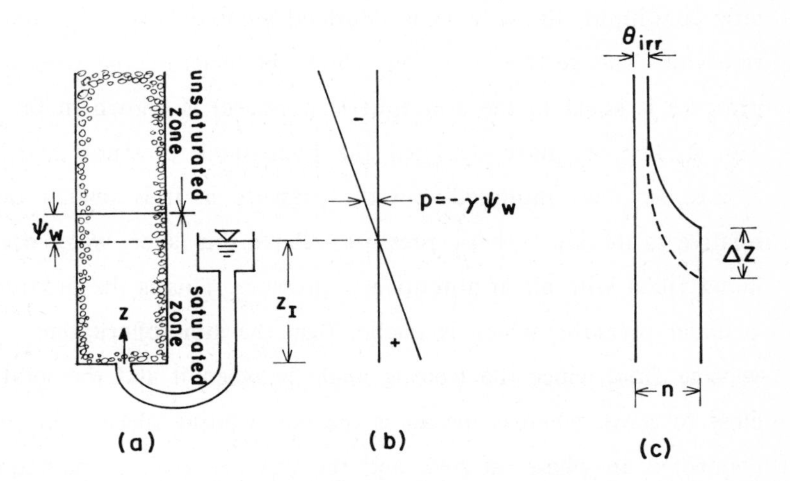

Fig. 2.13.6 Pressure and water content distributions beneath and above a free-surface under hydrostatical conditions : (a) definition sketch of vertical column; (c) water (moisture) content, full line for the initial stage (level at z_I) and dashed line for the final stage (level at $z_F = z_I - \Delta z$) and (b) pressure distribution.

physical terms, the liquid which leaves the porous medium spills in a thin layer along the slope and the pressure is atmospheric within it. The appropriate boundary condition is therefore

$$p = p_{at} \quad \text{i.e.} \quad \phi = \frac{p_{at}}{\gamma} + z = \frac{p_{at}}{\gamma} - x' \sin\theta \qquad (\text{ on } z'=0) \tag{2.13.8}$$

since the pressure is continuous at the boundary. Differentiation along the boundary $z'=0$ and Darcy's law yield in (2.13.8)

$$q_{x'} = - K \frac{\partial\phi}{\partial x'} = K \sin\theta \tag{2.13.9}$$

i.e. the tangential component of the specific discharge is constant. Its maximum occurs for a vertical seepage face for which $\theta=\pi/2$ and $q_{x'}=K$.

2.13.4 A free-surface (water-table, phreatic surface)

This is one of the most complex cases of fluids flow in a porous medium. To get acquainted with the phenomenon we shall consider first the static case. In Fig. 2.13.6a we have depicted a vertical porous column which is connected at its bottom with a constant level reservoir. Under hydrostatic conditions, the column is saturated up to a height ψ_w above the level $z=z_I$ of the water in the reservoir. The surface $z=z_I$ on which the macroscopic pressure is equal to zero (i.e. the absolute pressure is equal to the atmospheric pressure) is known in the literature as the phreatic surface. In Fig. 2.13.6b we have depicted the hydrostatic pressure distribution along the vertical. The layer $z_I<z<z_I+\psi_w$ is saturated, but the pressure in it is smaller than zero, i.e. water is under tension, relative to the atmospheric pressure. Above this layer, water occupies only part of the voids, the rest being filled with air at atmospheric pressure. This is the unsaturated zone and the water phase there is under pressure, which is smaller than the atmospheric one. The water constitutes in this case the wetting fluid, since the wetting angle between it and the solid is positive (as a matter of fact it is close to zero), whereas the air is the non-wetting phase. The pressure difference between the interconnected air phase, at $p=0$, and the water at $p<0$, is maintained by the surface tension acting on the menisci separating the two fluids (see Eq. 2.3.26). The pressure $p=-\gamma\psi_w$ acting at the free-surface (Fig. 2.13.6b) is known as bubbling-pressure or as air-entry-value. The latter nomenclature is related to an experimental setup in which air is pushed into a saturated sample and $-p$ is the extra pressure needed in order to penetrate into the largest pores.

On a macroscopic level, water in the unsaturated zone is characterized by the moisture content θ, defined as volume of water per unit volume of medium, by the pressure $p=-\gamma\psi$, where the positive

ψ is known as suction, and by the hydraulic conductivity K (it is generally assumed that Darcy's law is obeyed in the unsaturated zone as well). The relationships between these macroscopic variables are dealt with extensively by soil physics, and a succinct presentation may be found in the recent treatise by Bresler et al (1982). Such a presentation is beyond our scope and we shall limit ourselves only to discussing a few general concepts in a descriptive manner.

On both experimental and theoretical grounds it is assumed that there is a relationship between θ and ψ, well defined for a given soil, and the graph $\psi(\theta)$ is known as the retention curve. A physical explanation to such a relationship is that as suction increases, the menisci have to reduce their radii of curvature, and this is possible if they occur in smaller pores. Thus, an increase of the suction results in drainage of the largest pores filled with water. In Fig. 2.13.6c we have represented schematically the variation of θ along the column, assuming that at saturation $\theta=n$.

At high values of ψ, the water content tends to a constant value known as θ_{ir}, the irreducible saturation. This may be regarded as the volume of water trapped in pockets which are connected so weakly, that further increase of the suction cannot remove the water.

Returning now to the free-surface boundary condition, let us consider the following simple experiment: at t=0 the level in the external reservoir of Fig. 2.13.6a is lowered suddenly from its initial height z_I to z_F, i.e. by a drop of $\Delta z=z_I-z_F$. As a result, a transient drainage process takes place and after a sufficiently long time has elapsed, the water reaches again hydrostatic equilibrium, with the free-surface, phreatic surface and entire θ profile translated downwards by Δz.

The total volume drained ultimately from the column is given by

$$V_\infty = n_{ef}\, \Delta z\, A \quad ; \quad n_{ef} = n - \theta_{ir} \tag{2.13.10}$$

which is illustrated in Fig. 2.13.6c. In (2.3.10) n_{ef} stands for the effective drainable porosity, which is smaller than n, due to the water left at content θ_{ir} in the soil.

Vachaud (1968) has carried out laboratory experiments of this type. The soil was a graded river sand, of granulometry shown in Fig. 2.13.7. In the same figure we have reproduced the measured $\psi(\theta)$ and $K(\theta)$ curves. The unsaturated zone is characterized by a length scale related to the retention curve, namely the value of ψ for which $\theta\simeq\theta_{ir}$ and $K\simeq0$. Under the static conditions of Fig. 2.13.6, ψ is the height of the capillary fringe in which $\theta>\theta ir$ and in the example of Fig. 2.13.7 $\psi\simeq60$ cm.

The results of Vachaud's experiments are represented in Fig. 2.13.8 (reproduced from Dagan and Kroszynski, 1973) in a dimensionless form. Thus V is the volume of water drained through the column bottom up to time t and q_0 is the specific discharge at t=0. By Darcy's law applied to the saturated zone, we have $q_0=K_s(z_I-z_F)/(z_I+\psi_w)$, where K_s is the hydraulic conductivity of saturated soil. In Vachaud's study, experiments differed by the ultimate length of the saturated zone.

A complete theoretical solution of the same problem requires solving simultaneously the equation of saturated flow, i.e. Laplace equation, beneath the free-surface and the equation of unsaturated

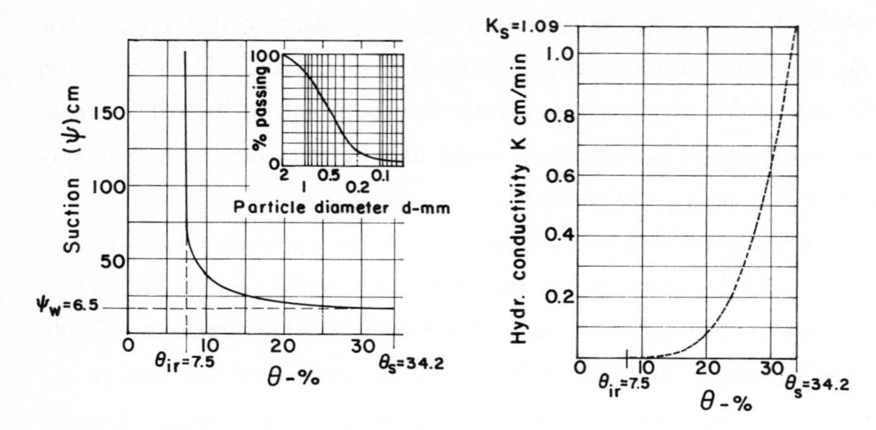

Fig. 2.13.7 Examples of retention and hydraulic conductivity curves of the river sand tested by Vachaud (1968) (reproduced from Dagan & Kroszinski, 1973)

flow (the so called Richard's equation) above the water table. The two zones are matched at the moving free-surface by the requirements of continuity of the pressure and of the specific discharge, which represent the complete boundary conditions at a free-surface. Such a solution, underlain by the experimental information embedded in the two curves of Fig. 2.13.7, can be achieved only numerically, because of the nonlinear nature of the equation of unsaturated flow.

A tremendous simplification of the flow problem and of the boundary condition is achieved if it is assumed that the unsaturated θ profile translates together with the free-surface, i.e. $\theta(z,t)=\theta(z-\eta)$ where $z=\eta(t)$ is the equation of the free-surface. Such an assumption can be justified if the free-surface moves downward so slowly, that the unsaturated zone adapts itself "instantaneously" to the changing position of the water table. By the same token, this requirement of slowness can be expressed with the aid of the ratio between the height of the capillary fringe $\psi(\theta_{ir})$ and that of the saturated zone z_I or z_F, i.e.

$$\frac{\psi(\theta_{ir})}{z_F} << 1 \tag{2.13.11}$$

If this assumption is adopted, the two conditions prevailing on the free-surface are as follows

$$p = -\gamma\,\psi_w \qquad ; \qquad q = n_{ef}\,\frac{d\eta}{dt} \tag{2.13.12}$$

The first condition, the dynamical one, is straightforward and expresses the constancy of the pressure, which is equal to the entry-air-value at the free-surface (Fig. 2.13.6).

The second condition, of a kinematical nature, stems from the conservation of mass. Indeed, let us consider a control volume bounded by a fixed plane at z, beneath the water table $z=\eta$, and the top of the column z=L. Since there is no flow at the top, conservation of mass of the water (both water and the solid matrix are regarded as incompressible) requires

$$q(z,t) = \frac{\partial}{\partial t}\int_z^L \theta(z',t)\, dz' \tag{2.13.13}$$

where the integral in the r.h.s. is the water volume per unit area contained in the control volume. The integral can be split in two parts, beneath and above the free-surface, as follows

$$\int_z^L \theta\, dz' = \int_z^\eta n\, dz' + \int_\eta^L \theta\, dz' \tag{2.13.14}$$

and time differentiation yields

$$\frac{\partial}{\partial t}\int_z^L \theta\, dz' = n_{ef}\frac{d\eta}{dt} + \frac{\partial}{\partial t}\int_\eta^L (\theta-\theta_{ir})\, dz' \tag{2.13.15}$$

Substituting (2.13.15) in (2.1.3.13) and letting $z\to\eta$, leads to the general relationship

$$q = n_{ef}\frac{d\eta}{dt} + \frac{\partial}{\partial t}\int_\eta^L (\theta-\theta_{ir})\, dz' \qquad (z=\eta) \tag{2.13.16}$$

Now, when the θ profile translates downwards, as depicted in Fig. 2.13.6c, the last integral in (2.13.16), which is equal to the water volume in the unsaturated zone, remains constant and its time derivative vanishes. Then, Eq. (2.13.12) is recovered, and the drainage from the unsaturated zone into the saturated one is instantaneous.

It is easy to solve the problem of saturated flow in the vertical column for an initial sudden drop and with the simplified boundary conditions (2.13.12). The solution for $V(t)$, whose derivation is left as an exercise, is given by

$$\frac{V(t)}{V_\infty} - \frac{z_F+\psi_w}{\Delta z}\,\ell n\left(1-\frac{V}{V_\infty}\right) = \frac{z_I+\psi_w}{\Delta z}\,\frac{q_0 t}{V_\infty} \tag{2.13.17}$$

This curve is represented in Fig. 2.13.8 for the extreme case of $z_F=0$. The validity of approximation (2.13.12) can be checked by comparing its outcome (2.13.17) with experiments. Comparison with experiments, as in Fig. 2.13.8, shows indeed that the actual volume is smaller than that of (2.13.17), i.e. there is a delayed yield from the unsaturated zone. It can be shown, however, that as z_F increases, the experimental results approach the solution (2.13.17) and they become very close for, say, $\psi(\theta_{ir})/z_F < 0.05$.

A comprehensive discussion of the validity of the simplified free-surface conditions (2.13.12), including the analysis of the role played by the matrix compressibility in the unsaturated zone, has been provided by Brutsaert and El-Kadi (1984). Comparison with a few experimental and numerical studies has led them to the conclusions that matrix compressibility is generally negligible in the unsaturated zone and that the assumption of instantaneous drainage is bound to be obeyed for the criterion given above.

In most cases of aquifer flow the thickness of the saturated part of the formation is indeed much larger than the length scale of the unsaturated zone, and the assumption of instantaneous drainage can be safely adopted. There are cases, however, of thin aquifers and clay soils, for instance, for which (2.13.11) is not obeyed, and the complete flow problem has to be solved to obtain q (2.13.16).

The analysis so far has been confined to the vertical flow in a porous column. We shall derive now the free-surface boundary condition under more general conditions of three-dimensional flow. To generalize the kinematical condition (2.13.16), we consider the conservation of mass of the liquid in a infinitesimal vertical column (Fig. 2.13.9) of horizontal cross-section $\Delta A = \Delta x . \Delta y$. The column extends from a horizontal plane z, within the saturated zone, to z=L, where the water content is constant and equal to θ_L. For generality we shall assume that vertical recharge from an external source occurs through the top. Its intensity (volume per unit area and unit time) R, is taken as positive for recharge and negative for uptake. Under this condition, θ_L, which prevails far above the free-surface, is larger than the irreducible moisture content θ_{ir}, and only for R=0, $\theta_L = \theta_{ir}$.

The equation of conservation of mass can be now written as follows

$$\Delta A \frac{\partial}{\partial t} \int_z^L \theta \, dz' - (q_z + R) \Delta A + \int (\mathbf{q}_H . \boldsymbol{\nu}) \, dA = 0 \tag{2.13.18}$$

where $\mathbf{q}_H$ is the specific discharge vector in the horizontal plane, i.e. of components q_x, q_y and 0, while $\boldsymbol{\nu}$ is a unit vector normal to the vertical sides of the infinitesimal column. The first term in (2.13.18) represents the change of water volume per unit time in the control volume, the second one is the flux through the top and the bottom and the last one represents the net lateral flux (Fig. 2.13.9). This latter term can be rewritten as follows

$$\int (\mathbf{q}_H . \boldsymbol{\nu}) \, dA = \Delta A \left[\frac{\partial}{\partial x} \int_z^L q_x \, dz' + \frac{\partial}{\partial y} \int_z^L q_y \, dz' \right] = \Delta A \, \nabla . \int_z^L \mathbf{q}_H \, dz' \tag{2.13.19}$$

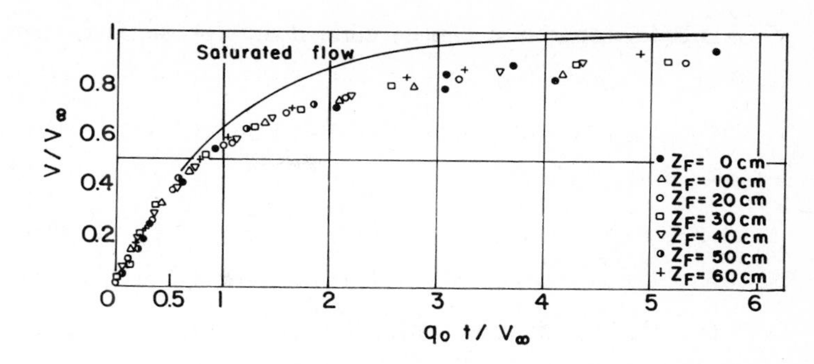

Fig. 2.13.8 Variation of the drained volume V with time for different lengths of the saturated zone z_F : experimental points of Vachaud (1968) and continuous curve based on instantaneous drainage approximation (2.13.17) (reproduced from Dagan & Kroszinski, 1973).

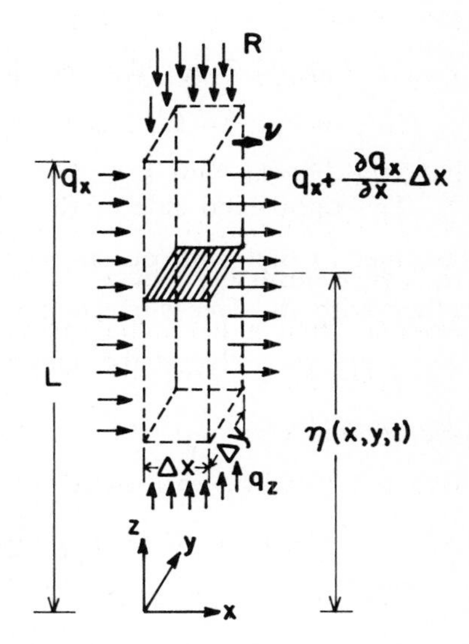

Fig. 2.13.9 Definition sketch for free-surface flow within an infinitesimal column of horizontal cross-section $\Delta A = \Delta x\ \Delta y$.

after expressing $\mathbf{q}_H$ explicitly with the aid of its Cartesian components and by taking into account the definition of the lateral envelope in Fig. 2.13.9. A further separation of the vertical interval into z to η and η to L, yields for the last term of (2.13.19)

$$\nabla.\int_z^L \mathbf{q}_H\ dz' = \nabla.\int_z^\eta \mathbf{q}_H\ dz' + \nabla.\int_\eta^L \mathbf{q}_H\ dz' = \mathbf{q}_H.\nabla\eta + \int_z^\eta \nabla.\mathbf{q}_H\ dz' + \nabla.\int_\eta^L \mathbf{q}_H\ dz' \qquad (2.13.20)$$

A similar separation can be carried out for the first term of (2.13.18), precisely like in (2.13.15), yielding

$$\frac{\partial}{\partial t}\int_z^L \theta \, dz' = n \frac{\partial}{\partial t}\int_z^\eta dz' + \frac{\partial}{\partial t}\int_\eta^L \theta \, dz' =$$

$$= n_{ef} \frac{\partial \eta}{\partial t} + \frac{\partial}{\partial t}\int_\eta^L (\theta - \theta_L) \, dz' \qquad \text{where} \qquad n_{ef} = n - \theta_L \tag{2.13.21}$$

Collecting now the results of (2.13.20) and (2.13.21) into (2.13.18), and letting z to tend to η, yields the kinematical free-surface boundary condition

$$n_{ef} \frac{\partial \eta}{\partial t} + q_x \frac{\partial \eta}{\partial x} + q_y \frac{\partial \eta}{\partial y} - q_z - R = \nabla . \int_\eta^L \mathbf{q}_H \, dz' - \frac{\partial}{\partial t}\int_\eta^L (\theta - \theta_L) \, dz' \qquad (z=\eta) \tag{2.13.22}$$

The right hand side of (2.13.22) represents the excess drainage from the unsaturated zone. If we assume, in line with our previous discussion, that the unsaturated profile translates downward together with the free-surface, i.e. that drainage is instantaneous, the terms in the r.h.s. are equal to zero, and we arrive at the common form of the free-surface condition

$$n_{ef} \frac{\partial \eta}{\partial t} - q_z + q_x \frac{\partial \eta}{\partial x} + q_y \frac{\partial \eta}{\partial y} = R \qquad (z=\eta) \tag{2.13.23}$$

Unlike (2.13.22), Eq. (2.13.23) does not include any variable related to the flow in the unsaturated zone, which does not have to be considered at all for solving the flow problem in the saturated zone.

The second condition, the dynamical one, is precisely (2.13.12) for the pressure, i.e.

$$p = - \gamma \psi_w \qquad \text{i.e.} \qquad \phi = \eta - \psi_w \qquad (z=\eta) \tag{2.13.24}$$

Eqs. (2.13.23) and (2.13.24) form the free-surface boundary conditions for saturated flow of an incompressible fluid. Both conditions are required, since not only ϕ has to be determined in the flow domain, but the position of the free-surface, i.e. the function $z=\eta(x,y,t)$ is unknown.

The above boundary conditions may be expressed in other various forms. Thus, by using Darcy's law for an isotropic medium and (2.13.23), we may eliminate $\mathbf{q}$ and η from (2.13.24) to obtain a boundary condition for ϕ solely

$$n_{ef}\frac{\partial\phi}{\partial t} + (K+R)\frac{\partial\phi}{\partial z} - K[(\frac{\partial\phi}{\partial x})^2 + (\frac{\partial\phi}{\partial y})^2 + (\frac{\partial\phi}{\partial z})^2] = R \qquad (z=\eta) \qquad (2.13.25)$$

Eq. (2.13.25) may serve for solving for ϕ, while (2.13.24) may be employed in order to determine the free-surface equation. These equations are not independent, however, since (2.13.25) is posed on $z=\eta$.

In a particular case of (2.13.25), for steady flow and in absence of recharge, the free-surface boundary conditions have the usual form

$$\frac{\partial\phi}{\partial z} - (\frac{\partial\phi}{\partial x})^2 - (\frac{\partial\phi}{\partial y})^2 - (\frac{\partial\phi}{\partial z})^2 = 0 \qquad (z=\eta) \qquad (2.13.26)$$

where ψ_w has been also neglected, consistent with condition (2.13.11), which underlies the assumption of instantaneous drainage.

Finally, it is worthwhile to mention here, that in the presence of the unsaturated zone, such that $\theta_{ir}>0$, the free-surface is not a material surface. In other words,the velocity of displacement of the free-surface along the normal ν (see Exercise 2.13.4) is not equal to the normal component of the fluid velocity $u_\nu=q_\nu/n$, but to q_ν/n_{ef}.

2.13.5 Boundary conditions for solute and heat transport

We shall analyze here only two types of boundaries, namely an impervious boundary and contact with free fluid.

In the case of a boundary between a porous medium and an impervious solid (Fig. 2.13.1) the solute satisfies the microscopic condition (2.3.23) with f=0, which leads to

$$q_{m,z} = 0 \qquad \text{i.e.} \qquad \frac{\partial C}{\partial z} = 0 \qquad (z=0) \qquad (2.3.27)$$

In the case of heat transport, conductive heat transfer through the impervious solid generally occurs. Then, the microscopic boundary conditions (2.3.21) and (2.3.22) give for the macroscopic temperature T

$$T = T_s \quad ; \quad q_{hz} = - K_{hs}\frac{\partial T_s}{\partial z} \qquad (z=0) \qquad (2.3.28)$$

where T_s and K_{hs} stand here for the temperature and the heat conductivity of the bounding solid, respectively, and q_{hz} is the component of the macroscopic heat flux in the porous medium, normal to the boundary which is given by (2.7.1), i.e. $q_{hz} = -K_h\,\partial T/\partial z$.

We consider now the case of contact between a saturated porous medium and a free fluid flowing normally, and towards, the boundary (Fig. 2.13.3). Let us assume that the fluid is at a constant concentration C_0 far from the boundary. These conditions are the usual ones in laboratory experiments at the entrance of porous columns. We wish to formulate now the boundary conditions satisfied by the concentration C at the boundary z=0.

The appropriate boundary conditions, needed in order to determine C(z,t) in the two zones are of continuity of macroscopic concentration and solute flux, i.e

$$C = C_{ext} \quad ; \quad q_{m,z} = - D_m \frac{\partial C_{ext}}{\partial z} + qC_0 \qquad (z=0) \qquad (2.3.29)$$

The term $q_{m,z}$ represents the solute flux in the porous medium. One may be tempted to write it in a form similar to that of the r.h.s. of (2.3.29), but with the diffusive flux expressed with the aid of the bulk dispersion coefficient (2.10.3). It is erroneous, however, to assume that the effective molecular diffusivity and the longitudinal dispersion coefficients, maintain their bulk values up to the boundary. D_m changes from the value in the free fluid to the bulk values throughout a thin boundary layer whose thickness is of the order of the pore-scale. The dispersion coefficient is a Lagrangian quantity (Chap. 2.10), which is attained only after the solute particles have travelled a few pore- scales through the medium. Thus, the concentration C for which the equality between the solute flux in the free fluid on one hand, and the one in the porous medium based on bulk effective properties, on the other hand, does not prevail at z=0 but at some distance downstream from the entrance. Application of the solute flux boundary condition, with effective coefficients up to the boundary, has led in the past to some erroneous conclusions, like the existence of a concentration jump or even propagation of solute by dispersion, against the average convective motion.

The molecular diffusion coefficient is very small and the diffusive terms in (2.3.29) can be safely neglected in presence of convection, i.e. for $q \neq 0$. Under these circumstances the second equation in (2.3.29) leads to the simplified boundary condition

$$C = C_0 \qquad (2.3.30)$$

replacing both equations (2.3.29). This was precisely the condition used in Chap. 2.10, Eq. (2.10.12), in order to obtain the expression of the breakthrough curve (2.10.13).

Thus, a complete solution of the solute transport problem requires to consider the macroscopic transport equation with variable effective properties and boundary condition (2.13.30). However, if the concentration is sought at a distance from the entrance which is large compared to the pore-scale, we can use the transport equation (2.10.3) with constant coefficients with a negligible error.

The same reasoning applies to the solution of the transport problem for the case depicted in Fig. 2.10.1, in which a solute body at constant concentration is inserted at t=0 in a porous medium. Only after the solute particles have travelled a distance of a few pore-scales, do the dispersion coefficients reach their constant, asymptotic, values of Chap. 2.10. Still, if we seek the concentration field after a sufficiently long travel time, a solution based on the transport equation (2.10.3) is quite accurate. This topic is of major consequences for flow in large porous formations and will be discussed in detail in Part 4.

Exercises

2.13.1 Derive equation (2.13.17) for the volume of water drained from a vertical porous column which is in hydrostatic equilibrium and at $t=0$ is subjected to a sudden drop of the water head at its bottom (see Fig. 2.13.6).

Generalize the solution for the case in which there is uniform recharge at the top and the initial and final states are of uniform downward flow.

2.13.2 Generalize boundary condition (2.13.25) for an anisotropic medium and in particular for the case of principal axes of the hydraulic conductivity tensor in the vertical z and in the x,y directions. Find the proper linear transformation of the variables t,x,y and z which render the equation of flow (2.11.1) and the free-surface boundary condition identical to those applying to an isotropic medium.

2.13.3 Rederive boundary condition (2.13.23) in terms of the velocity of the free-surface in a direction normal to itself. Interpret Eq. (2.13.23) as a material derivative in the case in which $n_{ef}=n$.

2.13.4 The hodograph plane, for two-dimensional flow, is known as the plane in which the components of the specific discharge q_x and $-q_y$ serve as axes (here y stands for a vertical coordinate). Depict the curve representing the free-surface condition (2.13.26) in the hodograph plane. Find also the image of a seepage face (Eqs. 2.13.8 and 2.13.9) and the confluence of the two types of boundaries, i.e. seepage face and steady free-surface.

PART 3. WATER FLOW AT THE LOCAL (FORMATION) SCALE

3.1 INTRODUCTION

Beginning with the present part, and in the rest of this book, we shall study flow and transport in natural porous formations. These formations differ considerably from the laboratory samples (considered in Part 2) in a few respects, but mainly in their dimensions. Thus, natural formations, or aquifers, appear generally as predominantly horizontal layers whose thickness may vary between a few to hundreds of meters. Their planar extent is generally much larger than the thickness, the horizontal size being as large as tens of kilometers. At these vast scales, as compared to the pore-scale, the properties and flow variables of interest are the macroscopic ones discussed in Part 2. In other words, measurements or computations are carried out for variables averaged over volumes or surfaces which are large enough to warrant the use of macroscopic quantities regarded as deterministic and attached to each point in space, in the spirit of the discussion of Chap. 2.1. The heterogeneity at pore-scale is, therefore, ignored and its presence manifests indirectly in the equations satisfied by macroscopic variables and in the coefficients (permeability, effective storativity, etc) which characterize the macroscopic properties of the porous medium. Along these lines, the porous medium and the fluid are regarded as continua and the macroscopic properties and variables are represented mathematically as continuous spatial scalar, vector or tensor fields. To illustrate the concept, let us consider the meaning of permeability k at a point $\mathbf{x}$: it is the one of a core surrounding the point, which is hypothetically or actually extracted from the formation and brought to the laboratory for measurement. The assumptions underlying the concept were clarified in Part 2: the core is large enough to warrant the validity of the ergodic hypothesis, on one hand, and the porous structure is statistically homogeneous in the neighborhood of $\mathbf{x}$. Summarizing this fundamental point, from now on we shall refer to macroscopic properties and variables solely, their relationship with the porous structure having been elucidated in Part 2.

The local (or formation) scale, the topic of this chapter, is of the order of the formation thickness in both vertical and horizontal directions, and flow and transport are generally of a three-dimensional nature at this scale. For the purpose of illustration, we have represented in Fig. 3.1.1 a few typical examples of flow at the local scale. Thus, a partially penetrating well, Fig. 3.1.1 a, either pumping or recharging, is one of the common man made devices. Similarly, the second example in Fig. 3.1.1 b, the merging of a vertical unsaturated plume with groundwater in an unconfined aquifer, is related to leaking repositories or recharging ponds. The third example, of flow toward the sea above a fresh-water salt-water interface (Fig. 3.1.1 c), occurs under natural conditions in coastal aquifers. The common, though not necessarily universal, feature of these flows is their three-dimensional character, i.e. the seepage velocity changes its magnitude and direction in space. Furthermore, the extent of the domain in which three-dimensional effects occur is of the order of the formation thickness.

The traditional approach to flows at local scale is to regard the medium properties as uniform

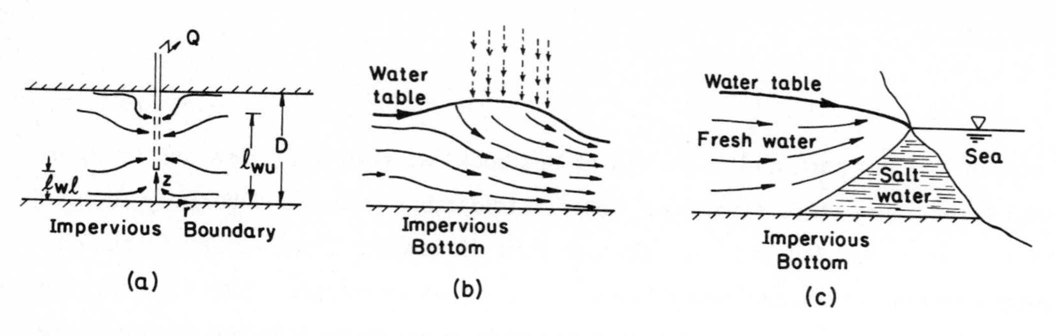

Fig. 3.1.1 Illustrative examples of flows at the local (formation) scale : (a) flow to a partially penetrating pumping well in a confined aquifer ; (b) the mound caused by a plume recharging through the unsaturated zone into a flowing, phreatic, aquifer ; (c) seaward flow of fresh water in a phreatic coastal aquifer above a salt water body. The two water bodies are separated by an idealized sharp interface.

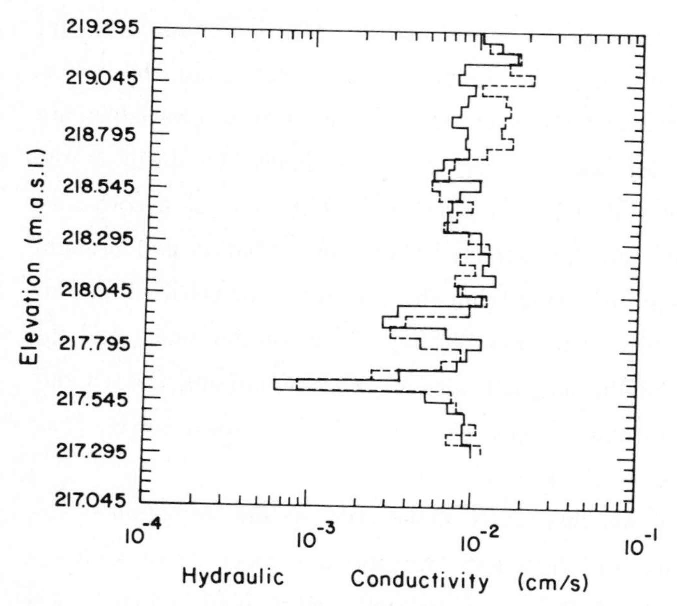

Fig. 3.2.1 Two examples of vertical hydraulic conductivity profiles determined from cores taken at one meter horizontal distance at the Borden tracer test site (from Sudicky, 1986).

over the entire domain, or in parts of it which are separated by well defined boundaries, i.e. to assume the formation to be homogeneous. Under these conditions, the equations of flow, summarized in Chap. 2.2, have been solved for a large variety of boundary and initial conditions, and such solutions can be found in various available texts on seepage and groundwater flow, e.g. Muskat (1937), Polubarinova Kochina (1962), Verrujt (1970), Freeze and Cherry (1979), Bear (1972, 1979), de Marsily (1986). Furthermore, the advent of electronic computers and the development of numerical methods, have expanded enormously the realm of complex problems of flow and transport which can be readily solved.

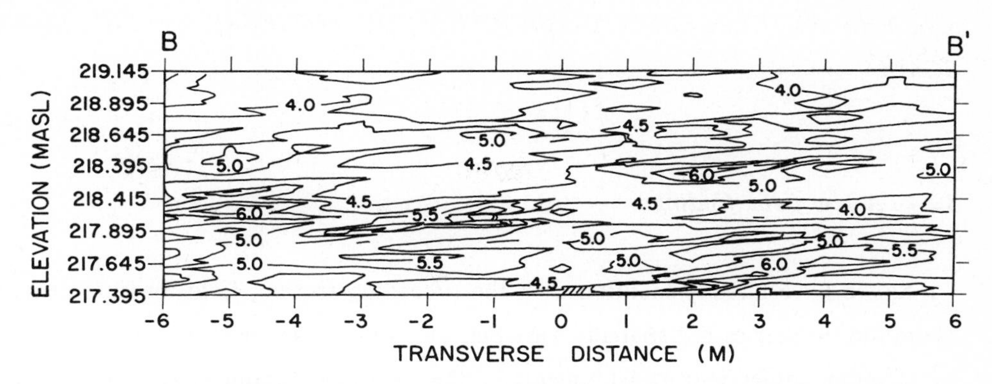

Fig. 3.2.2 Lines of constant -ln K (K in cm/sec) in a vertical cross-section at the Borden tracer site test (from Sudicky, 1986).

Still, the assumption of spatial homogeneity of aquifer properties is generally not supported by field findings. Referring, for instance, to the hydraulic conductivity, it is found as a rule that measurements at different points yield different values. Furthermore, K(**x**) varies in an irregular fashion in space, and these variations occur over distances which are much larger than the laboratory scale (the mathematical representation of these variations will be considered in Sect. 3.2.3). The spatial variability of K is vividly illustrated in Figs. 3.2.1, 3.2.2, which depict the distribution of measured K in a sandy aquifer. The apparent contradiction between this type of finding and the traditional approach can be conciliated if we regard properties, as well as flow variables, as space averages over volumes which are of much larger extent than the heterogeneity scale, by a reasoning similar to the one adopted when passing from pore to laboratory scale in Part 2. Such a passage, however, has to be carried out systematically and carefully, in order to assess its validity and limitations. In some important applications, like solute transport (see Part 4), accounting for spatial variability is essential for the understanding of field phenomena. In other cases, like unsteady flows in compressible aquifers (see Sect. 3.4.7), there are conceptual difficulties in defining effective properties. To deal with these problems, we shall adopt in this book an approach similar to that of Part 2 and different from the traditional one and we shall devote a major portion of this part to the study of spatial heterogeneity and its impact upon flow theory. In Chaps. 3.5 and 3.6 we shall recover the classical grounds and review succinctly a few problems and solutions for averaged flow variables. The transport problem is discussed separately in Part 4.

3.2 *THE HETEROGENEOUS STRUCTURE OF AQUIFERS AT THE LOCAL (FORMATION) SCALE*

3.2.1 *A few field findings*

As we have already mentioned in the preceding section, field measurements show as a rule that hydraulic properties are spatially variable, i.e. aquifers are heterogeneous. We shall start this section by recalling some field measurements of the hydraulic conductivity reported in the literature. Such measurements are neither easy to perform in unconsolidated formations, which are the ones of primary interest in the hydrological context, nor are they a matter of routine. One of the main obstacles stems from the difficulty of extracting undisturbed cores from unconsolidated formations. Furthermore, mapping K on a dense grid requires drilling a large number of wells, which is a costly operation. Nevertheless, such investigations are to be expected at an increased pace in the future, due to the interest in applications related to solute transport.

A systematic attempt to summarize field data on permeability from the literature has been carried out by Freeze (1975). His first finding was that $K(\mathbf{x})$ varies in an irregular manner in space and it should be characterized by a frequency distribution. Freeze's analysis was aimed at determining the type of p.d.f. (probability density function, see Chap. 1.1) which fits data best. The result was that such a best fit is achieved by a lognormal K distribution, i.e. for $Y = \ln K$ being $N[m_Y, \sigma_Y]$, where $m_Y = \langle Y \rangle$ is the expected value and σ_Y is the standard deviation (see Eq. 1.1.6). This finding has been strengthened by other studies and is accepted at present as a general tenet, as far as water bearing formations are concerned. A type of heterogeneity which might evade such a representation, is of thin impervious layers within a conductive matrix. We discuss an approach to represent their presence in Sect. 3.4.4 and in Chap. 4.7.

Most of the data collected by Freeze (1975) pertain to consolidated formations of low conductivity encountered mainly in reservoir engineering applications. We reproduce here only a few values of m_Y (for K in cm/sec) and σ_Y of a few formations, namely those of highest K_G (geometric mean) of Table 1 of Freeze

Since $K(\mathbf{x})$ is regarded as a random space function, its complete characterization is achieved with the aid of the various joint p.d.f. of its values at different points (Chap. 1.3). In particular, for Y multivariate normal, the statistical information is exhausted by m_Y, σ_Y and the auto-covariance $C_Y(\mathbf{x}_1,\mathbf{x}_2)$. Determining C_Y empirically, by statistical inference, requires measuring K at a large number of pairs of points, at various separation intervals. Such systematic measurements are costly and time-consuming and have been reported quite seldom in the literature.

An example of a thorough field investigation is the one carried out at the Borden site in Canada (Sudicky, 1986), related to a solute transport experiment (see Chap. 4.7), and for the purpose of illustration, we shall describe in some detail this instructive case. The aquifer is a sandy

TABLE 3.2.1 Statistical parameters of lognormal hydraulic conductivity (reproduced from Freeze, 1975; K in cm/sec, Y = ln K, K_G geometric mean).

Formation	K_G	m_Y	σ_Y
Sandstone	1.51×10^{-4}	-8.79	0.92
Sandstone	4.36×10^{-4}	-7.73	0.46
Sand and gravel	-		1.01
Sand and gravel	-		1.24
Sand and gravel	-		1.66
Silty clay	1×10^{-3}	-6.90	2.14
Loamy sand	2.09×10^{-3}	-6.16	1.98

one, and is comprised of primarily horizontal, discontinuous lenses of medium-grained, fine-grained and silty fine-grained sand. There are also infrequent silt, silty-clay and coarse sand layers. Thirty-two cores over a depth of two meters each have been extracted and have been subdivided into subsamples of 0.05 m length. The latter were quite homogeneous, and their properties were taken as representing point values at their centroid, in the sense of our preceding discussion. The hydraulic conductivity of each subsample has been measured in the laboratory after repacking, altogether such 1279 values being determined. The cores were collected along two vertical orthogonal planes, and an example of K profiles along two neighboring vertical profiles is reproduced in Fig. 3.2.1., while the distribution of the hydraulic conductivity in one of the planes is shown in Fig. 3.1.2.

The profiles in Fig. 3.2.1 illustrate convincingly the irregularity of the variation of K in space, typical for natural porous formations. The complex, and seemingly erratic, heterogeneous nature of the aquifer is displayed in Fig. 3.2.2. A striking feature is the presence of layering, predominantly horizontal, which seems to be a common property of sedimentary formations.

The statistical analysis of measurements (Sudicky, 1986) indicate that they fit quite accurately a lognormal distribution, with m_Y=-4.63 (corresponding to K_G=9.75×10^{-3} cm/sec). The total variance was found σ_Y^2+w=0.38, with σ_Y^2=0.28 and w=0.1. These values are in contrast with the smaller mean and larger variances of the formations catalogued in Table 3.2.1.

Finally, an analysis of the spatial correlation structure revealed that $Y(\mathbf{x})$ could be regarded as stationary, and the autocovariances of the correlated residuals in the horizontal and vertical directions could be approximated by the exponential, Poisson type, relationships

$$C_{Yh}(x,y,0) = \sigma_Y^2 \exp[-(x^2+y^2)^{1/2}/I_{Yh}] \; ; \; C_{Yv}(0,0,z) = \sigma_Y^2 \exp(-|z|/I_{Yv}) \qquad (3.2.1)$$

where x,y are the horizontal and z the vertical components of the separation vector $\mathbf{x}=\mathbf{x}_1-\mathbf{x}_2$ and I_{Yh}, I_{Yv} are horizontal and vertical integral scales, respectively. The inferred values were $I_{Yh}=2.8$ m and $I_{Yv}=0.1$ m, reflecting the layering visible in Fig. 3.2.2. An interesting point, of definite significance, is that the Y values cease to be correlated at a distance of a few meters in the horizontal plane, in contrast with some models of sedimentary formations in which continuous layering is assumed to persist for considerable distances. Although one may expect that layers continuity prevails at a larger extent in lacustrine formations than in alluvial ones, it still seems that spatial homogeneity has not been detected over distances exceeding tens of meters.

Sudicky (1986) suggests on the basis of (3.2.1) that $C_Y(x)$ can be represented as

$$C_Y(\mathbf{x}) = w[1-H(r)] + \sigma_Y^2 \exp[-(\frac{x^2+y^2}{I_{Yh}^2} + \frac{z^2}{I_{Yv}^2})^{1/2}] \quad ; \quad H(r)=1\ (r=0),\ H(r)=0\ (r>0) \qquad (3.2.2)$$

rendering (3.2.1), for the correlated residuals, as particular cases. Thus, the formation could be categorized as statistically homogeneous (stationary) and anisotropic (see Sect. 3.2.2).

A less detailed analysis for an unconsolidated, alluvial, aquifer (Moltyianer, 1985) indicate similar results for the horizontal and vertical correlations scales, namely $I_{Yv} \simeq 0.64$ m, $I_{Yh} \simeq 3$ m.

A few additional data regarding σ_Y^2 and correlation scales have been collected from the literature by Gelhar (1986). Excluding data pertaining to the regional scale (see Part 4) or to the upper soil layer, herewith a few data quoted by Gelhar (1986)

TABLE 3.2.2 A few field inferred values of variances and correlation scales of the logconductivity (from Gelhar, 1986 ; scales are in meters)

Formation	σ_Y	I_{Yh}	I_{Yv}
Sandstone	1.5-2.2		0.3-1.0
Outwash sand	0.8	>10	0.4
Fluvial sand	0.9	>3	0.1
Sand and gravel	1.9	17	0.5

The analysis of the relationship between spatial variability of hydraulic conductivity of natural formations and their geological structure is a subject of considerable interest, which is beyond the scope of this book. However, it is worthwhile mentioning here that field investigations quite often show the presence of thin, horizontal, clay layers of very low conductivity, embedded into the more conductive, heterogeneous, medium. Such lenses, which might not be captured by a core analysis of the type discussed above, may prevent flow and transport in the vertical direction.

We shall discuss now briefly some data pertaining to other hydraulic properties, namely the porosity n and the specific storativity s (see Sect. 2.11.3 for its definition). Unfortunately, even less field data are available for s. Thus, Freeze (1975, Fig. 9) suggests, on a somewhat speculative basis, the following linear correlation between n and Y

$$n \simeq 0.375 - 0.011\,Y \quad \text{(for K in cm/sec)} \qquad (3.2.3)$$

The two main points revealed by (3.2.3), generally accepted in the literature, are that : (i) the porosity may be described by a normal p.d.f. and (ii) the variance of n is much smaller than that of Y, by four orders of magnitude, i.e. the porosity is of a much lesser spatial variability than the hydraulic conductivity. It is reminded (Chap. 1.1) that for low variances the log-normal and normal distributions are close.

For instance, the analysis of cores at Borden site (Sudicky, 1986) led to σ_n=0.0165, to be compared to σ_Y=0.62.

The linear correlation between $\log_{10}\alpha$ and $\log_{10}K$ proposed by Freeze (1975) has been already given in Eq. (2.11.23). Starting from the definition of specific storativity s (2.12.22), and neglecting the effect of water compressibility (see Chap. 2.11), yields in (2.11.23)

$$\ln s = \ln(\rho g) + \ln\alpha = - 16.9 - Y/3 \ ; \quad \text{(for K in cm/sec and s in cm}^{-1}\text{)} \qquad (3.2.4)$$

Eq. (3.2.4) suggests that s, similarly to Y, is lognormal, and its variance is smaller by an order of magnitude than that of Y.

On the basis of the limited information in the literature, no conclusion could be drawn about the correlation structure of porosity and specific storativity, regarded as space random functions. However, because of their lesser variability as compared to that of K, we shall concentrate mainly on the effect of the latter, which is anyway the only hydraulic property affecting steady flow (see Chap. 2.12).

3.2.2 Statistical representation of heterogeneous formations and their classification

Taking it for granted that natural porous formations are heterogeneous, albeit at different possible degrees, and on the basis of the field data of the preceding section, we shall formulate now the mathematical frame to describe heterogeneity. To simplify matters we shall refer to the hydraulic conductivity K only, for reasons given before, but the definitions apply to other properties as well.

$K(\mathbf{x})$ is regarded as a random function of space, to model its irregular spatial variability, as displayed, for instance, by Figs. 3.2.1, 3.2.2. The general definition of such functions, in terms of the joint p.d.f. of K at N points, has been discussed in Chap. 1.3. We assume that K is a scalar, i.e. there is no *systematic* anisotropy of the cores which represent the point values of K. At the local scale, anisotropic heterogeneity manifests in the covariance structure, as it happens for instance in the case of stratified formations.

Starting with N=1, K is characterized by $f(K;\mathbf{x})$, the p.d.f. of its values at the point $\mathbf{x}$. In compliance with field data, we shall assume that f is a continuous function of K, in contrast with the depiction of the porous medium (Chap. 2.2) as a two-phase heterogeneous structure. Still, it is convenient for computational purposes and for some particular types of formations, to consider also a discrete distribution

$$f(K;\mathbf{x}) = \sum_{j=1}^{M} n_j(\mathbf{x})\, \delta(K-K_j) \tag{3.2.5}$$

where δ stands, as usual, for the Dirac distribution. Eq. (3.2.5) may be viewed as a representation of a collection of blocks of various conductivities, with n_j the volume fraction of blocks of $K=K_j$. In particular, a two-phase formation, i.e. M=2, may be a convenient representation of a formation of a relatively uniform matrix which contains lenses of a definite, significantly different, permeability. The continuous f(K) may be regarded as the limit of (3.2.5) for $M\to\infty$, so that $n_j \to f(K_j)\, dK_j$.

Another important distinction is between the general case of a statistically nonhomogeneous (or nonstationary) formation, and a statistically homogeneous (stationary) one, for which f is independent of $\mathbf{x}$. We shall assume that the latter case prevails at the local scale, this being a basic hypothesis which can be justified on empirical grounds and by statistical inference techniques. Its fulfillment is essential in order to be able to identify f(K) from a single available realization, by using ergodic arguments (see Chap. 1.10). This basic limitation can be relaxed, for instance, by allowing for a trend which is slowly varying relative to the correlation scale of K, or by dividing the formation into distinct blocks, of dimensions much larger than the latter scale, f(K) being different in each block.

Last, in line with the field findings recalled in the preceding section, we shall assume that f(K) is lognormal , i.e. $Y = \ln K$ is $N[m_Y, \sigma_Y]$. It is worthwhile to recall here the following relationships (see Eq. 1.1.15)

$$\langle K \rangle = K_A = \exp(m_Y + \sigma_Y^2/2) \;;\; \sigma_K^2 = \exp(2m_Y + \sigma_Y^2)\,[\exp(\sigma_Y^2)-1] \;;\; K_G = \exp(m_Y) \tag{3.2.6}$$

where, as usual, <> stands for the expectation (ensemble average), K_A is the arithmetic and K_G is the geometric mean, respectively. We consider now the two-point ($N = 2$) joint p.d.f. $f(Y_1,Y_2)$, where $Y_1=Y(\mathbf{x}_1)$ and $Y_2=Y(\mathbf{x}_2)$, $\mathbf{x}_2$ and $\mathbf{x}_2$ being the coordinate vectors of two arbitrary points in the formation domain. The correlation between Y_1 and Y_2 is characterized by the covariance $C_Y(\mathbf{x}_1,\mathbf{x}_2) = \sigma_Y^2\, \rho_Y(\mathbf{x}_1,\mathbf{x}_2)$. Furthermore, along the lines of the basic assumption formulated above, we assume that by stationarity C_Y depends on the separation vector $\mathbf{r} = \mathbf{x}_2-\mathbf{x}_1$, rather than on $\mathbf{x}_1$, $\mathbf{x}_2$ separately.

If nothing is known about stationarity of higher multi-point moments ($M > 2$), Y is said to be weakly or second-order stationary. In some applications, like stochastic interpolation by kriging, this might be sufficient, no use being made of higher-order moments. We shall, nevertheless, assume that $f(Y_1,Y_2,...,Y_N)$ is multivariate normal for any set of points $\mathbf{x}_j$ ($j=1,2,...,N$). This assumption is not easy to verify for field data, but it is compatible with field findings and has the advantage of simplifying considerably the computational effort. The p.d.f. (Eq.1.2.15) and properties of multivariate normal distributions have been discussed in Chap. 1.2 and it is reminded here that it is completely defined with the aid of m_Y and $\sigma_{ij}=C_Y(\mathbf{x}_i,\mathbf{x}_j)$. Thus, under this simplifying scheme, the entire statistical structure of the stationary $Y(\mathbf{x})$ is defined by m_Y and $C_Y(\mathbf{r})$, and we shall discuss now a few properties of the latter, along the lines of Chap. 1.3.

Starting with the behaviour of $C_Y(\mathbf{r})$ near the origin $\mathbf{r}=0$, we have $\nabla C_Y=0$ for $r=0$, for a continuous $Y(\mathbf{x})$, whereas for a discrete distribution (3.2.5), C_Y has a discontinuous derivative at $r=0$. Its value is related to the specific area (see Chap. 1.3). It can be shown that for (3.2.5) and for quite general conditions

$$\frac{\partial \rho_Y}{\partial r_i} = - \frac{\sigma_i}{2} \quad \text{for } \mathbf{r}=0 \qquad (i=1,2,3) \tag{3.2.7}$$

where the specific area, σ_i, is defined as the projection of the area between blocks of various Y upon a plane normal to the x_i axis, divided by the formation volume.

Still, we shall often adopt a function C_Y of discontinuous slope at the origin, even for continuous Y, for the sake of convenience. This is permissible if Y serves as an input for *integration*, the detailed behaviour of C_Y near the origin being then of little consequence upon results. Furthermore, in practice C_Y is inferred from measurements and the points are not that close to warrant an accurate description of C_Y near the origin, and the shape there is generally obtained by extrapolating from finite $\mathbf{r}$ toward $r=0$.

A more severe discontinuity at the origin is associated with the presence of a "nugget" (Chap. 1.7), i.e. a discontinuity of C_Y which jumps from σ_Y^2+w for $r=0$ to a lower value $\sigma_Y{}^2$ at $r=0^+$, the nugget being equal to w. The nugget is related to uncorrelated Y variations, like the ones resulting from measurement random errors. It may also arise from "microregionalization" in the geostatistical terminology (Journel and Huijbregts, 1978), which can be associated with the existence of blocks of uniform various Y whose dimensions are smaller than the minimal r for which measurements are available. Usually, the nugget effect appears in the extrapolation of data available at finite r toward r=0. Again, in processes in which Y serves as input for integration, and this is the case for flow and transport treated in this book, the nugget is irrelevant, since its measure is zero. We shall disregard its presence, unless specified, and will relate only to the correlated part of C_Y, i.e. $\sigma_Y^2=C_Y(0)$ stands in fact for the value at $r=0^+$.

The behaviour of C_Y for large $r=|\mathbf{r}|$ is of definite significance. Its drop to zero, is indicative of lack of correlation between the values of Y at points at sufficiently large separation distance. We shall assume that this is indeed the case and furthermore, that the drop is sufficiently rapid to warrant the existence of various integral scales (see Chap. 1.4), e.g. the linear ones defined by

$$I_{Yi} = \frac{1}{\sigma_Y^2}\int_0^\infty C_Y(\mathbf{r})\, dr_i = \int_0^\infty \rho_Y(\mathbf{r})\, dr_i \qquad (i=1,2,3) \tag{3.2.8}$$

I_Y is indicative of the distance over which the values of Y cease to be correlated and can be employed as the definition of the correlation scale of Y. In some circumstances, like when sampling from a volume whose extent is not large compared to I_Y, C_Y might not display a clear drop to zero. In such a case it is useful to employ the variogram $\Gamma_Y(\mathbf{r}) = 1/2\ [Y(\mathbf{x}+\mathbf{r}) - Y(\mathbf{x})]^2$ (Chap. 1.7), which can be inferred satisfactorily for r smaller than the domain extent. Obviously, for ordinary stationarity, C_Y and Γ_Y are interrelated by $\Gamma_Y = \sigma_Y^2 - C_Y$.

In some studies (e.g. Bakr et al, 1978) it has been assumed that C_Y may have negative values for large r, i.e. C_Y is a "hole" covariance, and furthermore, its integral scale is exactly zero. This choice has been motivated by some computational convenience when using spectral methods (Chap. 1.6). Its advocates suggest that measurements underlying the inference of the tail of the C_Y curve are imprecise and its extrapolation to negative values is to some extent at the discretion of the modeler. Although negative correlation may be associated with periodic phenomena, we shall refrain from employing the "hole" covariances for Y, which apparently does not display such a behaviour in field measurements. As for computational difficulties, they can be overcome by adequate mathematical techniques (see Chap. 1.6).

The ratio between the linear extent D of the formation and the integral scale I_Y plays a fundamental role in the modeling of heterogeneous formations (for a detailed discussion, see Dagan, 1986). Only in the case in which $D/I_Y >> 1$, ergodicity arguments can be invoked and stationarity may be employed in order to infer the statistical parameters of Y from a single realization (see Chap. 1.10). Otherwise, one may infer only the variogram Γ_Y for separation distances $r<<D$, but not m_Y, σ_Y^2 and I_Y separately.

We shall discuss now additional properties of the covariance C_Y. Its simplest form is attained for statistically isotropic formations, for which there is no preferential direction in space. Then, (see Chap. 1.3) C_Y is a function of $r=|\mathbf{r}|$ solely and can be represented by a unique curve. For a formation of a discrete structure (Eq. 3.2.5), a conceptually convenient model is one of a union of spherical blocks of various conductivities and diameters which fill the space.

In the case of anisotropy, some preferential directions exist in space. A simple first case is the one of axisymmetric anisotropy, i.e. only one preferential direction prevails. In this case, by virtue of invariance requirements (see Chap. 1.3) one may write

$$C_Y(\mathbf{r}) = \sigma_Y^2 \rho_Y(\mathbf{r}.\mathbf{i}\ ,\ r) \tag{3.2.9}$$

where $\mathbf{i}$ is a unit vector along the axis of symmetry. In a Cartesian system in which $\mathbf{i}$ is in the vertical direction, ρ_Y (3.2.9) can be also rewritten as $\rho_Y(r_h, r_v)$, where the arguments are the horizontal and vertical components of $\mathbf{r}$, respectively.

Based on dimensionality requirements, the isotropic autocorrelation may be written as $\rho_Y(r/I_Y)$. A major simplification of anisotropic covariances may be achieved if they can be reduced to an isotropic one by a linear scaling of the Cartesian components of $\mathbf{r}$ in three orthogonal, anisotropy directions (see, for instance, Gelhar and Axness, 1983). Thus, with I_{Yx}, I_{Yy} and I_{Yz} three length scales, which are taken for convenience equal to the linear integral scales in the same directions, we have

$$C_Y(\mathbf{r}) = \sigma_Y^2 \rho_Y(r') \quad \text{with} \quad r' = [(r_x/I_{Yx})^2 + (r_y/I_{Yy})^2 + (r_z/I_{Yz})^2]^{1/2} \tag{3.2.10}$$

In the case of axisymmetric anisotropy (3.2.10) degenerates into $r' = [(r_h/I_{Yh})^2 + (r_v/I_{Yv})^2]^{1/2}$.

For a discrete conductivity formation, a convenient model of a statistically anisotropic formation with covariance of type (3.2.10) is a collection of ellipsoidal blocks with the same ratios between their axes, which are also parallel. Other simplifications of the general anisotropic C_Y, e.g. assuming that $\rho_Y(\mathbf{r}) = \rho_{Yx}(r_x)\, \rho_{Yy}(r_y)\, \rho_{Yz}(r_z)$, are possible, and will be adopted occasionally in the sequel. As a matter of fact, field data are seldom of an extent and quality which permit one to identify the anisotropy directions as well as the general functional dependence on the three components of $\mathbf{r}$. Thus, (3.2.10) may be a convenient approximation accessible in practice, and this was precisely the covariance structure (Eq. 3.2.2) adopted for the Borden site (Sudicky, 1985).

A particular case, encountered quite often in the literature, is the one of a layered formation. Continuous horizontal layering is a particular case of axisymmetric anisotropy for which $I_h \to \infty$, i.e. Y values are perfectly correlated in the horizontal plane and

$$C_Y(\mathbf{r}) = \sigma_Y^2 \rho_Y(r_v/I_{Yv}) \tag{3.2.11}$$

Such perfect layering is quite improbable in natural formations, and the axisymmetric case, with $I_{Yh} > I_{Yv}$ and finite I_{Yh}, is more typical (a few values of field correlation scales are given in the preceding section).

Another anisotropic structure results from the existence of thin, parallel, horizontal layers of a very low permeability (e.g. clay), embedded in a matrix of much larger conductivity. If such layers are ideally of zero thickness, their presence will not show up in f(Y) or C_Y, but they may have a large impact on flow, which behaves as if the formation were of a two-dimensional nature in the horizontal plane, i.e.

$$C_Y(\mathbf{r}) = \sigma_Y^2 \rho_Y(r_h / I_{Yh}) \tag{3.2.12}$$

where I_{Yh} is the horizontal integral scale of the medium making up the matrix and Y stands for the vertical, spatial, average of Y over an interval of the order of the distance between the impervious lenses. This representation works better if the latter is small compared to the horizontal extent of the lenses, which in turn is large compared with I_{Yh}. Hence, modeling the heterogeneous structure as two-dimensional may be a convenient way to treat flow and transport in a special case of anisotropic three-dimensional structure (see Chap. 4.7).

3.2.3 A few examples of covariances $C_Y(\mathbf{r})$

We shall review now a few specific examples of logconductivity autocovariances that are encountered in the literature ; some of them will be used in the sequel. Most of these covariances are isotropic, but can be generalized to anisotropic ones by the transformation (3.2.10). Hence, to each isotropic covariance we can attach a general three-dimensional anisotropic one (3.2.10), an axi-symmetric (3.2.9), a two-dimensional or a layered one-dimensional (3.2.11) covariance, obtained by a linear transformation of the Cartesian components of the separation vector **r**.

It is also reminded that in the case of ordinary stationarity C_Y may be replaced by the variogram $\Gamma_Y = \sigma_Y^2 + w - C_Y$.

We shall make frequent use of the Fourier transform (FT) of C_Y, defined (Chap. 1.6) by $\hat{C}_Y(\mathbf{k}) = (2\pi)^{-m/2} \int C_Y(\mathbf{r}) \exp(i\mathbf{k}.\mathbf{r})\, d\mathbf{r}$, where m = 1,2,3 is the number of the space dimensions. It is reminded that in the case of an isotropic covariance, $\hat{C}_Y$ is a function of $k = |\mathbf{k}|$ and that $\hat{C}_Y(0) = (2\pi)^{-m/2} I_Y$, whereas the limit $k \to \infty$ is related to the behaviour of C_Y near the origin. Thus, for a C_Y of discontinuous slope, $\hat{C}_Y \doteq O(k^{-m+1})$ for $kI_Y >> 1$. Furthermore, in the anisotropic case of the particular form (3.2.10) we have

$$\hat{C}_Y(\mathbf{k}) = I_{Yx} I_{Yy} I_{Yz} \hat{C}_Y(k') \;;\; k' = \left[(k_x I_{Yx})^2 + (k_y I_{Yy})^2 + (k_z I_{Yz})^2\right]^{1/2} \tag{3.2.13}$$

where $\hat{C}_Y(k')$ is the FT of $C_Y(r')$ (3.2.10), leading to straightforward generalization of FT of isotropic covariances. It is also reminded (Chap. 1.6) that $\hat{C}_Y$ is equal to the spectrum times $(2\pi)^{-m/2}$, the difference stemming from the definition adopted here for the Fourier transform.

Herein are a few useful examples of covariances of finite integral scale and of their Fourier transforms. To any of these covariances a nugget $w[1-H(r)]$ may be added, but we shall omit it. In any case, it does not show up in the FT of C_Y, but appears as the FT of $wH(r)$, i.e. $(2\pi)^{m/2} w\,\delta(k)$ in the FT of Γ_Y.

(i) Exponential.

The exponential covariance

$$C_Y(r) = \sigma_Y^2 \exp(-r/I_Y) \tag{3.2.14}$$

has been already encountered in Sect. 2.2.1 as a model of a two-phase Poisson medium and in Sect. 3.2.1, in its anisotropic version, as a representation of field data. It has been employed quite often in models of flow and transport due to its simplicity. Although C_Y (3.2.14) has a discontinuous slope at the origin, it has been used to model continuous $Y(\mathbf{x})$ for reasons discussed in Sect. 3.2.2. The linear integral scale is I_Y and this C_Y is particularly advantageous for analytical studies, since it is an analytic function of r in the entire range $r>0$.

The FT of C_Y (3.2.14) is given by

$$\hat{C}_Y = \sqrt{\frac{8}{\pi}}\, \sigma_Y^2 I_Y^3 \frac{1}{(1 + k^2 I_Y^2)^2} \tag{3.2.15}$$

In the case of the two-dimensional version of (3.2.14), i.e. for $r^2=x^2+y^2$, the FT has the expression

$$\hat{C}_Y = \sigma_Y^2 I_Y^2 \frac{1}{(1 + k^2 I_Y^2)^{3/2}} \tag{3.2.16}$$

(ii) Semi-spherical.

$$C_Y(r) = \sigma_Y^2 \left(1 - \frac{3}{2}\frac{r}{\ell_Y} + \frac{1}{2}\frac{r^3}{\ell_Y^3}\right) \quad \text{for } \frac{r}{\ell_Y} < 1 \;; \quad C_Y = 0 \quad \text{for } \frac{r}{\ell_Y} > 1 \tag{3.2.17}$$

where ℓ_Y is known as the range. This is also a two-parameter covariance, of discontinuous first derivative at the origin and discontinuous second-derivative at $r=\ell_Y$. The linear integral scale is now

$$I_Y = \int_0^{\infty} \rho_Y(r)\, dr = \frac{3}{8}\ell_Y \tag{3.2.18}$$

The FT of (3.2.18) is given by

$$\hat{C}_Y = \sqrt{\frac{2}{\pi}}\, \frac{\sigma_Y^2 \ell_Y^3}{k'^2} \left[\cos k' \left(-2 + \frac{17}{k'^2} - \frac{36}{k'^4}\right) + \frac{6\sin k'}{k'}\left(1 - \frac{6}{k'^2}\right) + \frac{1}{k'^2} + \frac{36}{k'^4}\right] \;;\; k' = k\,\ell_Y \tag{3.2.19}$$

(iii) Linear.

$$C_Y(r) = \sigma_Y^2\left(1 - \frac{r}{\ell_Y}\right) \text{ for } \frac{r}{\ell_Y} < 1 \;;\; C_Y = 0 \text{ for } \frac{r}{\ell_Y} > 1 \tag{3.2.20}$$

which has discontinuous first derivatives at both $r=0$ and $r=\ell_Y$. The formation model leading to (3.2.20) is a collection of blocks of constant, but random lognormal conductivities, and of same dimensions. In contrast, (3.2.14) represents a formation made up of blocks of different, continuously varying sizes. The linear integral scale is now $I_Y = 0.5\,\ell_Y$. The variogram

$$\Gamma_Y(r) = \frac{\sigma_Y^2}{\ell_Y}\, r \tag{3.2.21}$$

can be employed for formations of stationary increments and unbounded variance. In this case σ_Y^2/ℓ_Y combines in an unique parameter which cannot be decomposed into its two components unless the extent of the formation is larger than ℓ_Y and (3.2.20) is valid.

The FT of of (3.2.20) is as follows

$$\hat{C}_Y = \sqrt{\frac{2}{\pi}}\, \frac{\sigma_Y^2 \ell_Y^3}{k'^3} \left[\frac{2(1-\cos k')}{k'} - \sin k'\right] \;;\; k' = k\,\ell_Y \tag{3.2.22}$$

(iv) Gaussian.

The Gaussian covariance

$$C_Y(r) = \sigma_Y^2 \exp(- r^2/\ell_Y^2) \qquad (3.2.23)$$

has the distinguished property of being analytical for any r, representing a continuous variation of Y(**x**). Its linear integral scale is $I_Y = \ell_Y \sqrt{\pi}/2$.

The FT of (3.2.23) has the following expression

$$\hat{C}_Y = \frac{\sqrt{8}\,\sigma_Y^2 \ell_Y^3}{\pi} \exp(-k^2\ell_Y^2/4) \qquad (3.2.24)$$

(v) The "separated" exponential.

$$C_Y(\mathbf{r}) = \sigma_Y^2 \exp(-\frac{|x|}{I_{Yx}}) \exp(-\frac{|y|}{I_{Yy}}) \exp(-\frac{|z|}{I_{Yz}}) \qquad (3.2.25)$$

This covariance function is again discontinuous at the origin and is intrinsically anisotropic, even if $I_{Yx} = I_{Yy} = I_{Yz} = I_Y$, when C_Y becomes

$$C_Y(\mathbf{r}) = \sigma_Y^2 \exp(- \frac{|x|+|y|+|z|}{I_Y}) \qquad (3.2.26)$$

The linear integral scale of (3.2.26) changes with the angle between **r** and one of the axes, and varies between I_Y and $\sqrt{3}I_Y$, in three dimensions, and I_Y to $\sqrt{2}I_Y$ in two dimensions, respectively. Still, (3.2.26) may prove itself as quite useful in analytical studies of flow or transport.

The FT of (3.2.26) is given by

$$\hat{C}_Y = \frac{8\,\sigma_Y^2 I_{Yx} I_{Yy} I_{Yz}}{(2\pi)^{3/2}} \frac{1}{1+k_x^2 I_{Yx}^2} \frac{1}{1+k_y^2 I_{Yy}^2} \frac{1}{1+k_z^2 I_{Yz}^2} \qquad (3.2.27)$$

The two-dimensional version of C_Y (3.2.26) is obtained by suppressing the variable z. Its FT has the following expression

$$\hat{C}_Y = \frac{2\,\sigma_Y^2 I_{Yx} I_{Yy}}{\pi} \frac{1}{1+k_x^2 I_{Yx}^2} \frac{1}{1+k_y^2 I_{Yy}^2} \qquad (3.2.28)$$

(vi) The "white noise".

The definition of a "white noise" process and its covariance have been given in Chap. 1.4. Accordingly,

$$\rho_Y(\mathbf{r}) = A\ \delta(\frac{r}{I_Y}) = A\ \delta(\frac{x}{I_Y})\ \delta(\frac{y}{I_Y})\ \delta(\frac{z}{I_Y}) \tag{3.2.29}$$

where δ is the Dirac distribution and A is a constant. As already mentioned in Sect. (1.1.4), this singular covariance may be used as a convenient representation of a covariance of finite integral scale whenever C_Y appears in a convolution integral in which the kernel varies slowly over the actual integral scale. In such a case the constant A is related to the exact C_Y by

$$A = (1/I_Y^3) \int \rho_Y(\mathbf{r}\ d\mathbf{r}) \tag{3.2.30}$$

The Fourier transform of (3.2.30) is given by

$$\hat{C}_Y = A\ \sigma_Y^2\ \ell_Y{}^3/(2\pi)^{3/2} \tag{3.2.31}$$

A few covariances and their Fourier transforms are represented, for the sake of illustration, in Fig. 3.2.3.

3.2.4 Statistical properties of the space average $\overline{Y}$

A point of major interest is the relationship between the statistical structure of $Y(\mathbf{x})$ and that of its space average, or "block" value, defined by

$$\overline{Y}(\mathbf{x}) = \frac{1}{V} \int_V Y(\mathbf{x}')\ d\mathbf{x}' = \frac{1}{V} \int_D Y(\mathbf{x}')\ \Omega(\mathbf{x} - \mathbf{x}')\ d\mathbf{x}' \tag{3.2.32}$$

where V is a line interval, an area or a volume whose centroid is at $\mathbf{x}$ and $\Omega(\mathbf{x}')$ is a step function equal to unity for $\mathbf{x}' \in V(0)$ and to zero for $\mathbf{x}'$ outside V. These definitions, as well as the properties of the random function $\overline{Y}$, have been discussed in a general manner in Chap. 1.9, and will be expanded herein.

The space averaging (3.2.32) has important applications, which will be explored in Sect. 3.7.6 (see, also, Dagan, 1986). For instance, in many cases, and rather as a rule, K is not measured by point sampling in the formation, but by some device which averages it over a volume, like a well.

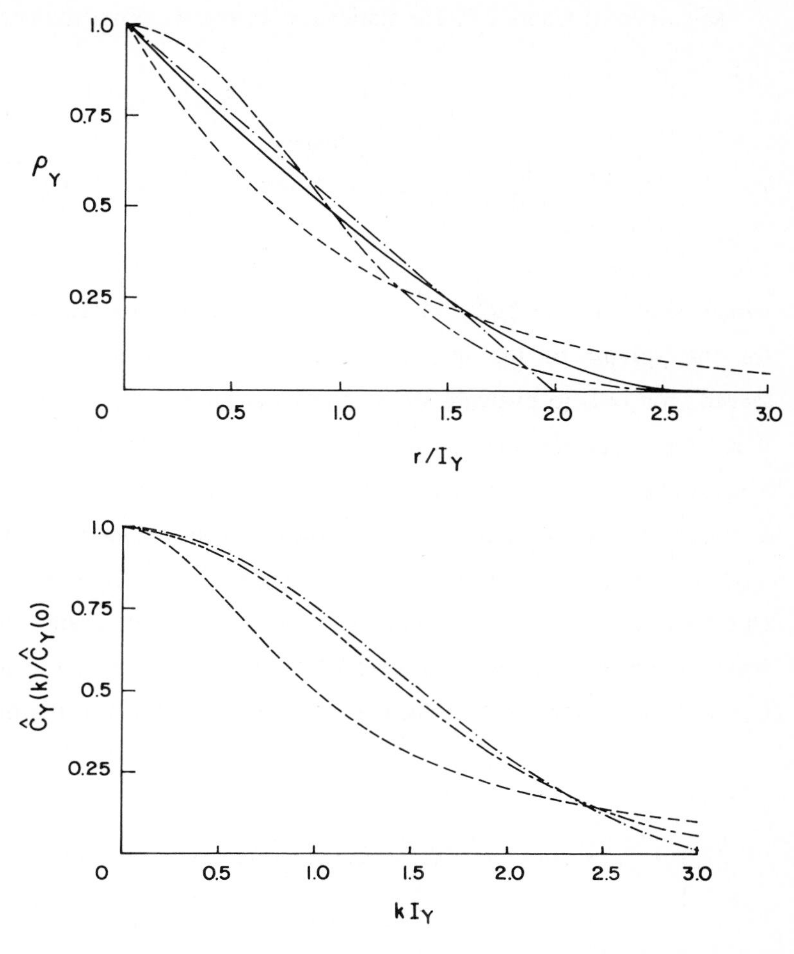

Fig. 3.2.3 Examples of (a) isotropic logconductivity covariances C_Y and (b) their Fourier transforms $\hat{C}_Y$: ------exponential (Eqs. 3.2.14, 3.2.15) ; ———semi-spherical (Eqs. 3.2.17, 3.2.19) ; -·-·- linear (Eqs. 3.2.20, 3.2.22) ; —·—·— Gaussian (Eqs. 3.2.23, 3.2.24). The separation distance and the wave-number have been made dimensionless with respect to the linear integral scale $I_Y = (1/\sigma_Y^2) \int_0^\infty C_Y(r)\, dr$.

As a matter of fact, when using slug, packer or pumping tests, K is averaged in a complex manner over a volume surrounding the well, which requires the generalization of (3.2.22) to include a weight function (see, for instance, Cushman,1984). In any case, the medium in the immediate neighborhood of the well has the largest weight and V in (3.2.32) may be adjusted to be a line or a cylinder. In a different context, when modeling flow or transport by numerical methods, one may be interested in the average of K or Y over the finite elements pertaining to the particular numerical scheme.

As shown in Chap. 1.9, for stationary Y, the two first moments of $\overline{Y}$ (3.2.32) are given by

$$\langle\overline{Y}\rangle = \langle Y\rangle \tag{3.2.33}$$

$$C_{\overline{Y}}(\mathbf{r}) = \frac{1}{V^2}\iint C_Y(\mathbf{x}')\,\Omega(\mathbf{x}'')\,\Omega(\mathbf{x}'+\mathbf{x}''-\mathbf{r})\,d\mathbf{x}'\,d\mathbf{x}'' = \frac{1}{V^2}\int C_Y(\mathbf{x}')\,H(\mathbf{x}'-\mathbf{r})\,d\mathbf{x}' \tag{3.2.34}$$

which stem from (3.2.22). In equation (3.2.34), known also as the Cauchy algorithm, H($\mathbf{x}$) stands for the volume of the intersection between $V(0)$ and its translation $V(\mathbf{x})$. This expression has served, for r=0, to evaluate the porosity variance in Sect. 2.2.2. Here, we wish to analyze in some detail the properties of $C_{\overline{Y}}$ and its dependence on the extent of V (for an extensive analysis, see Vanmarcke, 1983). It is important to mention first that for Y multivariate normal (see Chap. 1.2), $\overline{Y}$ is also MVN, since by (3.2.32) it results from a linear operation applied to Y. Hence, the two moments (3.2.33, 3.2.34) characterize completely $\overline{Y}$. To evaluate $C_{\overline{Y}}$ we need to know C_Y and H($\mathbf{x}$). The latter has closed form expressions for simple shapes of V, e.g. segment, circle or sphere, which have been given in Eq. (1.9.11). Alternative expressions can be obtained for the FT of $C_{\overline{Y}}(\mathbf{r})$ by applying the Faltung theorem (see Eq. 1.6.19) to the middle term of (3.2.34) leading to

$$\hat{C}_{\overline{Y}}(\mathbf{k}) = \frac{(2\pi)^3}{V^2}\,\hat{C}_Y(\mathbf{k})\,\hat{\Omega}(\mathbf{k})\,\hat{\Omega}^*(\mathbf{k}) \tag{3.2.35}$$

where $\hat{\Omega}^*$ is the complex conjugate of $\hat{\Omega}$. Again, $\hat{\Omega}$ has closed form expressions for simple V shapes. Thus, for a segment of length ℓ lying on the z axis we have

$$\hat{\Omega}(\mathbf{k}) = \frac{2}{(2\pi)^{3/2}}\,\frac{\sin(k_z\ell/2)}{k_z}\,\exp[i(k_x x+k_y y)] \tag{3.2.36}$$

whereas for a circle in the x,y plane or a sphere of radii d

$$\hat{\Omega} = \frac{1}{\sqrt{2\pi}}\,\frac{d\,J_1(kd)}{k}\,\exp(ik_z z)\ ;\quad k = (k_x^2+k_y^2)^{1/2}$$

$$\hat{\Omega} = -\sqrt{\frac{2}{\pi}}\,\frac{kd\cos kd - \sin kd}{k^3}\ ;\quad k = (k_x^2+k_y^2+k_z^2)^{1/2} \tag{3.2.37}$$

respectively.

A few general properties of $C_{\overline{Y}}(\mathbf{r})$, discussed also in Chap. 1.9, are recalled here :

(i) The variance $\sigma^2_{\overline{Y}}$ is given by (3.2.34) for $\mathbf{r}=0$. It tends to σ^2_Y for $V/I^3_Y \to 0$ and it tends to zero like I^3_Y/V for $V\to\infty$. Furthermore, $\sigma^2_{\overline{Y}}$ is a monotonic decreasing function of V.

(ii) The spatial integral scale of $\overline{Y}$ can be easily computed by noting that $H(0) = V$ and $\int H(\mathbf{x})\, d\mathbf{x} = V^2$. Hence,

$$I_{\overline{Y}} = \frac{1}{\sigma^2_{\overline{Y}}} \int C_{\overline{Y}}(\mathbf{x}')\, d\mathbf{x}' = \frac{\sigma^2_Y}{\sigma^2_{\overline{Y}}} \int \rho_Y(\mathbf{x}')\, d\mathbf{x}' \tag{3.2.38}$$

The ratio between the volume integral scales is, therefore, given by

$$I_{\overline{Y}}/I_Y = (\sigma^2_Y/\sigma^2_{\overline{Y}})^{1/m} \tag{3.2.39}$$

where m=1,2,3 is the space dimensionality. By using the estimate of $\sigma^2_{\overline{Y}}$ of (i), it is seen that for large V/I^3_Y we get $I_{\overline{Y}} \sim V^{1/m}$ in (3.2.38), i.e. the volume integral scale of $\overline{Y}$ is of the order of the radius of the averaging volume.

(iii) An uncorrelated component, "nugget" effect, of C_Y is wiped out under the space averaging procedure, i.e. $C_{\overline{Y}}$ becomes continuous. A similar smoothing effect occurs for a discontinuous derivative and for this reason the space averaging operation is termed as a "regularization" by Matheron (1967).

Summarizing, space averaging has a smoothing effect, it increases the correlation scale and it decreases the variance of the point variable. The latter property is crucial in ensuring ergodicity for $\overline{Y}$ (see Chap. 1.10).

To illustrate further the effect of space averaging, let us consider the particular C_Y (3.2.25), which lends itself to very simple calculations. We wish to examine the effect of space averaging of Y over a vertical segment of length ℓ, as presumably achieved by a well whose length is large compared to its diameter. By using the Fourier transforms (3.2.27) and (3.2.36) in Eq. (3.2.35), we immediately obtain

$$\hat{C}_{\overline{Y}}(\mathbf{k}) = \frac{8\, I_{Yz}}{\sqrt{(2\pi)}\,\ell^2}\; \frac{1}{1+k_z^{\,2} I_{Yz}^{\,2}}\; \frac{\sin^2(k_z \ell/2)}{k_z}\; \hat{C}_{Yh}(k_x, k_y) \tag{3.2.40}$$

where, for the sake of generality, C_{Yh} denotes the horizontal part of the covariance (3.2.25), i.e. $C_Y = \exp(-|z|/I_{Yz})\, C_{Yh}(x,y)$.

Inversion of (3.2.40) yields

$$C_{\overline{Y}}(\mathbf{x}) = \frac{2\,(\ell' - z' - e^{-z'} + e^{-\ell'} \cosh z')}{\ell'^2}\, C_{Yh}(x,y) \quad \text{for } |z|<\ell\, ;$$

$$C_{\overline{Y}}(\mathbf{x}) = \frac{4 \sinh^2(\ell'/2)}{\ell'^2} \exp(-|z'|)\, C_{Y_h}(x,y) \quad \text{for } |z|>\ell \tag{3.2.41}$$

where $z' = z/I_{Yz}$, $\ell' = \ell/I_{Yz}$

To illustrate the results, $C_{\overline{Y}}$ is represented as function of z′, for a few values of ℓ′, in Fig. 3.2.4. The effects of space averaging mentioned above manifest clearly in this figure. For instance, the variance is given by $\sigma^2_{\overline{Y}} = 2\,(\ell'-1+e^{-\ell'})\,\sigma^2_Y/\ell'^2$ and it decreases like $1/\ell' = I_{Yz}/\ell$ for large ℓ′.

3.2.5 Effect of parameter estimation errors and summarizing comments

In the preceding sections it was shown that the statistical structure of the logconductivity could be expressed in terms of a few parameters, e.g. w, m_Y, σ^2_Y and I_Y, once a particular form of the covariance function is selected. For brevity, we shall denote by $\boldsymbol{\theta}$ the vector of P parameters, i.e. $\theta_1 = m_Y$, θ_2 = w, $\theta_3 = \sigma^2_Y$, ... So far, it was implicit in our analysis that these parameters are known deterministically, i.e. that their "true" values are known. In field applications, however, they are inferred from a finite sample of measured values, and they are affected by estimation errors. Hence, only the estimates of the parameters, denoted by $\tilde{\boldsymbol{\theta}}$, are accessible, whereas the "true" values remain unknown. Thus, $\tilde{\boldsymbol{\theta}}$ is generally a random vector characterized by the joint p.d.f. $f(\tilde{\boldsymbol{\theta}})$ of its components, which tends to normality for a sufficiently large number of measurements. While the ensemble mean of $\tilde{\boldsymbol{\theta}}$ remain undetermined, their variances of estimation can be estimated. We shall denote by Θ_{ij} (i,j=1,...,P) the variance-covariance matrix of the estimates $\tilde{\boldsymbol{\theta}}$.

The discussion of the techniques of statistical inference, e.g. maximum likelihood, is beyond the scope of this book. In the hydrological context, they have been the object of a few recent investigations (see, for instance, Kitanidis and Vomvoris, 1983 and Kitanidis, 1986). The analysis of the impact of estimation errors upon uncertainty of Y is a much simpler than that of spatial variability. To demonstrate this, we consider a functional $W(\mathbf{Y})$, depending on the vector of values $\mathbf{Y}$ of Y at a few points, which may be the outcome of the solution of a flow or transport problem. Its expectation value, the ensemble average, cannot be evaluated since $\boldsymbol{\theta}$ in the p.d.f. $f(\mathbf{Y},\boldsymbol{\theta})$ is not known. W is estimated by the following formula

$$\tilde{W} = \int W(\mathbf{Y})\, f(\mathbf{Y}|\tilde{\boldsymbol{\theta}})\, d\mathbf{Y} = \langle W \rangle \Big|_{\boldsymbol{\theta}=\tilde{\boldsymbol{\theta}}} \tag{3.2.42}$$

where $f(\mathbf{Y}|\tilde{\boldsymbol{\theta}})$ is the p.d.f. of Y, conditioned on the estimates $\tilde{\boldsymbol{\theta}}$. Eq. (3.2.42) reads in words: the estimate of W is equal to the expected value of W in which the "true" values of the parameters are

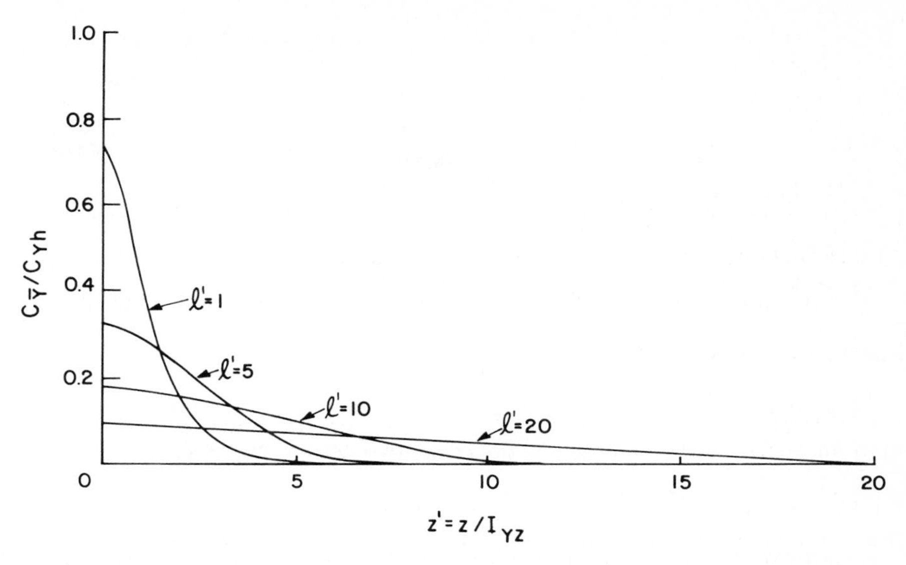

Fig. 3.2.4 Effect of vertical space averaging of logconductivity of covariance $C_Y = \exp(-|z|/I_{Yz})\, C_{Yh}(x,y)$ on an interval of length ℓ ($\ell' = \ell/I_{Yz}$), upon the covariance $C_{\bar{Y}}$ (Eq. 3.2.41).

replaced by their estimates, which is the straightforward definition in which the $\tilde{\theta}$ are regarded as "true". By the same token, we get for the variance of W

$$\sigma_W^2 = \langle[W - \langle W\rangle]^2\rangle = \langle[W - \tilde{W}]^2\rangle + \langle[\tilde{W} - \langle W\rangle]^2\rangle = \int (W-\tilde{W})^2 f(Y|\theta)\, dY + \sigma_{\tilde{W}}^2 \tag{3.2.43}$$

where $\sigma_{\tilde{W}}^2$ is the variance of estimation of $\tilde{W}$. The first term in the right-hand side of (3.2.43) cannot be computed exactly but only estimated, by replacing θ by its estimate. Thus, (3.2.43) is rewritten as follows

$$\tilde{\sigma}_W^2 \simeq \sigma_W^2\Big|_{\theta=\tilde{\theta}} + \sigma_{\tilde{W}}^2 \tag{3.2.44}$$

i.e. the estimate of the variance is equal to the variance in which the parameters are replaced by their estimates, *supplemented* by the variance of estimation of $\tilde{W}$. The last term can also be estimated only, by replacing again θ by $\tilde{\theta}$. Its evaluation simplifies considerably if the coefficients of variation of $\tilde{\theta}$ are small. Then, a first-order expansion of $\langle W\rangle$ in (3.4.40), i.e. $\langle W(\theta)\rangle = \tilde{W} + \Sigma_{i=1}^{P}$ $(\partial\tilde{W}/\partial\tilde{\theta}_i)\,(\theta_i - \tilde{\theta}_i)$ yields

$$\sigma^2_{\tilde{W}} = \langle(\tilde{W} - \langle W\rangle)^2\rangle \simeq \sum_{i=1}^{P}\sum_{j=1}^{P} \frac{\partial \tilde{W}}{\partial \tilde{\theta}_i} \frac{\partial \tilde{W}}{\partial \tilde{\theta}_j} \Theta_{ij} \tag{3.2.45}$$

The derivatives of the estimate $\tilde{W}$ with respect to $\tilde{\boldsymbol{\theta}}$ are known as sensitivity coefficients, since they reflect the impact of the uncertainty of the parameters estimates upon that of the function.

To illustrate these results we select the simple case W=Y. Then $\theta_1=m_Y$, $\theta_2=\sigma^2_Y$ and (3.2.42,43) lead to the classical relationship $\tilde{\sigma}^2_W = \tilde{\theta}_2 + \Theta_{11}$.

The main conclusion of this discussion is that the variance of a functional depending on a random space function is made up from two terms: one reflecting the uncertainty associated with spatial variability and the other one with the errors of estimation of the parameters characterizing the moments of the function. At any rate, we shall regard in the following sections, dealing with flow and transport at local scale, the parameters $\boldsymbol{\theta}$ as fixed and either deterministic, if the coefficient of variation of the estimation errors is much smaller than σ_Y, or random and subjected to the generalization implied by (3.2.45). The application of these concepts is discussed in Chap. 5.5.

Summarizing this section, we shall assume that the properties of heterogeneous formations affecting flow are represented at the local scale as random space functions, which are stationary and ergodic. In particular, the hydraulic conductivity, which displays the largest spatial variability, is assumed to be multivariate lognormal and characterized completely by the mean and the two-point covariance. The latter may be anisotropic, and is represented conveniently by a few analytical formulae, which in turn depend on a few parameters. In some applications the space average of the logconductivity may be of interest. The latter is also multivariate normal, of lesser variance and larger integral scale than the point variable.

The three length scales set forth so far, namely the heterogeneity correlation scale I_Y, the domain extent D and the extent of the averaging volume V (denoted above by ℓ or d) and their hierarchy, play a fundamental role in the theory of flow and transport (see Dagan, 1986). This point will be retaken in the following sections.

Exercises

3.2.1 Derive the expression of the two-point covariance of the hydraulic conductivity $K(\mathbf{x})$, if K is lognormal and stationary. The moments m_Y and $C_Y(\mathbf{x})$ are given. Is K stationary and isotropic if Y is so? What is the linear integral scale of K for C_Y linear?

Hint : use the characteristic function (1.2.18).

3.2.2 Derive the function $H(\mathbf{x})$ of (3.2.34) and its Fourier transform in the case in which the averaging volume V is a cube of sides d. Derive the variance of $\overline{Y}$ for C_Y of type "exponential separated" (3.2.25). What is the magnitude of d which renders $\sigma^2_{\overline{Y}}/\sigma^2_Y < 0.01$?

3.2.3 Derive and represent graphically the variogram of $\overline{Y}$ for Γ_Y linear and isotropic, in presence of a "nugget" and for a spherical V.

3.2.4 Assume that the covariance can be decomposed as follows : $C_Y(\mathbf{x}) = \rho_{Yv}(z)\, C_{Yh}(x,y)$, where $\rho_{Yv} = \exp(-|z|/I_{Yz})$. Furthermore, $I_{Yz} >> I_{Yh}$, the latter being the horizontal integral scale. Thin horizontal clay layers of extent much larger than I_{Yh} are present and the average vertical distance between them is equal to I_{Yv}. Derive the covariance $C_{\overline{Y}}$ for $\overline{Y}$ defined as a vertical average over strata of formation between the layers, i.e. over I_{Yv}.

3.3 GENERAL FORMULATION OF THE DIRECT PROBLEM AND OF EQUATIONS OF FLOW

3.3.1 General statement of the direct problem

The flow of water at the local scale is formulated in mathematical terms with the aid of the equations satisfied by the macroscopic variables, which were derived in Part 2. The formulation of the direct problem of flow is as follows : given the flow domain Ω and the formation properties K, n and s, derive the water head field $\phi(\mathbf{x},t)$. Once the latter is known, one may determine other variables of interest, like the specific discharge $\mathbf{q} = -K\, \nabla\phi$ and the velocity $\mathbf{u} = \mathbf{q}/n$. This is the main aim of the theory of flow at the local scale to be pursued here. The approach is to solve the partial differential equation satisfied by ϕ (see Chap. 2.11), with appropriate boundary and initial conditions (Chap. 2.13). The first case of primary interest is that of unsteady flow of homogeneous liquid in a compressible matrix, analyzed in Sect. 2.11.2, with ϕ satisfying Eq. (2.11.21), which is reproduced herein for the sake of convenience

$$s\, \frac{\partial\phi}{\partial t} = \nabla.(K\, \nabla\phi) \tag{3.3.1}$$

The second class of widely encountered problems is that of steady flows, or unsteady ones in which compressibility is negligible (Sect. 2.11.1), when Eq. (3.3.1) reduces to

$$\nabla.(K\, \nabla\phi) = 0 \tag{3.3.2}$$

These equations simplify considerably for homogeneous formations, of constant properties, for which (3.3.1) and (3.3.2) become the linear equations with constant coefficients of heat conduction and Laplace equation, respectively. To illustrate a typical problem of flow at local scale and its mathematical formulation, let us consider the example of Fig. 3.1.1a, i.e. a pumping well in a confined, elastic, aquifer. Assuming a homogeneous formation of given K and s, Eq. (3.3.1) has to be solved with the following boundary and initial conditions (see Chap. 2.13)

$$\frac{\partial \phi}{\partial z} = 0 \quad \text{on the impervious boundaries } z=0 \text{ and } z=D$$

$$\frac{\partial \phi}{\partial r} = 0 \quad \text{on the well impervious upper portion } r=r_w \ , \ \ell_{wu}<z<D$$

$$\phi=\phi_w(t) \ , \ \int_{\ell_{wl}}^{\ell_{wu}} \frac{\partial \phi}{\partial r} dz = \frac{Q}{2\pi r_w} \quad \text{on the well screen } r=r_w \ , \ \ell_{wl}<z<\ell_{wu} \tag{3.3.3}$$

$$\phi=\phi_0=\text{const at } t=0$$

where it was assumed that initially there was no flow and at t=0 the well starts to pump at a given discharge Q. Furthermore, the unknown head ϕ_w in the well was assumed to depend on time only. The solution of this problem, which will be retaken in Chap. 3.5, and of many similar ones, has been the object of numerous studies providing either exact or approximate analytical solutions (a few books on the subject have been quoted in Chap. 3.1).

As we have mentioned in the preceding sections, actual formations are generally heterogeneous, and the assumption of constant, or regularly varying, properties in space is not met. One of the main aims of this book is to extend the traditional approach to heterogeneous formations, relying on the recent literature on the subject (for recent reviews see Dagan, 1986, and Gelhar, 1986).

In line with the probabilistic approach of the preceding two sections, we shall consider Eqs. (3.3.1) and (3.3.2) as the starting point of the theory of flow, with K and s, however, regarded as random space functions (RSF). Consequently, these are equations with stochastic coefficients, and the solution $\phi(\mathbf{x},t)$ is also an RSF, and the same is true for the other variables, like $\mathbf{q}$ and $\mathbf{u}$, depending on ϕ. As shown in Chap. 1.1 and also discussed in the preceding sections, ϕ is defined in statistical terms by the joint probability density function of its values $\phi_j=\phi(\mathbf{x}_j,t)$ at an arbitrary set of N points $\mathbf{x}_j$ (j=1,2...,N). Of definite significance are the two first moments, the expectation (ensemble mean) $\langle\phi\rangle$ and the two-point covariance $C_\phi(\mathbf{x}_1,\mathbf{x}_2,t)$ and particularly the variance $\sigma_\phi^2(\mathbf{x},t) = C_\phi(\mathbf{x},\mathbf{x},t)$. In practice one may be satisfied with these two moments rather than a comprehensive representation of ϕ by its various moments. In particular, if ϕ_j happens to be multivariate normal, the second order representation is exhaustive and may serve for deriving any other higher order statistical moment (Chap. 1.2). As we have mentioned already in the preceding section in relation to K, in practice we are generally interested in some space average of ϕ. For instance, when measuring the water head by a piezometer, ϕ is averaged over the cylindrical surface of the screen. Such a space average is defined by

$$\bar{\phi}(\mathbf{x},t) = \frac{1}{V}\int_V \phi(\mathbf{x}',t)\,d\mathbf{x}' = \frac{1}{V}\int \phi(\mathbf{x}',t)\,\Omega(\mathbf{x}-\mathbf{x}')\,d\mathbf{x}' \tag{3.3.4}$$

where $\mathbf{x}$ is the centroid of V (line interval, surface or volume) and Ω is the step function equal to unity within V (see Chaps. 1.9 and 3.2). Obviously, $\bar{\phi}$ is an RSF whose moments can be derived from those of ϕ, by the procedure applied in Chap. 3.2 to Y. Furthermore, the point value ϕ may be viewed as a particular case of (3.3.4) for $V\to 0$, except for an uncorrelated, "nugget" effect, which is wiped out by space averaging.

We are now in a position to formulate the basic problem of flow through heterogeneous formations of random spatial structure : given the flow domain Ω and given the formation properties as random space functions, determine the RSF $\bar{\phi}$, with ϕ satisfying the flow equation and appropriate boundary and initial conditions. As already indicated, we may be satisfied with determining the first two moments of $\bar{\phi}$, namely $\langle\bar{\phi}\rangle$ and $C_{\bar{\phi}}$. Furthermore, we adopt for Y the stationary, multivariate normal, distribution discussed at length in the preceding section, whereas s and n have been shown to be of much lesser variability and may be taken as constant. Under these conditions, the flow problem can be stated in a restricted form as follows : for given domain Ω and for Y = ln K stationary and multivariate normal, defined by m_Y and $C_Y(\mathbf{x})$, determine $\langle\bar{\phi}\rangle$ and $C_{\bar{\phi}}(\mathbf{x})$, with ϕ satisfying the flow equation and boundary and initial conditions. The solution of this problem constitutes the object of the following sections.

It is worthwhile to mention here a few points of principle which will be discussed in detail later, in the frame of Part 5, which refers to flow at the regional scale. First, the problem formulated above has been termed as the direct one, being of the type studied by classical mathematical physics. We shall discuss in Part 5 the "inverse" problem, which essentially is the one in which the water head $\phi(\mathbf{x},t)$ is given and the formation properties are the ones to be identified. Second, the input RSF, as well as ϕ, are represented in terms of their unconditional probability distributions. Again, we shall study in Part 5 the effect of conditioning, i.e. of using the p.d.f. of Y and ϕ conditioned on their measured values at a few points. Third, we have regarded the heterogeneity as random, whereas the boundary and initial conditions were considered to be deterministic. In fact, the latter are also in practice subjected to uncertainty, but it is much easier to deal with these uncertainties than with those related to spatial variability of properties. To illustrate this point, let us assume that in (3.3.3) Q, the well discharge, is a random variable, subjected to uncertainty because of errors of measurement, fluctuations in the pump performance, etc. Its impact on the head field is easily accounted for by solving the deterministic problem (3.3.1), (3.3.3) conditioned on a fixed value of Q, and by using subsequently the approach of Sect. 3.2.5 for θ=Q to investigate the statistics of ϕ.

3.3.2 A few general observations on the stochastic problem

The basic partial differential equations (PDE) (3.3.1) and (3.3.2) with constant or deterministic coefficients are classical equations of the mathematical physics and have been studied intensively in the literature. Less is known about the solution of these equations when the coefficients, i.e. K and/or s, are random space functions (for a review related to problems of continuum theories one may consult Beran, 1968).

We shall discuss here only a few general approaches in solving stochastic PDE, skipping the more fundamental mathematical aspects of existence, uniqueness and stationarity of the solution.

(i) Monte Carlo simulations

A powerful technique to be briefly reviewed first is that based on Monte Carlo simulations, which is conceptually simple and of increasing availability with the advent of new generations of computers. In Monte Carlo simulations the PDE of flow, like (3.3.1) and (3.3.2), are first cast in a numerical form, e.g. by finite differences or finite elements, and the solution is sought in the form of a vector of values of the water head ϕ at the nodes of a spatial grid. In the numerical version the variable coefficients representing the properties are also represented by vectors of their values at a finite number of nodes, say the hydraulic conductivity $K_j = K(\mathbf{x}_j)$ ($j=1,2,...,N$). In the first stage of the simulation, a realization of the random vector K_j, say $K_{j,1}$ is computer generated by well known techniques. Thus, for $Y_j = \ln K_j$ multivariate normal of given mean and covariance, the generation of Y_j is a standard operation. In the second stage $\phi_{j,1}$ is derived numerically in the given realization by solving the flow problem as a deterministic one. This operation is repeated M times, with $M>>1$, to obtain the set of solutions $\phi_{j,k}$ ($j=1,2,...,N;k=1,2,...,M$), which are viewed as the set of realizations of the vector ϕ_j. Based on this information any moment of interest of $\phi(\mathbf{x},t)$ can be computed approximately. It is emphasized that unlike the solution of deterministic problems, two types of approximations are implied in this process : (i) replacement of the continuous functions Y and ϕ by a discrete set of dimension N, which is the usual approximation. However, since a random function may fluctuate violently, the problem of selecting a grid size which ensures a desired accuracy is a delicate one ; (ii) the simulation of the random variables Y_j and ϕ_j by finite samples of dimension M, $Y_{j,k}$ and $\phi_{j,k}$, respectively.

The Monte Carlo simulations have been applied to one-dimensional flow through heterogeneous formations by Freeze (1975) and to two-dimensional flow by Smith and Freeze (1979). The method is conceptually simple and powerful, in the sense that it can be used to solve problems with complex boundaries and large variances of the input variables. Nevertheless, it has serious limitations which have restricted its use to a few special cases. Herein are a few such limitations: (i) the computer time can be exceedingly large and in fact no systematic and controlled simulations have been carried out for general three-dimensional flows so far (for an early attempt, see Warren and Price,

1961 and a recent one, Desbarats, 1987). The requirements for large computer memory and time stem from the need to operate on dense grids in order to ensure an accurate representation of highly oscillatory random functions and from the large number of realizations necessary in order to achieve a representative sample. Both these requirements become more stringent as the variance of the input variable, i.e. σ_Y^2, increases, and its integral scale I_Y decreases; (ii) as a matter of fact, even in the case of two-dimensional flow (Smith and Freeze, 1979) it has been found that the numerical scheme does not converge for large σ_Y^2. Verification of convergence is a difficult matter in itself; (iii) by its nature, the Monte Carlo simulation may yield valuable solutions in some particular cases, but its usefulness in providing insight and in drawing general properties and conclusions is quite limited. For these reasons, and especially for the last one, we shall regard Monte Carlo simulations as numerical experiments rather than a general tool.

(ii) Small perturbation expansion.

Another approach of wide applicability, which lends itself to general analytical results or to much simpler numerical schemes, is the small perturbation approximation. This line of attack has been employed frequently in the last few years and appropriate references will be quoted in the following chapters, a general outline being given in Beran (1968).

Before describing the approach, it is worthwhile to mention that ideally, one would like to be able to derive partial differential equations satisfied by the various statisitical moments of ϕ and to solve them separately. Unfortunately, this is not possible in a general manner, since the equation which is derived from (3.3.1,2) for a moment includes higher-order moments (see Beran, 1968). Thus, a closure approximate condition is needed in order to truncate the moment equations.

The small perturbation asymptotic approximation of the head field ϕ, which evades this difficulty, may be presented in a few ways. A simple approach is to refer to the general dependence of the moments of ϕ on the various parameters of the problem. Thus, for the sake of illustration, let us assume that the statistical structure of the input property, the logconductivity Y, is characterized by the parameters m_Y, σ_Y^2 and the length scale I_Y, in line with the models presented in Chap. 3.2. The moments of interest, i.e. the mean $\langle\phi\rangle$, the two-point cross-covariance $C_{Y\phi}$ and the covariance C_ϕ, as well as higher-order moments, are functions of the parameters representing the boundary conditions (which drive the flow), of the domain geometry, of the dimensionless coordinate $\mathbf{x}/I_Y$ and of m_Y and σ_Y^2. We may write, for instance

$$\langle\phi\rangle = \langle\phi(\mathbf{x}/I_Y, tK_G/I_Y^2; m_Y, \sigma_Y^2, ...)\rangle \; ; \; C_\phi = C_\phi(\mathbf{x}_1/I_Y, \mathbf{x}_2/I_Y, tK_G/I_Y^2; m_Y, \sigma_Y^2, ...) \qquad (3.3.5)$$

and similar expressions for other moments. In the limit case in which $\sigma_Y^2=0$, $\phi=\phi^{(0)}$ becomes a deterministic solution for flow in a homogeneous formation of constant $Y=\langle Y\rangle=m_Y$,

$K=K_G=\exp(m_Y)$. Hence, C_ϕ and all other higher moments vanish for $\sigma_Y^2=0$. Consequently, assuming that $\langle\phi\rangle$, $C_{Y\phi}$, C_ϕ ... may be expanded in an asymptotic sequence for small σ_Y^2, we may write

$$\langle\phi\rangle = \langle\phi\rangle^{(0)} + \langle\phi\rangle^{(1)} + \langle\phi\rangle^{(2)} + ...$$

$$C_{Y\phi}(\mathbf{x}_1,\mathbf{x}_2) = \langle Y(\mathbf{x}_1)\, \phi(\mathbf{x}_2)\rangle = C_{Y\phi}^{(1)} + C_{Y\phi}^{(2)} + ... \tag{3.3.6}$$

$$C_\phi(\mathbf{x}_1,\mathbf{x}_2) = \langle \phi(\mathbf{x}_1)\, \phi(\mathbf{x}_2)\rangle = C_\phi^{(1)} + C_\phi^{(2)} + ...$$

where $\langle\phi\rangle^{(j+1)}/\langle\phi\rangle^{(j)} \to 0$ for $\sigma_Y^2 \to 0$, and similarly for $C_{Y\phi}^{(j)}$, $C_\phi^{(j)}$,... . It is possible now to derive PDE for these approximations by starting from the equations of flow (3.3.1,2). We shall illustrate here the procedure for the steady-state flow, by rewriting first (3.3.2) as follows

$$\nabla^2\phi + \nabla Y.\nabla\phi = 0 \qquad (Y = \ln K) \tag{3.3.7}$$

Now, we expand *formally* ϕ in an asymptotic sequence similar to (3.3.6), i.e.

$$\phi = \phi^{(0)} + \phi^{(1)} + \phi^{(2)} + ... \tag{3.3.8}$$

where, say, the variances of each term of (3.3.8) form an asymptotic sequence with respect to σ_Y^2. Substituting (3.3.8) into (3.3.7) and solving it iteratively we obtain the following sequence of equations

$$\nabla^2\phi^{(0)} = 0 \;;\; \nabla^2\phi^{(1)} = -\nabla Y.\nabla\phi^{(0)} \;;\; \nabla^2\phi^{(2)} = -\nabla Y.\nabla\phi^{(1)} \;... \tag{3.3.9}$$

Since we assume that the boundary conditions are deterministic (though it is relatively easy to generalize the procedure for random ones), $\phi^{(0)}$, satisfying (3.3.9), is seen to be deterministic. Hence, we have for the expectation, from (3.3.6) and (3.3.9) and for constant m_Y

$$\langle\phi^{(0)}\rangle = \phi^{(0)} \;;\; \langle\phi^{(1)}\rangle = 0 \tag{3.3.10}$$

For the crosscovariance we get at first-order

$$C_{Y\phi}^{(1)}(\mathbf{x}_1,\mathbf{x}_2) = \langle Y(\mathbf{x}_1)\, \phi^{(1)}(\mathbf{x}_2)\rangle \tag{3.3.11}$$

and the PDE satisfied by $C_{Y\phi}^{(1)}$ is readily obtained by multiplying the second equation of (3.3.9) by $Y(\mathbf{x}_1)$ and averaging. Hence

$$\nabla_2^2 C_{Y\phi}^{(1)}(\mathbf{x}_1,\mathbf{x}_2) = -\nabla\phi^{(0)}(\mathbf{x}_2).\langle Y(\mathbf{x}_1)\ \nabla Y(\mathbf{x}_2)\rangle = -\mathbf{J}^{(0)}(\mathbf{x}_2)\cdot\nabla_2 C_Y(\mathbf{x}_1,\mathbf{x}_2) \qquad (3.3.12)$$

where ∇_1 and ∇_2 stand for the gradient with respect to $\mathbf{x}_1$ and $\mathbf{x}_2$, respectively, and $\mathbf{J}^{(0)} = -\nabla\phi^{(0)}$ is the first approximation of the head gradient. Eq. (3.3.12) is a Poisson equation for $C_{Y\phi}^{(1)}$ whose right-hand side is expressed in terms of the zero-order approximation $\phi^{(0)}$, which satisfies Laplace equation with deterministic boundary conditions, and of the given C_Y. Furthermore, it is seen that $C_{Y\phi}^{(1)}$ is $O(\sigma_Y^2)$.

By a similar procedure we obtain for the head two-point covariance at first-order

$$C_\phi^{(1)}(\mathbf{x}_1,\mathbf{x}_2) = \langle\phi^{(1)}(\mathbf{x}_1)\ \phi^{(1)}(\mathbf{x}_2)\rangle$$

$$\nabla_2^2\nabla_1^2\ C_\phi^{(1)} = -\sum_j\sum_k J_j^{(0)}(\mathbf{x}_1)\ J_k^{(0)}(\mathbf{x}_2)\ \frac{\partial^2}{\partial x_{1j}\ \partial x_{1k}}\ C_Y(\mathbf{x}_1,\mathbf{x}_2) \qquad (3.3.13)$$

where again $C_\phi^{(1)}$ satisfies a linear nonhomogeneous equation and it is $O(\sigma_Y^2)$.

Similar, though more cumbersome, equations may be obtained for the higher-order approximations of the various statistical moments (see, e.g., Dagan, 1985b). An important point is that the RSF $\phi^{(1)}$, of zero mean, is obtained from (3.3.9) by a linear operator applied to Y. Hence (see Chap. 1.1) if Y is multivariate normal, the same is true for $\phi^{(1)}$. Consequently, the various statistical moments of Y and $\phi^{(1)}$ are determined entirely by C_Y, $C_{Y\phi}^{(1)}$ and $C_\phi^{(1)}$. The first-order approximation will be used extensively in the following sections.

The question of validity of the perturbation approximation, i.e. of the convergence of (3.3.6) and of the limiting value of σ_Y^2 which ensures the accuracy of the first-order approximation, is a difficult one. Some partial results have been given by Dagan (1985b) and will be discussed in Sect. 3.7.3.

The small perturbation technique has been applied extensively to the study of flow through heterogeneous formations, e.g. Schwydler, 1962, Matheron, 1967, Bakr et al, 1978, Sagar, 1978, Gutjahr et al, 1978, 1981, Dettinger and Wilson, 1981, Dagan, 1982a, and these studies will be quoted in relation to specific subjects in the sequel.

(iii) The self-consistent or renormalization technique

This approach has been applied to determining the effective properties of multi-phase, heterogeneous, media (for reviews, see Beran, 1968, Landauer, 1978, and for applications to porous formations Dagan,1981, and Sect. 3.4.3 here). Let us consider a multi-phase medium of p.d.f. of the type (3.2.5), which represents a formation made up from a collection of blocks of different conductivities K_j. In the first-order perturbation approximation, the head field $\phi^{(1)}$ can be determined as the sum of the head residuals caused by flow in the formation of uniform conductivity equal to K_G in which each time a block of logconductivity Y_j is inserted. At next order, two blocks have to be considered simultaneously and the procedure can be extended in principle to any order. In the self-consistent approximation the head field is the one caused by an isolated block in a formation of constant, but unknown, conductivity K_{ef}, which is determined by the requirement that K_{ef} is the effective conductivity of the composite medium.

The great advantage of the self-consistent approximation over the first-order perturbation approximation is its applicability to large variances σ_Y^2. It has been applied to deriving the effective conductivity and it has been shown (Dagan, 1981) to be accurate for a "completely random" medium, i.e. one in which there is no correlation between the conductivities of neighboring blocks, and for a large number of phases, for which this is a plausible assumption.

This powerful technique will be applied to flow through heterogeneous formations in the following section.

Exercises

3.3.1 A Monte Carlo simulation is carried out for one-dimensional flow. The conductivity field is generated by random, uncorrelated, values of $Y = \ln K$ at equidistant nodes on the x axis. Y at intermediate points is obtained by linear interpolation between the values at adjacent nodes. What is the corresponding $C_Y(x_1,x_2)$?

3.3.2 Derive the partial differential equation satisfied by $\langle\phi\rangle^{(1)}$, the first order approximation of the average head. Show that in case of stationary Y, $\langle\phi\rangle^{(1)} \equiv 0$.

3.3.3 Rewrite Eqs. (3.3.9) in terms of Fourier transforms of ϕ,Y for an infinite domain, for Y stationary and for $\mathbf{J}^{(0)}$ constant.

3.4 THE EFFECTIVE HYDRAULIC CONDUCTIVITY

3.4.1 Steady uniform flow : general statement and absolute bounds

In this chapter we shall derive the relationships satisfied by the expected values (ensemble averages) of the water head $\langle\phi\rangle$ and of the specific discharge $\langle\mathbf{q}\rangle$ in formations of heterogeneous structure, i.e. we generalize Darcy's law. This is the most important step toward a theory of flow for the following reasons : first, the expected value is the primary statistical information and second, in the case in which space averages $\bar{\phi}$ and $\bar{\mathbf{q}}$ are sought and the ergodic hypothesis is obeyed, the ensemble and space averages are interchangeable. Hence, in the latter case the expectations exhaust the information we need, precisely like in the case of averaging over the pore scale, discussed in Part 2. No wonder that this subject has been studied extensively in various branches of physics and engineering dealing with heterogeneous media (see, e.g. Beran, 1968). The same problem, at the laboratory scale, has been already discussed in Part 2 of this book and the developments here will follow a similar line of attack.

We consider a heterogeneous porous formation of stationary random logconductivity $Y = \ln K$ and steady water flow satisfying the equations of continuity and Darcy's law at each point (see Part 2)

$$\nabla.\mathbf{q} = 0 \quad ; \quad \mathbf{q} = - K\ \nabla\phi \qquad (\mathbf{x} \in \Omega) \tag{3.4.1}$$

where Ω is the flow domain. We shall assume that the boundary conditions on $\partial\Omega$, the domain boundary, are deterministic. Averaging of the continuity equation in (3.4.1), yields for the expectation of the specific discharge $\nabla.\langle\mathbf{q}\rangle=0$. One would expect that the averaging of Darcy's law would yield a linear equation of a similar type, i.e. $\langle\mathbf{q}(\mathbf{x})\rangle = -\mathbf{K}_{ef}\ \nabla\langle\phi(\mathbf{x})\rangle$, with the tensor $\mathbf{K}_{ef}$ representing a property of the medium. Such a representation is not warranted, however, under general conditions, and $\mathbf{K}_{ef}$ may depend, for instance, on the values of the gradient of ϕ at other points than $\mathbf{x}$ or on the distance to the boundary. To simplify matters, in line with the approach adopted in Part 2, we shall start by considering for ϕ the following boundary condition

$$\phi = - \mathbf{J}.\mathbf{x} \qquad (\mathbf{x} \in \partial\Omega) \tag{3.4.2}$$

where $\mathbf{J}$ is a constant vector. In a homogeneous formation $\nabla^2\phi = 0$ and ϕ is identically equal to (3.4.2) in the entire domain, whereas $\mathbf{q} = K\ \mathbf{J}$ is constant. A simple example of domain and boundary condition is shown in Fig. 3.4.1, for which $J_x=(\phi_A - \phi_B)/L$ and two impervious boundaries are present. Consider now a heterogeneous formation of large extent compared to the heterogeneity scale, i.e. with $L/I_Y>>1$. By a reasoning similar to that of Sect. 2.7, at the limit of an unbounded

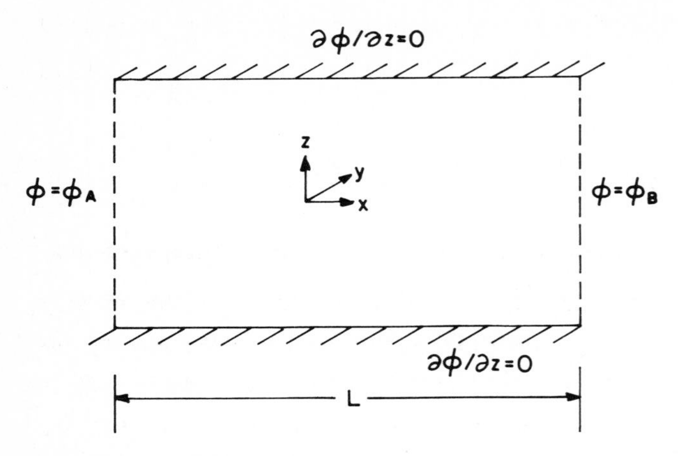

Fig. 3.4.1 Definition sketch for average uniform flow in a porous formation.

formation, $\nabla\langle\phi\rangle$ and $\langle \mathbf{q}\rangle$ are constant and furthermore, by the linearity of (3.4.1), they are given by

$$\langle\phi\rangle = -\mathbf{J}.\mathbf{x} \quad ; \quad \langle \mathbf{q}\rangle = -\mathbf{K}_{ef}\ \nabla\langle\phi\rangle = \mathbf{K}_{ef}.\mathbf{J} \qquad (3.4.3)$$

where the constant tensor $\mathbf{K}_{ef}$ is usually defined as the effective conductivity tensor (the generalization of these definitions for nonuniform average flows, bounded domains and unsteady flows will be discussed in Sects. 3.4.5., 3.4.6 and 3.4.7, respectively). From dimensional considerations we may write that $\mathbf{K}_{ef}/K_G$ is a function of σ_Y^2 and of the ratios between the length scales characterizing C_Y. The main objective of this chapter is to derive this function.

Toward this aim and along the lines of Part 2, Sects. 2.7 and 2.9, we may employ conveniently the dissipation function E, defined as the energy per unit weight of fluid dissipated by friction, i.e. $E = -\mathbf{q}.\nabla\phi$. With residuals defined by $\phi' = \phi - \langle\phi\rangle$ and $\mathbf{q}' = \mathbf{q} - \langle\mathbf{q}\rangle$, is it easily found from (3.4.1-3) that for stationary ϕ and $\mathbf{q}$, we have

$$\langle E\rangle = -\nabla.\langle \mathbf{q}\ \phi\rangle = \langle \mathbf{q}\rangle.\mathbf{J} - \nabla.\langle\phi'\mathbf{q}'\rangle = \langle\mathbf{q}\rangle.\mathbf{J}$$
$$\langle E\rangle = \langle K\ \nabla\phi.\nabla\phi\rangle = \mathbf{J}.\mathbf{K}_{ef}.\mathbf{J}^T = \langle\mathbf{q}\rangle.\mathbf{K}_{ef}^{-1}.\langle\mathbf{q}\rangle^T \qquad (3.4.4)$$

where $\mathbf{J}^T$ and $\mathbf{q}^T$ are the transposed of $\mathbf{J}$ and $\mathbf{q}$, respectively.

The last equations may serve as alternative definitions of $\mathbf{K}_{ef}$. Furthermore, it is seen from the second equation in (3.4.4) that $\langle E\rangle \geq 0$, i.e. $\mathbf{K}_{ef}$ is positive definite and invertible. By a slight extension of (3.4.4) for two uniform flows of different $\mathbf{J}$ (see Sect. 2.9.2) it can be shown that it is symmetrical as well. It can be reduced, therefore to a diagonal matrix, with three positive principal values in orthogonal directions K_{efI} ,K_{efII} and K_{efIII} , respectively.

The most general theoretical results for the dependence of $\mathbf{K}_{ef}$ on the heterogeneous structure were found for its bounds. The argument, following Batchelor (1974), has been given already in

Chap. 2.7. It was shown by simple manipulations of the flow equations (3.4.1) and by stationarity that

$$\langle E \rangle = K_A \ \mathbf{J}.\mathbf{J} + \langle K \ \nabla\phi'.\nabla\phi' \rangle = K_H^{(-1)} \ \langle \mathbf{q} \rangle.\langle \mathbf{q} \rangle - \langle \frac{\mathbf{q}'.\mathbf{q}'}{K} \rangle \tag{3.4.5}$$

where $K_A = \langle K \rangle$ and $K_H = (\langle K^{(-1)} \rangle)^{(-1)}$ are the arithmetic and harmonic means, respectively. From (3.4.4) and (3.4.5) it is immediately seen that K_A and K_H are upper and lower bounds of $\mathbf{K}_{ef}$. Furthermore, by using the expressions of K_A and K_H for lognormal conductivity (see Eq. 1.1.15), we may write

$$\exp(-\sigma_Y^2/2) \le \mathbf{K}_{ef}/K_G \le \exp(\sigma_Y^2/2) \tag{3.4.6}$$

These can be shown to be best bounds for anisotropic formations, since they are effectively attained for a perfectly stratified medium of covariance (3.2.11), for which K_A and K_H are the effective conductivities for flow parallel and perpendicular to the layers, respectively (the proof is left as an exercise).

Narrower bounds have been obtained by Hashin and Shtrikman (1962) for isotropic media (see Chap. 2.7) for a multiphase medium, but they do not provide an improvement over (3.4.6) for a continuous distribution of K between zero and infinity. An application of their technique to a two-phase formation is left for Sect. 3.4.3.

The two bounds (3.4.6) become widely separated for large σ_Y^2 and are, therefore, of limited use in such cases. In the following two sections we shall obtain values of $\mathbf{K}_{ef}$ by using the approximate techniques described in Sect. 3.3.2.

3.4.2 Small perturbation, first-order approximation of $\mathbf{K}_{ef}$

We shall employ now the small perturbation approximation (Sect. 3.3.2) to derive the expressions relating $\mathbf{K}_{ef}$ to the heterogeneous structure, the latter being represented by the moments of the RSF Y.

Starting with Darcy's law $\mathbf{q} = - K \ \nabla\phi = - \exp(m_Y + Y') \ \nabla\phi$, we formally expand both exp(Y') and ϕ (3.3.8) to obtain

$$\mathbf{q} = - K_G \ (1 + Y' + \frac{Y'^2}{2} + ...) \ (\nabla\phi^{(0)} + \nabla\phi^{(1)} + \nabla\phi^{(2)} + ...) \quad \text{i.e.} \tag{3.4.7}$$

$$\mathbf{q}^{(0)} = -K_G \ \nabla\phi^{(0)} \ ; \ \mathbf{q}^{(1)} = -K_G (Y'\nabla\phi^{(0)} + \nabla\phi^{(1)})$$

$$\mathbf{q}^{(2)} = -K_G (\frac{Y'^2}{2}\nabla\phi^{(0)} + Y'\nabla\phi^{(1)} + \nabla\phi^{(2)})$$

We first observe that $\phi^{(0)} \equiv -\mathbf{J}.\mathbf{x}$, by virtue of (3.3.9) and (3.4.2). Consequently, each $\phi^{(j)}$ (j=1,2,...) satisfies homogeneous boundary conditions on $\partial\Omega$ and is determined by (3.3.9). By ensemble averaging the latter and by stationarity arguments we easily find out that

$$\nabla^2 \langle\phi^{(1)}\rangle = 0 \quad \text{i.e.} \quad \langle\phi^{(1)}\rangle \equiv 0$$

$$\nabla^2 \langle\phi^{(2)}\rangle = -\nabla.\langle(Y'\nabla\phi^{(1)})\rangle + \langle Y'\nabla^2\phi^{(1)}\rangle = -\nabla.\langle Y'\nabla\phi^{(1)} - \mathbf{J}\, Y'^2/2\rangle = 0 \quad \text{i.e.} \quad \langle\phi^{(2)}\rangle \equiv 0 \tag{3.4.8}$$

Hence, ensemble averaging of (3.4.7), with (3.4.8) taken into account, yields at first-order

$$\langle\mathbf{q}^{(0)}\rangle = K_G\, \mathbf{J} \quad ; \quad \langle\mathbf{q}^{(1)}\rangle = 0 \quad ; \quad \frac{\langle\mathbf{q}^{(2)}\rangle}{K_G} = \frac{\mathbf{J}\,\sigma_Y^2}{2} - \langle Y'\nabla\phi^{(1)}\rangle$$

$$\langle\nabla\phi\rangle = \nabla\phi^{(0)} + O(\sigma_Y^4) \tag{3.4.9}$$

Consequently, the expectation of $\mathbf{q}$ at first-order, $\langle\mathbf{q}\rangle = \langle\mathbf{q}^{(0)}\rangle + \langle\mathbf{q}^{(2)}\rangle$, is determined by known expressions, except the last term of (3.4.9), which we shall denote for brevity by $\alpha_j = \langle Y'(\mathbf{x})\, \partial\phi^{(1)}(\mathbf{x})/\partial x_j\rangle$. We shall concentrate on the derivation of this last term, by using the Green function and Fourier transform methodology.

The Green function for Laplace equation in an infinite domain and its Fourier transform are defined by

$$\nabla^2 G_\infty = -\delta(\mathbf{x}-\mathbf{x}')$$

$$G_\infty(\mathbf{x},\mathbf{x}') = -\frac{1}{4\pi r} \quad ; \quad r = [(x-x')^2+(y-y')^2+(z-z')^2]^{1/2} \tag{3.4.10}$$

$$\hat{G}(\mathbf{k}) = \frac{1}{(2\pi)^{3/2}} \int G(\mathbf{r})\, e^{i\mathbf{k}.\mathbf{r}}\, d\mathbf{r} = \frac{1}{(2\pi)^{3/2}} \frac{1}{k^2}$$

where δ is the Dirac distribution. In the case of a finite domain, $G=G_\infty+g$, where g is a regular function satisfying Laplace equation in Ω, whereas G satisfies homogeneous boundary conditions on $\partial\Omega$. Thus, for the domain of Fig. 3.4.1 G=0 for $\mathbf{x}$ on the boundaries of constant head and $\partial G/\partial z = 0$ on the impervious boundaries, whereas for the domain of Fig. 3.4.2 G=0 on $\partial\Omega$.

The solution of Eq. (3.3.9) for $\phi^{(1)}$, with $\nabla\phi^{(0)} = -\mathbf{J} = \text{const}$, is given by

$$\phi^{(1)}(\mathbf{x}) = -\int_{\Omega} [\mathbf{J}.\nabla Y(\mathbf{x}')]\, G(\mathbf{x},\mathbf{x}')\, d\mathbf{x}' = \mathbf{J}.\int_{\Omega} Y(\mathbf{x}')\, \nabla' G(\mathbf{x},\mathbf{x}')\, d\mathbf{x}' \qquad (3.4.11)$$

This is a well known identity, which can be easily verified by applying the operator ∇^2 to (3.4.11) and using the first equation of (3.4.10), satisfied by G (see, e.g. Carslaw and Jaeger, 1965).

Multiplication of $\phi^{(1)}$ by Y′ and averaging yields for α_j the expression

$$\alpha_j = \mathbf{J}.\int_{\Omega} C_Y(\mathbf{x},\mathbf{x}')\, [\frac{\partial}{\partial x_j} \nabla' G(\mathbf{x},\mathbf{x}')]\, d\mathbf{x}' \qquad (3.4.12)$$

The computation in (3.4.12) can be carried out directly, by letting Ω to expand and by using the expression of G in an infinite domain. It is more convenient, however, to employ the Fourier transform. Switching to the coordinate $\mathbf{r}=\mathbf{x}'-\mathbf{x}$, applying the Faltung theorem (Chap. 1.6) to the integral in (3.4.12), using the expression of $\hat{G}$ (3.4.10) and inverting, yield

$$\alpha_j = \frac{1}{(2\pi)^{3/2}} \int \frac{(\mathbf{J}.\mathbf{k})k_j}{k^2} \hat{C}_Y(\mathbf{k})\, d\mathbf{k} \qquad (3.4.13)$$

Similar expressions have been obtained by Gutjahr et al (1978) by direct application of the FT to Eq. (3.3.9). As we shall show in the sequel, the Green function formalism has definite advantages in other cases.

Hence, for a given C_Y and its FT, we can derive (3.4.11) by integration in (3.4.13). Toward this aim we shall consider a formation of anisotropic structure of type (3.2.10), which is sufficiently general to cover most conceivable applications. We select a Cartesian system attached to the principal axes of C_Y, i.e. for which it can be written by transformation (3.2.10) as an isotropic C_Y. Furthermore, a similar transformation is carried out in the wave-number space, i.e. $k_j' = k_j I_{Yj}$ (j=1,2,3) where I_{Yj} are the three length scales characterizing C_Y, rendering $\hat{C}_Y(k')$ as the FT of the isotropic covariance $C_Y(r')$. Switching from **k** to **k′** in (3.4.13) yields

$$\int \frac{(\mathbf{J}.\mathbf{k})k_\ell}{k^2} \hat{C}_Y(\mathbf{k})\, d\mathbf{k} = \sum_{j=1}^{3} J_j L_{j\ell}$$

$$L_{j\ell} = \int m_{j\ell}\, \hat{C}_Y(k')\, dk' \quad ; \quad m_{j\ell} = \frac{(k_j' k_\ell')/(I_{Yj} I_{Y\ell})}{(k_1'/I_{Y1})^2 + (k_2'/I_{Y2})^2 + (k_3'/I_{Y3})^2} \qquad (3.4.14)$$

The next step toward the computation of (3.4.14) is to switch to spherical coordinates in the $\mathbf{k}'$ space, i.e. $k'_1 = k'_x = k' \sin\theta \cos\Phi$, $k'_2 = k'_y = k' \sin\theta \sin\Phi$, $k'_3 = k'_z = k' \cos\theta$, $d\mathbf{k}' = k'^2 \sin\theta \, dk' d\theta \, d\Phi$. Substitution into (3.4.14) shows immediately that $m_{j\ell}$ is independent of k′ and the integral can be separated into $L_{j\ell} = M_{j\ell} \int \hat{C}_Y \, k'^2 dk'$, with $M_{j\ell} = \int\int m_{j\ell} \sin\theta \, d\theta \, d\Phi$.

The integral over $\hat{C}_Y$ has a simple expression resulting from the definition of the Fourier transform (see Chap. 1.6), namely

$$\int_0^\infty \hat{C}_Y(k') \, k'^2 \, dk' = \sqrt{\frac{\pi}{2}} \, \sigma_Y^2 \tag{3.4.15}$$

As for $M_{j\ell}$, it is easy to ascertain that it depends only on the ratios between I_{Yj} and that its off-diagonal elements are equal to zero, i.e. $M_{j\ell}=0$ for $j \neq \ell$. Summarizing the results obtained so far we have for the effective conductivity at first-order

$$\langle q_\ell \rangle = \sum_{j=1}^{3} K_{ef,j\ell} \, J_j \quad ; \quad K_{ef,j\ell} = K_G [\delta_{j\ell} + \sigma_Y^2 (\frac{\delta_{j\ell}}{2} - \frac{1}{4\pi} M_{j\ell})] + O(\sigma_Y^4) \quad (j.\ell=1,2,3) \tag{3.4.16}$$

where $\delta_{j\ell}$ is the Kronecker identity matrix, and $M_{j\ell}$ was found to be diagonal. A few important conclusions may be derived from (3.4.16) with regard to the first-order effective conductivity of an anisotropic formation of type (3.2.10): (i) $\mathbf{K}_{ef}$ is *independent* of the detailed expression of C_Y, which manifests only through σ_Y^2 and the scaling factors I_{Yj} ; (ii) the principal directions of the tensor $\mathbf{K}_{ef}$ are parallel to those of the covariance matrix C_Y and (iii) its components depend on the ratios between I_{Yj}, say I_{Yy}/I_{Yx} and I_{Yz}/I_{Yx}.

Closed form expressions can be derived for the important case of an axisymmetric C_Y, of the type (3.2.9), encountered in field applications. With $I_{Yx}=I_{Yy}=I_{Yh}$ the horizontal integral scale, $I_{Yz}=I_{Yv}$ the vertical one and $e=I_{Yv}/I_{Yh}$, we have from (3.4.14)

$$M_{11} = M_{22} = 2\pi\lambda$$

$$\lambda = \frac{1}{2\pi} \int_0^\pi \int_0^{2\pi} \frac{\sin^3\theta \cos^2\Phi}{e^2 \sin^2\theta + \cos^2\theta} \, d\Phi \, d\theta = \frac{e^2}{1-e^2} [\frac{1}{e\sqrt{1-e^2}} \tan^{-1} \sqrt{\frac{1}{e^2} - 1} - 1]$$

(3.4.17)

$$M_{33} = \frac{1}{e^2} \int_0^\pi \int_0^{2\pi} \frac{\cos^2\theta \sin\theta \sin^2\Phi}{e^2 \sin^2\theta + \cos^2\theta} \, d\Phi \, d\theta = 4\pi - 2\pi\lambda$$

These results lead to the final form of the horizontal and vertical effective conductivities

$$K_{efh}/K_G = 1 + \sigma_Y^2(\frac{1}{2} - \frac{\lambda}{2}) \quad ; \quad K_{efv}/K_G = 1 + \sigma_Y^2(-\frac{1}{2} + \lambda) \tag{3.4.18}$$

with $\lambda(e)$ given explicitly by (3.4.17). The dependence of K_{ef}/K_G upon σ_Y^2 and e is represented graphically in Figs. 3.4.4 and 3.4.5 and will be discussed in the next section. A few important limit cases can be obtained from (3.4.18) : (i) for an isotropic formation, i.e. e=1, $\lambda=2/3$ and consequently $K_{ef}/K_G = 1 + \sigma_Y^2/6$, result obtained by Gutjahr et al (1978) and Dagan (1979b), represented also in Fig. 3.4.5; (ii) for $e \to \infty$, i.e. for a formation of two-dimensional isotropic structure, $\lambda=1$ and $K_{ef}=K_{efh}=K_G$, i.e. the effective conductivity is equal to the geometric mean, result shown to be of general validity by Matheron (1967), who relies on the earlier results obtained with the aid of perturbation theory by Schwydler (1962) and (iii) for $e \to 0$, i.e. for a stratified formation we get $K_{efh}/K_G = 1 + \sigma_Y^2(1/2-\pi e/4) + O(e^2)$, $K_{efv}/K_G = 1 - \sigma_Y^2(1/2-\pi e/2) + O(e^2)$. It is seen that for e=0 these results coincide with the exact ones, i.e. K_A and K_G (3.4.6), of course at the order $O(\sigma_Y^2)$.

We discuss next the self-consistent approximation, which is supposedly free from the requirement of small σ_Y^2.

3.4.3 The self-consistent approach

We have employed already this technique in Chap. 2.7 for the effective coefficients of heat conduction in a two-phase medium. It has been used in different fields of physics and engineering (see, e.g., Beran, 1968 and Landauer, 1978). More recently, termed also as a renormalization approach, it has been applied, for instance, to deriving properties of dilute (Hinch, 1977) and concentrated (Kim and Russel, 1985) suspensions of spheres. The self-consistent model (also termed as the effective medium approximation) has served for evaluating the effective hydraulic conductivity of heterogeneous, isotropic, formations by Dagan (1979b, 1981), and is extended here to anisotropic formations. We shall rely here merely on an intuitive approach, a more formal derivation being given in Dagan (1981).

We consider an N-phase formation, i.e. a collection of blocks of conductivities K_j (j=1,2,...,N), within a domain Ω, say a sphere whose diameter will eventually tend to infinity (Fig. 3.4.2a). Each block is characterized by its shape, size and location of centroid $\bar{\mathbf{x}}_\ell$ (ℓ=1,2,...,M). We assume that the medium is completely random, in the sense that the blocks are set at random, with no correlation between the size and property of two different blocks. By a proper choice of the p.d.f. of block sizes, one can approximate quite accurately a given covariance C_Y (Dagan, 1981). Hence, the

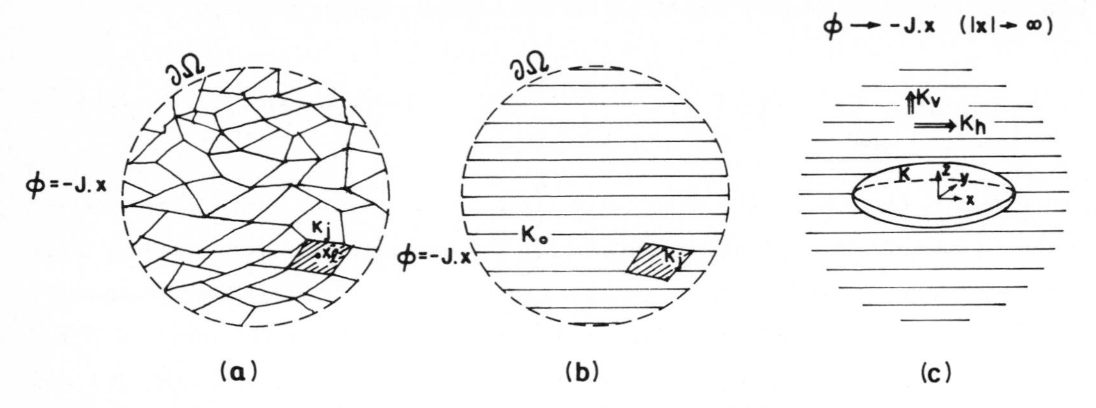

Fig. 3.4.2 The schematical representation of the model of formation serving for the self-consistent approach: (a) A multiphase formation made up from blocks of conductivity K_j with centroids at x_ℓ; (b) an individual block submerged in a uniform matrix of conductivity K_0 and (c) Representation of a block of conductivity K by a spheroid in an unbounded domain.

assumption of complete randomness is not too limitative, although it may prohibit the modeling of a formation of arbitrarily given higher order spatial correlations of K.

The head ϕ satisfies Laplace equation in each block and continuity of head and of normal component of specific discharge $q_\nu = -K\, \partial\phi/\partial\nu$ at the interface between blocks (see Chap. 2.13). At the boundary $\partial\Omega$, ϕ satisfies Eq. (3.4.2), i.e. $\phi = -\mathbf{J}.\mathbf{x}$, such that for a homogeneous formation the flow is uniform. In a given realization, the solution for ϕ can be represented as distributions of singularities, of "source" type, on the interfaces between blocks. Let now condition ϕ on a fixed block of $K=K_j$ and $\mathbf{x}_\ell$ (Fig. 3.4.2b), whereas it is ensemble averaged for the setting of all other blocks. In this averaging process the source distribution on the interface between the fixed block and the surrounding embedding matrix becomes the one between the block and a uniform medium of conductivity K_0, for an isotropic structure. Otherwise, K_0 is a tensor, as shown in the sequel. In the zero-order, small perturbation, approximation, K_0 is taken equal to K_G, while the Hashin and Shtrikman (1962) bounds can be shown to result from taking K_0 equal to the lowest or highest K_j, respectively. In the self-consistent approach K_0 is left unknown at this stage, and the head $\phi_{j\ell}$ and specific discharge $\mathbf{q}_{j\ell}$, and in particular the fluctuation $\mathbf{q}'_{j\ell} = \mathbf{q}_{j\ell} - K_0\, \mathbf{J}$, fields are computed for the configuration of Fig. 3.4.2b and for boundary condition (3.4.2). Now, $\mathbf{q}'_{j\ell}$ is averaged for all possible sizes of similar blocks, for all K_j and locations x_ℓ of the hitherto fixed block, to obtain the average specific discharge as follows

$$\langle \mathbf{q} \rangle = K_0\, \mathbf{J} + \langle \mathbf{q}'_{j\ell} \rangle \quad ; \quad \langle \mathbf{q}'_{j\ell} \rangle = \int_V\!\!\int \mathbf{q}'_{j\ell}\, f(K)\, f(x_\ell)\, dK\, dx_\ell \qquad (3.4.19)$$

where f(K) is the discrete p.d.f. (3.2.5) of conductivities and $f(\boldsymbol{x}_\ell)$ is the p.d.f. of block centroids. By the assumption of complete randomness the latter is supposed to be uniform, i.e. $f(\boldsymbol{x}_\ell)=1/V$, where V is the total volume. Hence, the averaging over $\boldsymbol{x}_\ell$ is a simple spatial integration and since $\mathbf{q}'_{j\ell}$ depends on $\mathbf{x}-\boldsymbol{x}_\ell$, it is identical to an integration over $\mathbf{x}$ in Ω with $\boldsymbol{x}_\ell$ kept fixed. At this last step the self-consistency argument is invoked, namely that $K_0 = K_{ef}$, the latter being given by $\langle\mathbf{q}\rangle = K_{ef}\ \mathbf{J}$. Hence, $K_0=K_{ef}$ is rendered by the condition in (3.4.19)

$$\langle\mathbf{q}'\rangle = \lim_{V\to\infty} \frac{1}{V} \int_V \int \mathbf{q}'(\mathbf{x},K,K_0)\, f(K)\, dK\, d\mathbf{x} = 0 \tag{3.4.20}$$

where $\boldsymbol{x}_\ell$, which is immaterial for an infinite domain, is taken at the origin.

This sequence of steps has been carried out in Dagan (1979b, 1981) for a collection of spheres or cylinders, to model an isotropic three- or two-dimensional formation, respectively. The procedure is generalized here for formations of anisotropic, axisymmetric, structures along the lines of Stroud (1975). Toward this aim we model the individual block as an oblate spheroid (Fig. 3.4.2c) of conductivity K and volume V_0, submerged in an infinite homogeneous matrix, of a tensorial conductivity. To represent an anisotropic medium, all spheroids are taken with parallel principal axes. Hence, the principal directions of the matrix conductivity are x,y, termed horizontal, and z, the "vertical", and the conductivity tensor is diagonal, of components K_h, K_h and K_v, replacing K_0 (Fig. 3.4.2c). The problem of flow can be formulated now explicitly as follows. We denote by S_0 the surface of the spheroid, of equation

$$\frac{x^2+y^2}{a_h^2} + \frac{z^2}{a_v^2} = 1 \quad ; \quad e = \frac{a_v}{a_h} < 1 \quad ; \quad V_0 = \frac{4\pi a_v a_h^2}{3} \tag{3.4.21}$$

where e is the eccentricity and V_0 is the volume. We denote by ϕ_{ext} and ϕ_{int} the head in the matrix and in the spheroid, respectively. They satisfy the following set of equations

$$\phi_{ext} = -\mathbf{J}.\mathbf{x} \quad (\text{on } \partial\Omega) \quad ; \quad K_h\left(\frac{\partial^2\phi_{ext}}{\partial x^2} + \frac{\partial^2\phi_{ext}}{\partial y^2}\right) + K_v \frac{\partial^2\phi_{ext}}{\partial z^2} = 0 \quad (\text{outside } S_0)$$

$$\phi_{ext} = \phi_{int} \quad ; \quad K_h\left(\frac{\partial\phi_{ext}}{\partial x}\nu_x + \frac{\partial\phi_{ext}}{\partial y}\nu_y\right) + K_v \frac{\partial\phi_{ext}}{\partial z}\nu_z = K\, \nabla\phi_{int}.\boldsymbol{\nu} \quad (\text{on } S_0) \tag{3.4.22}$$

$$\nabla^2\phi_{int} = 0 \quad (\text{inside } S_0)$$

As usual, $\boldsymbol{\nu}$ denotes a unit vector normal to S_0, of Cartesian components $\nu_x = x/(a_h^2\ F)$, $\nu_y = y/(a_h^2\ F)$, $\nu_z = z/(a_v^2\ F)$, $F = [(x^2+y^2)/a_h^4 + z^2/a_v^4]^{1/2}$.

Once the problem defined in a unique manner for ϕ by Eqs. (3.4.22) is solved, $\mathbf{q}_{ext} = -\mathbf{K}.\nabla\phi_{ext}$ and $\mathbf{q}_{int} = -K\nabla\phi_{int}$ can be evaluated, where $\mathbf{K}$ (K_h, K_h, K_v). Subsequently, the above steps of the self-consistent procedure lead to the effective conductivities $K_{efh}=K_h$ and $K_{efv}=K_v$. Since ϕ is linear in $\mathbf{J}$, we solve (3.4.22) first for a gradient in the x direction, i.e. for $\mathbf{J}$ (J,O,O). The difficulty of the problem, as compared with the one of a medium of isotropic heterogeneity, is in the presence of K_h and K_v in (3.4.22). To overcome this we switch to new coordinates defined by

$$x''=x \ , \ y''=y \ , \ z'' = z(K_h/K_v) \tag{3.4.23}$$

for which the spheroid S_0 is transformed into S_0'', of equation

$$\frac{x''^2+y''^2}{a_h''^2} + \frac{z''^2}{a_v''^2} = 1 \quad ; \quad a_h'' = a_h \ , \ a_v'' = a_v(K_h/K_v)^{1/2}$$

$$e''=a_v''/a_h''=e(K_h/K_v)^{1/2} \quad ; \quad V_0''=V_0(K_h/K_v)^{1/2} \tag{3.4.24}$$

Before writing the equations satisfied by $\phi_{ext}(\mathbf{x}'')$ and $\phi_{int}(\mathbf{x}'')$ outside and inside S_0'', respectively, we shall assume that ϕ_{int} represents simply a uniform flow in the x direction, i.e. $\phi_{int} \equiv -J_{int}x$. This is a substitution which will be proved to be exact a-posteriori, by solving the problem, which has a unique solution. Obviously, ϕ_{int} satisfies identically the Laplace equation or its transformed by application (3.4.24). With these preparatory steps, in the new coordinates system and after applying (3.4.23), Eqs. (3.4.22) become

$$\phi_{ext} = -J\,x'' \quad (\text{on } \partial\Omega'') \quad ; \quad \nabla^2\phi_{ext} = O \quad (\text{outside } S_0'')$$

$$\phi_{ext} = \phi_{int} \quad ; \quad K_h\nabla\phi_{ext}.\boldsymbol{\nu}'' = K\,\nabla\phi_{int}.\boldsymbol{\nu}'' \quad (\text{on } S_0'') \tag{3.4.25}$$

In (3.4.25) $\partial\Omega''$ denotes the transformed boundary of the flow domain. Since it is left to depart to infinity later on, its shape is immaterial. Similarly, $\boldsymbol{\nu}''$ is a unit vector normal to S_0'' and it is emphasized that the last equation in (3.4.25) is obtained only for $\phi_{int} \equiv -J_{int}x$, otherwise the transformation (3.4.24) is not effective. Inspection of (3.4.25) shows that the problem in the new coordinate system is identical to the one corresponding to a spheroid of conductivity K submerged in a homogeneous and *isotropic* matrix of conductivity K_h. It is emphasized, however, that the transformed spheroid S_0'' depends on the ratio K_v/K_h. We proceed now with the solution of flow in an isotropic matrix.

The head field can be derived explicitly by using ellipsoidal coordinates (e.g. Lamb,1932), and this derivation is given succinctly herein.

Let the equation of the oblate spheroid, with the origin at its center, i.e. $\mathbf{x}_\ell = 0$, be rewritten as

$$\frac{x''^2+y''^2}{1+\zeta_0^2} + \frac{z''^2}{\zeta_0^2} = 1 \quad \text{with} \quad e'' = \frac{\zeta_0}{\sqrt{1+\zeta_0^2}} \; ; \; \zeta_0 = \frac{e''}{\sqrt{1-e''^2}} \tag{3.4.26}$$

where ζ_0 is fixed. The coordinates here are normalized, for simplicity, with respect to the half-distance between the foci of the spheroid. At the limit $\zeta_0 \to 0$ the spheroid degenerates into a disc of unit radius in the x",y" plane, whereas for $\zeta_0 \to \infty$ it tends to a sphere of radius much larger than unity.

The head ϕ is now expressed in terms of ellipsoidal coordinates ζ,μ,Φ which are related to Cartesian ones by

$$x'' = \sqrt{(1-\mu^2)(1+\zeta^2)}\cos\Phi \; ; \; y'' = \sqrt{(1-\mu^2)(1+\zeta^2)}\sin\Phi \; ; \; z'' = \mu\zeta$$

$$0 < \zeta < \infty \; ; \; -1 < \mu < 1 \; ; \; 0 < \Phi < \pi \tag{3.4.27}$$

such that ζ=const represent confocal spheroids. The head field ϕ, solution of Laplace equation with the condition $\phi_{ext} = -Jx$ at infinity, is given by

$$\phi_{int} = B\, P_1^1(\mu)\, p_1^1(\zeta)\cos\Phi \; ; \; P_1^1(\mu)=\sqrt{1-\mu^2}\, , \; p_1^1(\zeta)=\sqrt{1+\zeta^2} \quad (0<\zeta\leq\zeta_0)$$

$$\phi_{ext} = -Jx - A\, P_1^1(\mu)\, q_1^1(\zeta)\cos\Phi \; ; \; q_1^1(\zeta)=\frac{\lambda(\zeta)}{\zeta\sqrt{1+\zeta^2}} \quad (\zeta_0\leq\zeta<\infty) \tag{3.4.28}$$

$$\text{with} \quad \lambda(\zeta) = \zeta(1+\zeta^2)\cot^{-1}\zeta - \zeta^2$$

For the definition of the tesseral harmonics and the derivation of (3.4.28) see Lamb (1932). The two arbitrary constants A and B are determined by the matching conditions (3.2.25) on S_0'', i.e. for $\zeta=\zeta_0$. This is possible for (3.4.28), in which ϕ_{int} represents indeed a uniform flow in the x direction, verifying a-posteriori our assumption. Subsequently, the specific discharge fluctuation associated with the ϕ field is integrated over the entire space, to account for the uniform probability of finding the centroid in space, according to the above observation about $f(\mathbf{x}_\ell)$. The final result of this sequence of computations is as follows

$$q' = \frac{1}{V''}\int (q - K_h\, J)\, dV = -J\, \frac{2K_h(K_h-K)}{(K-K_h)\,\lambda(e'')+2K_h}\, \frac{V_0''}{V''} \tag{3.4.29}$$

where $\lambda(e'')$ is the function in (3.4.28) in which $\zeta=\zeta_0$ is replaced by $\zeta_0(e'')$ (3.4.26). The result is precisely the expression (3.4.17) of λ with e replaced by e".

We are now in a position to apply the self-consistent argument, by ensemble averaging $\mathbf{q}'$ first for all spheroids of same conductivity K and subsequently for all K's. The first operation is equivalent with replacing the volume ratio V_0''/V'' in (3.4.29) by n(K), the volume fraction of phase of conductivity K. The second operation is tantamount to integrate over f(K) dK, either for the discrete N phases (3.2.25), or, by a limit process, for a continuous variation of K. The final result is the following equation for $K_h = K_{efh}$, the self-consistent approximation

$$K_{efh} = \frac{1}{2}\left[\int \frac{f(K)\,dK}{(K-K_{efh})\lambda(e'') + 2K_{efh}}\right]^{-1} \tag{3.4.30}$$

We have to repeat now this sequence of computations for a gradient applied in the vertical direction, i.e. for $\mathbf{J}$(O,O,J). By a new coordinates transformation $x''=x(K_v/K_h)^{1/2}$, $y''=y(K_v/K_h)^{1/2}$, $z''=z$, the problem formulated in (3.4.22) is reduced to one of flow past a stretched spheroid S_0'', surrounded by a homogeneous matrix of conductivity K_v this time. The important point, however, is that the eccentricity e'' is the same as in the case of horizontal gradient and given by (3.4.24). By a similar chain of computations it is found that the specific discharge fluctuation for a gradient applied in the z direction is given by

$$q_z' = J\,\frac{K_v(K_v-K)}{(K-K_v)\,\lambda(e'')-K)}\,\frac{V_0''}{V''} \quad ; \quad q_x = q_y = 0 \tag{3.4.31}$$

Subsequently, the self-consistent argument leads now to the following equation rendering $K_{efv} = K_v$

$$K_{efv} = \left[\int \frac{f(K)\,dK}{K + \lambda(e'')(K_{efv}-K)}\right]^{-1} \tag{3.4.32}$$

We can summarize now the results embodied by the final equations (3.4.30) and (3.4.32). The input information is: f(K), the conductivity p.d.f, the principal directions of the effective conductivity tensor and e, the eccentricity. The latter can be equated with the anisotropy ratio between the vertical and horizontal scales. Indeed, in a medium made up from spheroids of various sizes and conductivities, but of same eccentricities and orientations, the correlation scales in the principal directions are given by e, from geometrical similarity considerations. Furthermore, this ratio is also equal to that between the integral scales of Y, since the logarithmic transformation of K preserves this ratio. With this information, the principal components of $\mathbf{K}_{ef}$ $(K_{efh}, K_{efh}, K_{efv})$ can be determined from (3.4.30) and (3.4.32) by two quadratures and by an iterative process. It is emphasized that the unknown K_{efh} and K_{efv} appear indirectly in the above equations through the parameter

$e''=e(K_{efh}/K_{efv})^{1/2}$ in the function $\lambda(e'')$ (3.4.17). We shall illustrate the results for the particular, but important case, of K lognormal, but first we discuss a few limiting cases of a more general nature.

Furthermore, substitution of $K=K_G \exp(Y')$, of $f(K)\,dK = f(Y')\,dY'$, with Y' $N[0,\sigma_Y]$, in (3.4.31) leads to K_{ef}/K_G = function(e,σ_Y^2). Before embarking on the discussion of this function under general conditions, we shall consider a few limiting cases which lend themselves to simple calculations.

The first particular case is for e=1, i.e. for spheres representing a model of an isotropic formation, with $e=e''=1$ and $\lambda(1) = 2/3$. The result is for $K_{efh}=K_{efv}$

$$K_{ef} = \frac{1}{3}\left[\int \frac{f(K)}{K + 2K_{ef}}\, dK\right]^{-1} \qquad (3.4.33)$$

which has been obtained by Dagan (1979b) by solving the problem for spheres from the outset (see also Chap. 2.7).

A second class of solutions is obtained by a small perturbation expansion of (3.4.30) and (3.4.32), by substituting $K=K_G \exp(Y')$, expanding in a power series in Y' and retaining terms of zero and first-order in σ_Y^2. The result is found to coincide with that obtained by a general procedure in the preceding section (Eqs. 3.4.18) for a medium of covariance (3.2.10). This result strengthens the confidence in the validity of the self-consistent approach.

Finally, simple expressions can be obtained for small e, i.e. for highly anisotropic formations for which $I_{Yv}/I_{Yh} \ll 1$. At this limit the oblate spheroids degenerate into thin lenses and in (3.4.26) $\zeta_0 \to 0$, $\lambda(\zeta_0) \to (\pi e''/2)$. Expanding (3.4.30) and (3.4.32) for $\lambda \to 0$ yields the explicit expressions

$$K_{efh} \to K_A\left[1-\frac{\lambda}{2}\frac{\sigma_K^2}{K_A^2} + O\left(\frac{\lambda^2\sigma_K^4}{K_A^4}\right)\right]\ ;\ K_{efv} \to K_H\left[1+\lambda\frac{\sigma_{1/K}^2}{K_H^2} + O\left(\frac{\lambda^2\sigma_{1/K}^4}{K_H^4}\right)\right] \qquad (3.4.34)$$

It is seen that these approximations depend on the smallness of the product $e''\sigma_Y^2$ rather than e'', i.e. the asymptotic expansion (3.4.34) is not uniform in e. Obviously, if an additional expansion is carried out in (3.4.34) for small σ_Y^2, with $\sigma_K^2/K_A^2 = \sigma_{1/K}^2/K_H^2 = \exp(\sigma_Y^2)-1$ for lognormal K, the results coincide with those obtained by the expansion of (3.4.18) for small e. Another result of interest is that for $e=\lambda=0$, (3.4.34) degenerate into the exact formulae for a stratified formation, which constitutes another check of the self-consistent model.

To better grasp these results we have computed the effective conductivities for K lognormal, by substituting in (3.4.30) and (3.4.32) $K=K_G \exp(Y')$ and $f(K)dK=f(Y')dY'$, with $f(Y')$ the normal distribution $N[0,\sigma_Y]$. K_{efh}/K_G (3.4.30) and K_{efv}/K_G (3.4.32) as functions of e and for a few

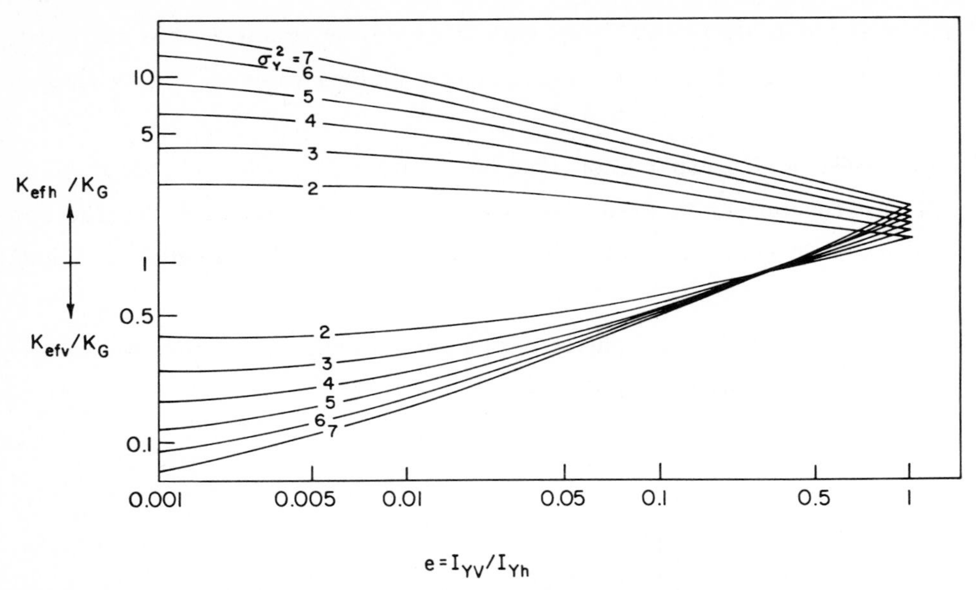

Fig. 3.4.3 The dependence of the principal values, horizontal and vertical, of the effective conductivity upon e and upon $\sigma_Y^2 \geq 2$, for a formation of anisotropic axisymmetric heterogeneous structure. The principal directions of C_Y are vertical and horizontal, respectively, and the anisotropy ratio $e=I_{Yv}/I_{Yh}$ is smaller than unity. The results have been obtained by numerical integration in Eqs.(3.4.30) and (3.40.32), i.e. the self-consistent approach.

values of $\sigma_Y^2 \geq 2$ are represented in Fig. 3.4.3. The results have been obtained by solving iteratively (3.4.30,32) and by computing the quadrature numerically.

The same results are represented in Fig. 3.4.4 for $\sigma_Y^2 < 1$. On the same figure we have also represented the first-order small perturbation approximation (3.4.18). It is seen that the range of validity in σ_Y^2 of this approximation depends also on e and it increases as e grows. The most severe limitation occurs for e=0, when in the exact solution $K_{efh}=K_A$ and $K_{efv}=K_H$ (3.4.6), $\exp(\pm\sigma_Y^2/2)$ is replaced by $1\pm\sigma_Y^2/2$, respectively. Thus for $\sigma_Y^2=1$ and at this limit, the first-order approximation for K_{efh} underestimates the exact result by 9% and K_{efv} by 18%.

Finally, we have represented in Fig. 3.4.4 K_{ef}/K_G (3.4.33) for an isotropic formation, as well as the first-order perturbation approximation $K_{ef}/K_G = 1+\sigma_Y^2/6$ (3.4.18). Even for $\sigma_Y^2=7$ the perturbation approximation exceeds only by 22% the complete self-consistent solution.

It is worthwhile to mention here that in the case of two-dimensional flow in an isotropic formation, in the self-consistent model the blocks are represented by circular cylinders of infinite height. The effective conductivity has been given by Dagan (1979) and its derivation is left as an exercise here. The final result is that K_{ef} satisfies the equation, similar to (3.4.32)

$$K_{ef} = \frac{1}{2}\left[\int \frac{f(K)}{K + K_{ef}}\, dK\right]^{-1} \tag{3.4.35}$$

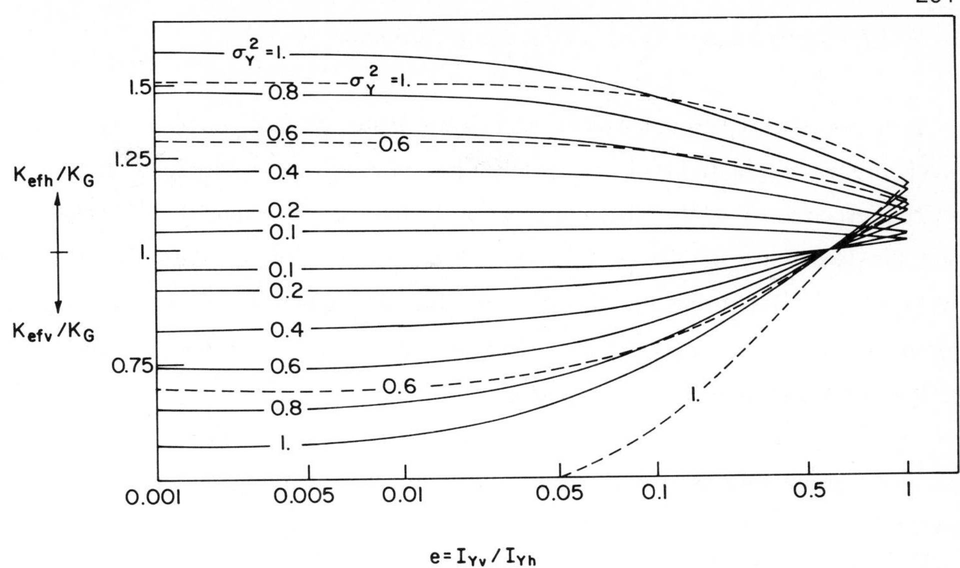

Fig. 3.4.4 Same as Fig. 3.4.3 (full line) for $\sigma_Y^2 \leq 1$ and the first-order small σ_Y^2 approximation (3.4.18) (broken lines).

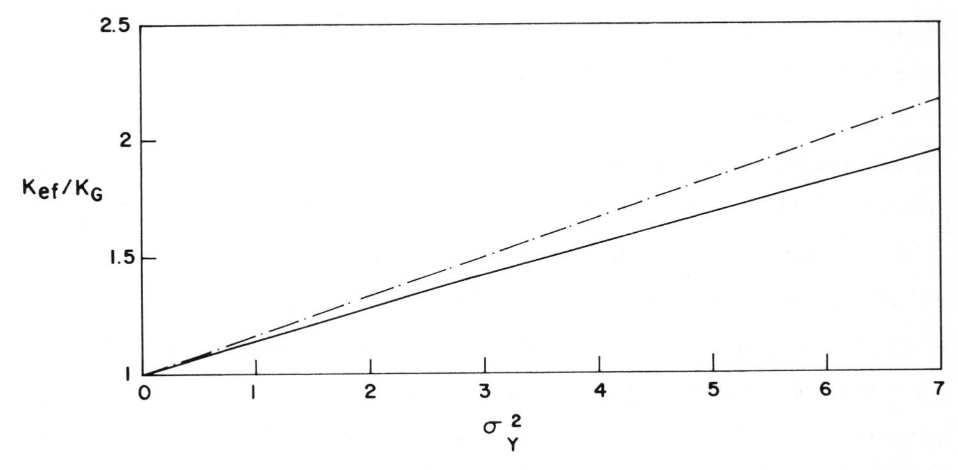

Fig. 3.4.5 The dependence of the effective conductivity of a formation of isotropic three-dimensional heterogeneous structure upon σ_Y^2. The numerical solution for the self-consistent approximation (3.4.33) (full line) and the first-order small σ_Y^2 approximation $K_{ef}/K_G = 1+\sigma_Y^2/6$ (broken line).

The numerical solution of this equation for K lognormal yielded with a high degree of accuracy $K=K_G$ for any σ_Y^2 . It is reminded that a similar result has been obtained at first-order in Sect. 3.4.1 and has been proved to be of general validity for K lognormal by Matheron (1967). Again, this agreement strengthens the confidence in the self-consistent model as a representation of a multiphase formation of random structure.

3.4.4 Effective conductivity of a two-phase formation

In some porous formations one encounters lenses of low conductivity embedded in a medium of much higher conductivity. Even if the latter displays some degree of heterogeneity, the large conductivities contrast is the dominant factor influencing the overall effective conductivity. Assuming that the lenses are spread in a random fashion, in terms of their extent, shape and location, the problem is to determine the effective conductivity given the two conductivities, say K_2 (effective conductivity of the matrix) and K_1 (lenses), their volume fractions n_1 and n_2, and the covariance C_h of the index function h (see Chap. 2.1).

We have discussed a similar problem in Chap. 2.7 about the effective thermal conductivity of a porous medium. We have emphasized that the self-consistent model is a poor approximation in such a case because it presumes that blocks of the two phases are set in an independent manner in space, whereas the picture here is that of phase two surrounding the blocks of phase one. We suggested in Chap. 2.7 to employ, for an isotropic medium, the ingenious model of Hashin and Shtrikman (1962) of a swarm of composite spheres, which was shown to yield results agreeing with experiments. We are going to generalize this model here for anisotropic media of the type discussed in the preceding sections, along the lines of Bergman (1982), who dealt with the dielectric constant of two-component composites.

We consider a composite oblate spheroid as depicted in Fig. 3.4.6, i.e. a spheroid of conductivity K_1 surrounded by a confocal spheroid, the layer between the two being of conductivity K_2. This composite body is submerged in a medium of tensorial conductivity $\mathbf{K}_0$ extending to infinity. In this model the lenses are idealized as oblate spheroids of revolution, of eccentricity e. Since the medium is regarded as a collection of such similar spheroids of different sizes, it can represent a formation of given anisotropic C_h of type (3.2.10). The idea forwarded by Hashin and Shtrikman (1962) and generalized by Bergman (1982) is to solve for ϕ obeying Laplace equation, matching conditions of continuity of ϕ and of the normal flux at the interfaces between the phases and boundary condition (3.4.2) at infinity, and to determine $\mathbf{K}_0$ from the requirement that the field in the surrounding matrix is *undisturbed*, i.e. $\phi\equiv-\mathbf{J}.\mathbf{x}$ outside the composite spheroid. If such a value can be found, one can fill the entire space with composite spheroids of different sizes, without affecting the flow and then it is easy to show that by its definition (3.4.3) $\mathbf{K}_{ef}=\mathbf{K}_0$.

It is emphasized that the matrix exterior to the composite spheroid in Fig. 3.4.6 can be taken as homogeneous and of K_0 equal to K_h or K_v, depending on the direction of the mean gradient, without carrying out the coordinates stretching of the previous section. Indeed, since in the present case the field is *uniform* in the surrounding matrix, it is only the principal component of $\mathbf{K}_{ef}$ in the direction of $\mathbf{J}$ which counts.

To carry out this program we switch to the ellipsoidal coordinates (3.4.27) and consider first lateral flow, i.e. parallel to the major axis x or y. The three heads, in the three regions of Fig. 3.4.2, can be written as follows (Lamb, 1932)

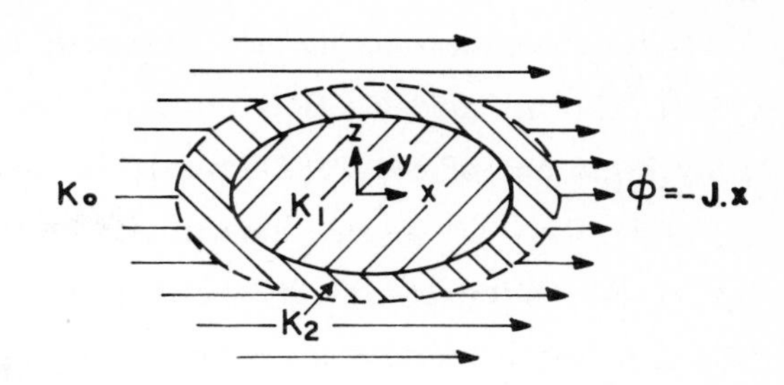

Fig. 3.4.6 A composite spheroid made up from an interior one of conductivity K_1, surrounded by a confocal spheroid with shell of conductivity K_2, submerged in an unbounded matrix of conductivity K_0.

$$\begin{aligned} \phi_1 &= A\, P_1^1(\mu)\, p_1^1(\zeta) \cos\Phi \quad (0<\zeta/\leq\zeta_0) \\ \phi_2 &= [B\, P_1^1(\mu)\, q_1^1(\zeta) + C\, P_1^1(\mu)\, p_1^1(\zeta)] \cos\Phi \quad (\zeta_0/\leq\zeta/\leq\zeta_1) \\ \phi_{ext} &= -J_x\, x = -J_x\, P_1^1(\mu)\, p_1^1(\zeta) \cos\Phi \quad (\zeta_1/\leq\zeta<\infty) \end{aligned} \tag{3.4.36}$$

where $\zeta=\zeta_0$ and $\zeta=\zeta_1$ define the two spheroids. The tesseral harmonics p_1^1 , P_1^1 and q_1^1 have been defined in (3.4.28).

The values of ζ_0 and ζ_1 are determined by the given volume ratio between the interior and exterior spheroid, i.e. the volume fraction of the lenses, and by the given eccentricity of spheroids, i.e.

$$n_1 = \frac{\zeta_0(1+\zeta_0^2)}{\zeta_1(1+\zeta_1^2)} \quad ; \quad \zeta_0 = \frac{e}{\sqrt{1-e^2}} \tag{3.4.37}$$

The constants A, B, C and K_0 are determined by the matching conditions between ϕ_1,ϕ_2 and ϕ_{ext}. The final result for $K_{efx}=K_{efy}=K_{efh}=K_0$ is as follows

$$\frac{K_{efh}}{K_2} = \frac{\Delta_1\, \dot{q}_1^1(\zeta_1) + \Delta_2\, \dot{p}_1^1(\zeta_1)}{\Delta\, \dot{p}_1^1(\zeta_1)} \quad \text{with}$$

$$\Delta_1 = -p_1^1(\zeta_1)\, [(K_1 - K_2)p_1^1(\zeta_0)\, \dot{p}_1^1(\zeta_1)] \;\; ; \;\; \Delta_2 = p_1^1(\zeta_1)[K_1\, \dot{p}_1^1(\zeta_1)\, q_1^1(\zeta_0) - K_2\, p_1^1(\zeta_0)\, \dot{q}_1^1(\zeta_1)] \tag{3.4.38}$$

$$\Delta = K_2\, p_1^1(\zeta_0)[\dot{p}_1^1(\zeta_1)q_1^1(\zeta_1) - p_1^1(\zeta_1)\dot{q}_1^1(\zeta_1)] - K_1\, \dot{p}_1^1(\zeta_1)[p_1^1(\zeta_0)q_1^1(\zeta_1) - p_1^1(\zeta_1)q_1^1(\zeta_0)]$$

the symbol $\dot{}$ standing for differentiation. A similar result is attained for the vertical effective conductivity with the same equations (3.4.38) in which p_1^1 and q_1^1 are replaced by p_1 and q_1, respectively, and the latter are defined by

$$p_1(\zeta) = \zeta \;\; ; \;\; q_1(\zeta) = 1 - \zeta\cot^{-1}\zeta \tag{3.4.39}$$

These theoretical results are compared in Fig. 3.4.7 with numerical simulations by Desbarats (1987), who has modeled flow through a formation containing lenses of shale of very low permeability $K_1 \simeq 0$ and of e=1/15. Desbarats has modeled the lenses by using random generation of K on a grid for a given anisotropic covariance C_K of exponential form, $C_K(r') = n_1(1-n_1)(K_1-K_2)^2 \exp(-r')$, $r'^2=(x^2+y^2)/I_h^{\,2}+z^2/I_v^{\,2}$. In view of these differences in defining the lenses, of the numerical errors and of the effect of the finite volume in the simulation, the comparison can be considered quite satisfactory.

3.4.5 Influence of boundary on effective conductivity

The effective conductivity has been calculated so far for flow in unbounded domains, which simplified considerably the computations, allowing for instance to employ the Fourier transform methodology. In applications, however, formations are of bounded extent, and the question is whether the effective conductivity, defined by Eq. (3.4.3), is influenced by the presence of boundaries. We shall limit the investigation of this point to the first-order approximation of K_{ef} in a perturbation expansion for small σ_Y^2 , along the lines of Sect. 3.4.2.

Again, we consider the condition of uniform flow (3.4.3), applying this time to the boundary $\partial\Omega$ of a bounded domain Ω. By its definition (3.3.9), the zero order approximation is given by $\phi^{(0)} \equiv -\mathbf{J}.\mathbf{x}$, whereas by (3.3.9) and by the use of the Green function (3.4.10), $\phi^{(1)}$ is given by the same expression (3.4.11), which is valid in a bounded domain as well.

The first simple case we choose to consider is a half-space bounded by an impervious lower boundary z=0. The Green function for this domain is easily written down by using the method of "images" as follows

$$G(\mathbf{x},\mathbf{x}') = G_\infty(x-x',y-y',z-z') + G_\infty(x-x',y-y',z+z') \quad ; \quad (z>0,\ z'>0) \tag{3.4.40}$$

It is easy to ascertain that $\partial G/\partial z=0$ for z=0 and the same is true for $\phi^{(1)}$ (3.4.11). We shall apply now the general equation (3.4.11) to this particular configuration, i.e. for the upper half-space and G (3.4.40). Furthermore, we assume that Y is stationary, i.e. $C_Y=C_Y(\mathbf{x}-\mathbf{x}')$ and has principal axes parallel to the Cartesian system. As a result α, as well as K_{ef}, is a function of the coordinate z only, i.e. of the distance to the boundary. Subsequently, we carry out the change of variables x"=x'-x, y"=y'-y, and assume that the vector **J** is parallel to the impervious boundary z=0, to obtain

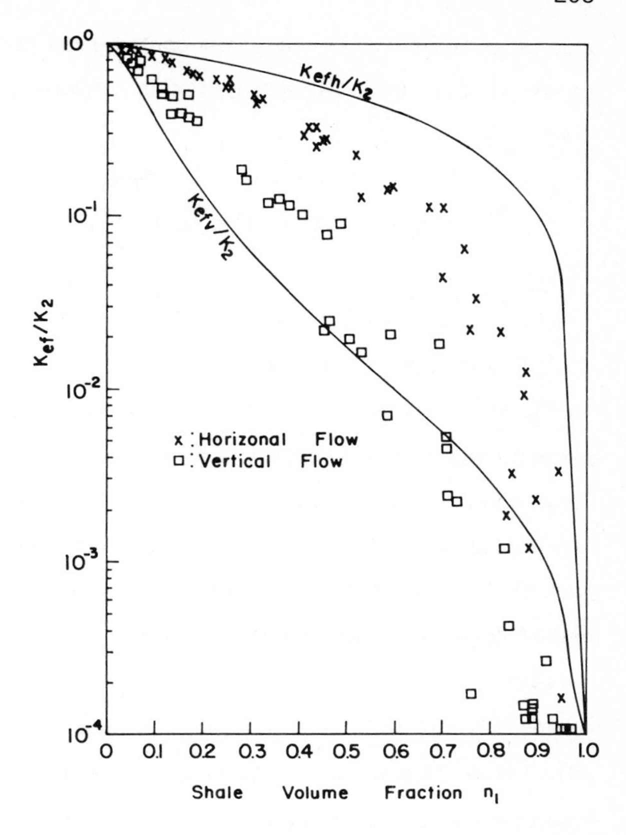

Fig. 3.4.7 The effective conductivities (horizontal and vertical) of a two-phase formation made up from a matrix of conductivity K_2 and shale lenses of very low conductivity. The full-line curves are based on (3.4.38, 3.4.39) and the points were obtained by numerical simulations by Desbarats (1987).

$$\alpha_x/J_x = \alpha_y/J_y =$$

$$-\int_{-\infty}^{\infty}\int_{-\infty}^{\infty}\int_{0}^{\infty} C_Y(x'',y'',z'-z)\left\{\frac{\partial^2}{\partial x''^2}[G_\infty(x'',y'',z'-z)+G_\infty(x'',y'',z'+z)]\right\}dz'dy''dz'' \qquad (3.4.41)$$

It is easy to ascertain that for z=0, following a change of variable z"=-z′ in the second term of the integral, α becomes identical to its expression for an unbounded domain, i.e. the specific discharge and the effective conductivity are not affected at first-order along the impervious boundary itself. The computation of K_{ef} for $z\neq 0$ is not as simple and is deferred to the case of two-dimensional flows in Part 5. It is relatively easy, however, to assess the impact of the boundary for large z compared to the integral scales of Y. Indeed, in this case the first term of the integral is approximately equal to α in an unbounded domain, whereas the presence of the boundary manifests in the second term which can be evaluated by adopting the "white noise" approximation of C_Y (3.2.29).

This is permitted since the main contribution results from C_Y near $z'=z$, where $G_\infty(x'',y'',z'+z)$ is regular. The result in (3.4.41) is, for an isotropic formation,

$$\frac{\alpha(z)}{J} = \frac{\alpha(\infty)}{J} - \sigma_Y^2 (I_Y)^3 \frac{\partial^2}{\partial x''^2} G_\infty(x'',y'',z'-z)\Big|_{x''=0,y''=0,z'=z} = \frac{\alpha(\infty)}{J} - \frac{\sigma_Y^2}{32\pi} \frac{1}{(z/I_Y)^3} \quad (3.4.42)$$

It is seen that α, and the related K_{ef}, relaxes quite rapidly to its value in an unbounded domain. Furthermore, the value of the correction, expressed by the last term in (3.4.42), is quite small compared to the term $\sigma_Y^2/6$ (Eq. 3.4.19). Hence, the influence of the presence of an impervious boundary upon K_{ef} is quite small and K_{ef} can be approximated accurately by its value in an unbounded domain. It is emphasized that in the case of an anisotropic formation, with layering parallel to the boundary, the effect is even smaller, since at the limit of a stratified formation (3.2.11) the boundary does not affect K_{ef} at all .

Different results are obtained for flow normal to a boundary of constant head. Indeed, since stationarity prevails on planes parallel to the boundary, $\langle q_x \rangle = \langle q_y \rangle = 0$ and by the continuity equation $d\langle q_z \rangle/dz=0$. Hence, $\langle q_z \rangle$ is constant and the same is true for the effective conductivity, which is therefore equal to its bulk value.

Concluding, on the basis of these assessments it was found that for uniform flow in a formation of stationary random structure, the presence of an impervious boundary has a small effect upon the effective conductivity, which is slightly different from its bulk value in a thin layer near the boundary.

3.4.6 The influence of nonuniformity of average flow

In the preceding sections we have derived the effective conductivity of a heterogeneous formation under conditions of applied uniform head gradient on the boundary (3.4.2). Furthermore, we have seen that K_{ef} evaluated in an unbounded formation applies to a bounded one, if the dimensions are large compared to the log-conductivity integral scales. In most applications, however, the average flow is nonuniform (see, for instance, Fig. 3.1.1) and the question is whether one can use the same Eq. (3.4.3) in the case in which $\nabla\langle\phi\rangle = -\mathbf{J}$ is no more constant. Furthermore, if the answer is affirmative, is it possible to adopt the constant K_{ef} of uniform flow in this case as well? We have examined briefly this problem, at the pore scale, when deriving Darcy's law, in Chap. 2.9, and we have emphasized the findings which confirmed the robustness of Darcy's law. Although of fundamental nature, this problem has received little attention in the literature on heterogeneous materials, which takes for granted that nonuniformity of the average field does not affect the constitutive equations and K_{ef}. This assumption can be easily defended when the length scales involved in the problem are much larger than the heterogeneity scale, which is indeed the case in most applications. This is not always the case, however, for flow through heterogeneous formations, especially at the regional scale (Part 5). Hence, this difficult point deserves more attention in the present context than in other similar fields.

We consider again the basic equations (3.4.1) and their averages, for a domain Ω, but this time we do not assume that (3.4.2) prevails, but only that ϕ or its normal derivative are given deterministically on the boundary $\partial\Omega$. In each realization of given $Y(\mathbf{x})$ this is a well posed problem (see Eq. 3.3.7), which can be solved in principle to derive ϕ and subsequently $\mathbf{q}$, as well as their averages. The general linear dependence of $\langle\mathbf{q}\rangle$ upon $\nabla\langle\phi\rangle$ can be written, following a reasoning similar to that of Saffman (1971), as follows

$$\langle q_j(\mathbf{x})\rangle = \int_\Omega \Big[\sum_{\ell=1}^{3} \frac{\partial\langle\phi(\mathbf{x}')\rangle}{\partial x'_\ell}\, \omega_{j\ell}(\mathbf{x},\mathbf{x}')\Big]\, d\mathbf{x}' \qquad (j=1,2,3) \tag{3.4.43}$$

where ω is a kernel which expresses the dependence of the flux at $\mathbf{x}$ on the head gradient not just at $\mathbf{x}$, but at other points as well. The kernel is generally a function of the statistical parameters of Y, of coordinates and of the shape of Ω. Such an operator does not qualify, of course, for a definition of an effective property which presumably depends on the structure solely. It is reasonable to assume that ω tends to zero as $\mathbf{x}-\mathbf{x}'\to\infty$ and furthermore, this happens on a distance of the order I_Y. A major simplification of (3.4.43), leading to meaningful effective properties, can be achieved if it is assumed that $\langle\phi\rangle$ is continuous and varies slowly at the I_Y scale. We may then expand $\langle\phi\rangle$ in a Taylor series, i.e. $\langle\phi(\mathbf{x}')\rangle = \langle\phi(\mathbf{x})\rangle + \Sigma(x'_j-x_j)\,\partial\phi/\partial x_j$ +... and substitute in (3.4.43) to obtain

$$\langle q_j(\mathbf{x})\rangle = -\sum_\ell K_{ef,j\ell}\frac{\partial\langle\phi(\mathbf{x})\rangle}{\partial x_\ell} + \sum_k\sum_\ell A_{jk\ell}\frac{\partial^2\langle\phi(\mathbf{x})\rangle}{\partial x_k\,\partial x_\ell} + \sum_k\sum_\ell\sum_p B_{jk\ell p}\frac{\partial^3\langle\phi(\mathbf{x})\rangle}{\partial x_k\,\partial x_\ell\,\partial x_p} + \ldots \tag{3.4.44}$$

where the tensorial coefficients result from the integration of the kernel ω times various products of the components of $\mathbf{x}'-\mathbf{x}$. If these coefficients are functions of the structure only, (3.4.44) constitutes a generalization of (3.4.3) for nonuniform flows. Unlike (3.4.43), Eq.(3.4.44) has a local character, but the average flux does not depend anymore on the average head gradient, but on its derivatives as well. Furthermore, since $A=0(I_Y)$, $B=0(I_Y^2)$,... the successive terms of (3.4.44) have decreasing orders of magnitude if $\langle\phi\rangle$ is slowly varying, and (3.4.44) can be truncated after a few terms.

We shall attempt here to investigate the first few coefficients in (3.4.44) under additional simplifying assumptions, namely of large, in fact unbounded Ω, as compared to I_Y, and of stationary Y of small σ_Y^2, warranting the application of the perturbation analysis of Sect. 3.4.2. Under these conditions we may identify immediately the kernel ω in (3.4.44) at first-order by recalling Eqs. (3.4.11) for $\langle \mathbf{q} \rangle$, which are valid for nonuniform flows as well if $\mathbf{J}$ is replaced by $-\nabla\phi^{(0)}$, and Eqs. (3.3.9) which define $\phi^{(0)}$ and $\phi^{(1)}$. We shall first concentrate, similarly to the preceding section, on $\alpha_j = \langle Y' \partial\phi^{(1)}/\partial x_j \rangle$, which determines $\langle \mathbf{q} \rangle$ to first-order. By using again the Green function to solve (3.3.9) for $\phi^{(1)}$, we have after an integration by parts

$$\alpha_j = \langle Y'(\mathbf{x})\, \nabla\phi^{(1)}(\mathbf{x}) \rangle = -\int_\Omega C_Y(\mathbf{x},\mathbf{x}') \frac{\partial^2 G(\mathbf{x},\mathbf{x}')}{\partial x_j \partial x'_\ell} \frac{\partial \phi^{(0)}}{\partial x'_\ell}\, d\mathbf{x}' \tag{3.4.45}$$

Furthermore, since $\phi^{(0)} = \langle\phi\rangle + O(\sigma_Y^2)$ we may replace $\phi^{(0)}$ by $\langle\phi\rangle$ in Eq. (3.4.45), which in combination with (3.4.11) permits one to identify the kernel ω as follows

$$\frac{\omega_{j\ell}(\mathbf{x},\mathbf{x}')}{K_G} = -\left(1+\frac{\sigma_Y^2}{2}\right) \delta(\mathbf{x}-\mathbf{x}') - C_Y(\mathbf{x},\mathbf{x}') \frac{\partial^2 G(\mathbf{x},\mathbf{x}')}{\partial x_j \partial x'_\ell} + 0(\sigma_Y^4) \tag{3.4.46}$$

Eq. (3.4.46) may serve as the starting point for future studies of the effective conductivity for various Ω and $\phi^{(0)}$. We limit here the investigation to the coefficients of (3.4.46) for an unbounded domain. It is easy to ascertain that $A_{jk\ell} = \int (x_k - x'_k)\, \omega_{j\ell}(\mathbf{x},\mathbf{x}')\, d\mathbf{x}'$ is identically zero, by the symmetry properties of G (3.4.10) and C_Y. Hence, the first important conclusion is that a "straining" type of motion, represented by a second-order polynomial in the expression of $\phi^{(0)}$, does not influence the effective conductivity. The next coefficients in (3.4.46) are different from zero and we shall examine them briefly. Rather than a general case, we consider a two-dimensional average flow in the x,y plane, which are also principal axes for C_Y. Since $\phi^{(0)}$ satisfies Laplace equation, the four derivatives $\partial^3\phi^{(0)}/\partial x_k \partial x_\ell \partial x_p$ (k,ℓ,p = 1,2), can be expressed in terms of two of them, namely $\partial/\partial x_\ell[\partial^2\phi^{(0)}/\partial x^2 - \partial^2\phi^{(0)}/\partial y^2]$ (ℓ=1,2). Furthermore, for the sake of illustration, let us assume that C_Y is isotropic, i.e. it is a function of $r=|\mathbf{x}-\mathbf{x}'|$. The computations reduce then to the unique coefficient of (3.4.44)

$$B_{1111} - B_{1122} = \frac{1}{4}\int (r_x^4 - r_x^2 r_y^2)\, C_Y(r) \frac{\partial^2 G(r)}{\partial r_x^2}\, dr = \frac{45}{64}\int_0^\infty r C_Y(r)\, dr \tag{3.4.47}$$

where the passage to the last expression results from replacing G by (3.4.10) and integration in a spherical coordinate system over the angles θ and Φ. We obtain, therefore, the final form for the generalized Darcy's law in a two-dimensional average flow in the x,y plane as follows

$$\langle \mathbf{q} \rangle = - K_{ef} \ \nabla\phi^{(0)} + [\frac{45}{64} \int_0^\infty rC_Y \ dr] \ \nabla(\frac{\partial^2\phi^{(0)}}{\partial x^2} - \frac{\partial^2\phi^{(0)}}{\partial y^2}) \qquad (3.4.48)$$

where $K_{ef} = 1+\sigma_Y^2/6$ is the uniform flow value.

To further illustrate the results let us adopt the exponential covariance (3.2.14) and to assume that $\phi^{(0)}$ represents a radial flow directed towards the origin between two cylinders of radii which are much larger than I_Y. At a certain $R=\sqrt{x^2+y^2}=R_0$, we may write by a Taylor expansion of $\phi^{(0)}=(D/2\pi)\ \ln R$ around R_0 and substitution in (3.4.48)

$$\frac{\langle q_r(R_0)\rangle}{K_G} = \frac{D}{2\pi R_0} [-(1+\frac{\sigma_Y^2}{6}) + \sigma_Y^2 \frac{45}{16} \frac{I_Y{}^2}{R_0{}^2}] \qquad (3.4.49)$$

where D is a constant. In Eq. (3.4.49), the last term is the correction to the relationship prevailing in uniform flow. It is seen first that the effect of nonuniformity, i.e. of convergence of streamlines, is to reduce the average flow as compared to that resulting from the application of (3.4.3) with K_{ef} pertinent to uniform flow. Furthermore, this correction is of the order $\sigma_Y^2(I_Y{}^2/R_0{}^2)$ which was supposed to be small to warrant the truncation of the series in (3.4.44).

The above derivations definitely show that the usual generalized Darcy's law (3.4.3) can be applied to an average nonuniform flow in a heterogeneous formation, provided that it is slowly varying, i.e. that the length scale L associated with the third spatial derivatives of $\langle\phi\rangle$ is large compared to I_Y, the error incurred to K_{ef} being of order $\sigma_Y^2(I_Y/L)^2$. The generalization of Darcy's law when this condition is not obeyed or when σ_Y^2 is much larger than unity, requires further investigation.

3.4.7 Effective conductivity and storativity in unsteady flow through compressible formations

In the preceding sections we have examined the effective conductivity in steady flow. The results are also valid for unsteady flow in incompressible formations as well, when both matrix and fluid compressibility are neglected. Indeed, unsteadiness may stem then only from time changes in boundary conditions, which are propagated instantaneously through the medium, such that the flow adapts itself instantaneously to the boundary conditions, and time is only a parameter.

In contrast, in the case in which compressibility is not negligible, e.g. for thick formations in which the flow starts from rest, the basic equations are (see Chap. 2.11 and Eq. 3.3.1)

$$\nabla.\mathbf{q} = -s \frac{\partial\phi}{\partial t} \quad ; \quad \mathbf{q} = - K \ \nabla\phi \quad \text{i.e.} \quad s \frac{\partial\phi}{\partial t} = \nabla.(K \ \nabla\phi) \qquad (3.4.50)$$

where s is the specific storativity. In heterogeneous formations, s and K are RSF which have been analyzed in Chap. 3.1, being assumed that they are stationary and lognormal. The basic question is, first, whether the expectations of ϕ and $\mathbf{q}$ satisfy similar equations , i.e.

$$\nabla.\langle\mathbf{q}\rangle = -s_{ef}\frac{\partial\langle\phi\rangle}{\partial t} \quad ; \quad \langle\mathbf{q}\rangle = -K_{ef}\nabla\langle\phi\rangle \quad \text{i.e.}$$

$$s_{ef}\frac{\partial\langle\phi\rangle}{\partial t} = \nabla.(K_{ef}\nabla\langle\phi\rangle) \tag{3.4.51}$$

If the answer is affirmative, the next question is whether s_{ef} and K_{ef} depend only on the statistical structure of s and K or they depend also on t, $\mathbf{x}$ or ϕ. Only in the first case they do qualify as effective properties in the usual sense, otherwise they are of limited usefulness in solving (3.4.51) to derive $\langle\phi\rangle$.

In the literature on heterogeneous materials this point has not received attention, being assumed that (3.4.6) apply, with K_{ef} identical with its steady-state value. As we shall see, this assumption can not be taken for granted for flow in heterogeneous formations.

A first attempt to tackle this problem was presented by Freeze (1975), who has solved numerically, by Monte Carlo simulations, one-dimensional flow through blocks set in series. Since we are interested here mainly in three-dimensional flows, we shall not dwell upon Freeze's results. The main conclusion was, however, that one cannot generally define meaningful effective properties, since they were shown to be time dependent. An attempt to attack the problem in a more general manner, by using analytical approximations has been presented by Dagan (1982 c). We follow here succinctly this study and the interested readers may find details in the original article.

The mathematical problem for the RSF $\phi(\mathbf{x},t)$ is stated as follows : ϕ satisfies (3.4.50) in a domain Ω, with initial and boundary conditions

$$\phi(\mathbf{x},0) = \phi_{in}(\mathbf{x}) \quad ; \quad \phi = \phi_b(\mathbf{x},t) \quad \text{or} \quad \frac{\partial\phi}{\partial\nu} = \frac{\partial\phi_b}{\partial\nu} \qquad (\mathbf{x} \in \partial\Omega) \tag{3.4.52}$$

where ϕ_{in}, ϕ_b and $\partial\phi_b/\partial\nu$ are deterministic, given functions. By assuming that the initial state is deterministic, we refer in fact to a flow which starts from rest. This can be considered as an extreme case, in which the effect of unsteadiness is particularly important. Furthermore, in (3.4.51) K and s are lognormal, stationary RSF. As shown in Chap. 3.2, s is of much smaller variance than Y, and its lognormal distribution can be approximated by a normal one quite accurately, and this was indeed assumed in the past. Under these conditions Y and s are statistically characterized by the expectations m_Y and $\langle s\rangle$ and by the covariances C_s, C_{Ys} and C_Y, which are assumed to be given (if Eq. 3.2.4 is adopted, these covariances are expressed in terms of C_Y solely). Our aim at this stage is to derive the equation satisfied by $\langle\phi\rangle$ and to find out particularly whether, and under what conditions, it is of type (3.4.51). As we have already seen in the simpler case of steady flow, the problem is a difficult one and there is no general answer. To simplify matters we are adopting here again the small perturbation approximation, which provides general results and insight, in spite of its limitation to sufficiently small σ_Y^2. Following the procedure of Chap. 3.2, ϕ is formally expanded in an asymptotic sequence (3.3.8) and the equations satisfied by the different terms are obtained by expanding (3.4.50) for small Y′ and s′. For the sake of generality we shall assume at this stage that they are of same order, i.e. that σ_Y^2 and σ_s^2 are of same order, in spite of their disparity. The straightforward expansion of (3.4.50) yields at zero-order

$$L[\phi^{(0)}] \equiv \frac{1}{\kappa}\frac{\partial(\phi^{(0)})}{\partial t} - \nabla^2\phi^{(0)} = 0$$

$$\phi^{(0)}(\mathbf{x},0)=\phi_{in} \quad ; \quad \phi^{(0)} = \phi_b \quad (\mathbf{x} \in \partial\Omega) \tag{3.4.53}$$

where $\kappa=K_G/\langle s\rangle$ and L stands for the linear operator of the equation of heat conduction (see, e.g. Carslaw and Jaeger, 1965 for its properties). For the next term we similarly get

$$L[\phi^{(1)}] = \rho_1(\mathbf{x},t) \quad \text{with} \quad \rho_1 = -\frac{s'}{\langle s\rangle}\frac{\partial\phi^{(0)}}{\partial t} + \nabla.(Y'\nabla\phi^{(0)})$$

$$\phi^{(1)}(\mathbf{x},0) = 0 \quad ; \quad \phi^{(1)}(\mathbf{x},t) = 0 \quad (\mathbf{x} \in \partial\Omega) \tag{3.4.54}$$

i.e. $\phi^{(1)}$ satisfies the nonhomogeneous heat equation with a random forcing term and with homogeneous initial and boundary conditions, and similar equations are obtained for the higher-order approximations. Following an approach similar to that of steady state, we shall express the solutions of (3.4.53,54) with the aid of the appropriate Green function $G(\mathbf{x},\mathbf{x}',t-\tau)$. The latter, representing an instantaneous "heat source" at $t=\tau$ (see Carslaw and Jaeger, 1965), has the following properties

$$L[G] = \frac{1}{\kappa}\frac{\partial G}{\partial t} - \nabla'^2 G = 0 \quad \text{for } t > \tau \;;\; \lim_{\tau\to t}\int_\Omega G\, d\mathbf{x} = \lim_{\tau\to t}\int_\Omega G\, d\mathbf{x}' = 1$$

$$G(\mathbf{x},\mathbf{x}',t-\tau) = 0 \quad (\mathbf{x} \in \partial\Omega) \tag{3.4.55}$$

The Green function and its FT have the following close form expressions for an infinite domain Ω and for a space of m dimensions (Carslaw and Jaeger, 1965)

$$G(\mathbf{r},t') = \frac{1}{2^m(\pi\kappa t')^{m/2}}\exp[-\frac{r^2}{4\pi\kappa t'}] \quad \text{with} \quad \mathbf{r}=\mathbf{x}'-\mathbf{x} \;,\; t'=t-\tau$$

$$\hat{G}(\mathbf{k},t') = \frac{1}{(2\pi)^{m/2}}\exp(-\kappa t' k^2) \quad (m=1,2,3) \tag{3.4.56}$$

The solution for $\phi^{(1)}$ (3.4.54) can be written down as follows

$$\phi^{(1)}(\mathbf{x},t) = K\int_0^t\int_\Omega \rho_1(\mathbf{x}',\tau)\, G(\mathbf{x},\mathbf{x}',t-\tau)\, d\mathbf{x}'\, d\tau \qquad (3.4.57)$$

The specific discharge $\mathbf{q}$ is given by the same equation (3.4.9) of steady flow, which stems from Darcy's law. Hence, the computation of K_{ef} boils down to deriving the vector $\alpha_j = \langle Y'\, \partial\phi^{(1)}/\partial x_j\rangle = \alpha_{Ys,j}+\alpha_{Y,j}$. The latter are obtained from (3.4.57), after an integration by parts, as follows

$$\alpha_{Ys,j} = -\frac{1}{\langle s\rangle}\int_0^t\int_\Omega C_{Ys}(\mathbf{x},\mathbf{x}')\, \frac{\partial\phi^{(0)}(\mathbf{x}',\tau)}{\partial\tau}\, \frac{\partial G}{\partial x_j}\, d\mathbf{x}'\, d\tau$$

$$\alpha_{Y,j} = \int_0^t\int_\Omega C_Y(\mathbf{x},\mathbf{x}')\, \frac{\partial\phi^{(0)}(\mathbf{x}',\tau)}{\partial x'_\ell}\, \frac{\partial^2 G}{\partial x'_j\, \partial x'_\ell}\, d\mathbf{x}'\, d\tau \qquad (3.4.58)$$

The expressions (3.4.58) may constitute the starting point toward deriving the effective properties under quite general conditions, except for the presumed smallness of σ_Y^2. To obtain meaningful and simple results, we shall assume, similarly to the steady flow case, that $\phi^{(0)}$ is slowly varying both in space and with time and expand it in a Taylor expansion as follows

$$\frac{\partial\phi^{(0)}(\mathbf{x}',\tau)}{\partial x'_\ell} = \frac{\partial\phi^{(0)}(\mathbf{x},t)}{\partial x_\ell} + \sum_j (x'_j - x_j)\, \frac{\partial^2\phi^{(0)}(\mathbf{x},t)}{\partial x_\ell\, \partial x_j} +$$

$$(\tau - t)\, \frac{\partial^2\phi^{(0)}(\mathbf{x},t)}{\partial x_\ell\, \partial\tau} + \ldots \quad ; \quad \frac{\partial\phi^{(0)}(\mathbf{x}',\tau)}{\partial\tau} = K\, \nabla^2\phi^{(0)}(\mathbf{x},t) + \ldots \qquad (3.4.59)$$

The expansions (3.4.59) are substituted now in (3.4.58) and the domain Ω is left to become unbounded, such that G can be replaced by (3.4.56). Both operations are permitted because the kernels in the integrals of (3.4.58) vanish for large $|\mathbf{x} - \mathbf{x}'|$ and $t-\tau$. It is easily found that due to the anti-symmetry of ∇G, the first integral, namely $\alpha_{Ys,j}$, is equal to zero. Hence, the cross-correlation between Y and s does not influence at the leading order the effective conductivity. The remaining term, $\alpha_{Y,j}$, has been computed in a close form for C_Y isotropic by Dagan (1982c) and the final result is presented herein. First, only the first term of the expansion of (3.4.59) contributes to $\alpha_{Y,j}$ in (3.4.58), again for reasons of symmetry of C_Y and G. Furthermore, for isotropic C_Y only the term with $j=\ell$ survives in $\alpha_{Y,j}$ and $\partial^2 G/\partial x'^2_\ell$ can be replaced by $(1/m)\, \nabla^2 G$ (m=1,2,3). After these preparatory steps we can write

$$\alpha_{Y,\ell} = \kappa \frac{\partial\phi^{(0)}(\mathbf{x},t)}{\partial x_\ell} \int_0^t \int_\Omega C_Y(\mathbf{r}) \frac{\nabla^2 G(\mathbf{r},t-\tau)}{m} \, d\mathbf{r} \, d\tau \tag{3.4.60}$$

Making use of (3.4.56) in (3.4.60) yields the final result for $\alpha_{Y,\ell}$

$$\alpha_{Y,\ell} = \frac{\partial\phi^{(0)}(\mathbf{x},t)}{\partial x_\ell} \sigma_Y^2 \frac{b_m(t)-1}{m} \qquad (m = 1,2,3)$$

$$b_m(t) = 4\pi \int_0^\infty \rho_Y(r) \, G(r,t) \, r^2 \, dr \quad ; \quad t = \frac{t K_G}{\langle s\rangle I_Y^2} \tag{3.4.61}$$

The last step towards obtaining the expression of K_{ef} is to substitute α (3.4.61) into (3.4.7), i.e. to evaluate $K_{ef} = -\langle q_\ell\rangle/\langle\partial\phi^{(0)}/\partial x_\ell\rangle$. The result is, for flow in a formation of m-dimensional structure, as follows

$$\frac{K_{ef}}{K_G} = 1 + \sigma_Y^2 \left[\frac{1}{2} - \frac{1}{m} + \frac{b_m(t)}{m}\right] \qquad (m = 1,2,3) \tag{3.4.62}$$

From the properties of G (3.4.56) it is seen that $b_m(0)=1$, whereas $b_m(\infty)\to 0$. Furthermore, it is reminded that in the small perturbation analysis (see Eq. 1.1.15) $K_A/K_G = 1+\sigma_Y^2/2$, whereas the effective conductivity in steady flow is given by $K_{efs}/K_G = 1+\sigma_Y^2(1/2-1/m)$, m=1,2,3. Under these conditions, (3.4.62) can be rewritten at first order in the equivalent and elegant form

$$\frac{K_{ef}(t) - K_{efs}}{K_A - K_{efs}} = b_m(t) \tag{3.4.63}$$

Summarizing the results obtained so far, it was shown that for an unsteady flow in a formation of compressible matrix and for a flow starting from rest, the effective conductivity depends on both structure, namely on C_Y, and on the dimensionless time $t = t K_G/\langle s\rangle I_Y^2$. Furthermore, K_{ef} is initially equal to the arithmetic mean K_A and relaxes subsequently to its steady state value. The relaxation time t_r depends on b_m, which in turn can be evaluated by (3.4.61) for a given covariance C_Y. For illustration we reproduce in Fig. 3.4.8 b_m as a function of t for the isotropic exponential covariance (3.2.14) (Dagan, 1982c). In this simple case b_m could be determined analytically ; for instance, $b_3 = -2\sqrt{t/\pi} + (1+2t)\exp(t)\,\mathrm{erfc}(\sqrt{t})$. Inspection of Fig. 3.4.8 shows convincingly that t_r is largest for one-dimensional structures (m=1) and drops considerably for m=3. The one-

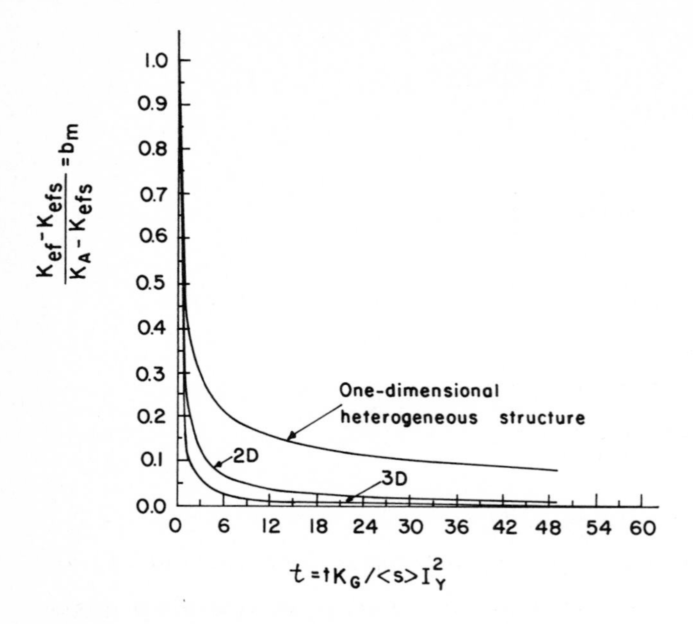

Fig. 3.4.8 The dependence of the effective conductivity upon time (Eq. 3.4.62) for unsteady flow starting from rest in heterogeneous formations of isotropic one-, two- or three-dimensional structures (reproduced from Dagan 1982 c).

dimensional flow corresponds to the configuration investigated by Freeze (1975) or to flow normal to layers in a stratified formation. This result explains the results of Freeze, namely that K_{ef} in unsteady flow may not tend to its steady state value after considerable time. In contrast, for an isotropic three-dimensional structure $b_3=0.05$ for $t_r \simeq 3$, i.e. $t_r \simeq 3\langle s\rangle I_Y{}^2/K_G$. This time is of order of minutes to seconds for some conceivable values of parameters.

We are now in a position to define quantitatively the notion of slowly time-varying average gradient, to justify retaining only the leading terms in the Taylor expansion (3.4.59) : indeed, the time scale defined with the aid of $\partial\nabla\phi^{(0)}/\partial t$, has to be large compared to t_r. Furthermore, if this is the case the steady state value of K_{ef} can be adopted. This is even more so in an anisotropic formation in which flow is parallel to layering.

As for the effective storativity, defined by $s_{ef} = -\nabla.\langle q\rangle/(\partial\langle\phi\rangle/\partial t)$, it has been shown by Dagan (1982c) under the same approximations that it also relaxes to its steady state value $s_{ef} = \langle s\rangle$ during a time of order t_r, but altogether it differs very little from $\langle s\rangle$ due to the smallness of the ratio $\sigma_s^2/\langle s\rangle^2$.

A different unsteady regime examined by Dagan (1982c) is the one in which the average flow is periodic, e.g. as caused by seasonal recharge variations. This type of flow will be discussed in Part 5, dealing with regional scale.

Summarizing, it has been found that for flow through formations of three-dimensional heterogeneous structures, the effective conductivity in steady flow and the specific storativity arithmetic average can be adopted in the equation satisfied by the expected values of the head and specific discharge, provided that they are slowly varying in space, at the I_Y scale, and in time, at the t_r scale. Otherwise, the effective conductivity is changing with time and its precise dependence upon time is related to the history of the flow. This transient stage seems to be quite small for the usual range of parameter values at local scale.

Exercises

3.4.1 Prove that the arithmetic mean K_A and the harmonic mean K_H are best bounds for one dimensional flow in a stratified formation of random structure.

3.4.2 Derive the expressions of the effective conductivity for flow, uniform in the average, through isotropic formations of one-, two- or three-dimensional heterogeneous structures by the first-order perturbation approximation and for lognormal K. Hint : use the Green function and take advantage of indifference to direction in space.

3.4.3 Derive the equation defining the effective conductivity, by using the self-consistent approach, for flow in an isotropic formation of two-dimensional structure. Apply the result to a two-phase medium.

3.4.4 Prove, by invariance arguments, that the relationship between the average specific discharge and head gradient in an anisotropic, axisymmetric, formation can be written in an elegant form as follows

$$\langle \mathbf{q} \rangle = A\ \mathbf{J} + B\ (\mathbf{J}.\boldsymbol{\nu})\ \boldsymbol{\nu} \qquad (3.4.64)$$

where A and B are two constants, $\boldsymbol{\nu}$ is a unit vector in the direction of the symmetry axis and $\mathbf{J} = -\nabla\langle\phi\rangle$. Relate A and B to the principal values of the effective conductivity tensor. Use (3.4.64) in order to derive the dependence of the Cartesian components of $\mathbf{K}_{ef}$ upon its principal values for an coordinate system of arbitrary orientation.

3.4.5 The directional effective conductivity can be defined either by the ratio between $q=|\mathbf{q}|$ and the projection of $\mathbf{J}$ upon it or by the ratio between the projection of $\mathbf{q}$ upon $\mathbf{J}$ and J. Investigate the change of the directional conductivity with the direction of $\mathbf{J}$ for an anisotropic axi-symmetric formation. Examine the dependence of the angle between $\mathbf{q}$ and $\mathbf{J}$ upon the direction of $\mathbf{J}$.

3.4.6 Derive the first-order approximation of the effective conductivity in unsteady flow through a compressible formation if the initial condition is one of steady uniform flow rather than rest.

3.5 SOLUTIONS OF THE MEAN FLOW EQUATIONS (EXAMPLES OF EXACT SOLUTIONS)

3.5.1 General

The aim of the present chapter is to present a few solutions of the partial differential equations satisfied by the expected values of the head $\langle\phi(\mathbf{x},t)\rangle$ and of the specific discharge $\langle\mathbf{q}\rangle$. This is the primary knowledge to be obtained with the aid of the statistical theory. Furthermore, it exhausts the information about the flow field either if the formation is homogeneous and of properties known with certainty, or if space averages (see Chap. 3.3) are sought. Since averaging domains are usually of large dimensions compared to the local heterogeneity scale, the variance of the space averages is generally quite low. Consequently, most of the literature on heterogeneous materials deals primarily with the equations of the mean field and their solutions. Nevertheless, we shall discuss briefly the derivation of higher statistical moments in Chap. 3.7, while a major portion of Parts. 4 and 5 are devoted to second-order moments. We limit here the discussion to flow of water, i.e. of a homogeneous liquid, the presence of a solute being considered in Part 4. Nevertheless, we shall analyze briefly the flow of two liquids separated by a sharp interface, an approximation which renders the problem similar to that of free-surface flow of one fluid.

The field equations to be considered here have been discussed in the preceding sections. First, the averaged Darcy's law is

$$\langle\mathbf{q}\rangle = -\,\mathbf{K}_{ef}\,\nabla\langle\phi\rangle \tag{3.5.1}$$

where the meaning of the effective conductivity has been analyzed at length in Chap. 3.4. In a formation of stationary random structure, $\mathbf{K}_{ef}$ is of constant components (or may have a slow trend), provided that the requirements of slowly varying average flow in space and time, discussed in Chap. 3.4, are satisfied. We shall assume that this is the case, although (3.5.1) will be applied to particular problems in which it is not obvious that $\mathbf{K}_{ef}$ is constant. Incorporating the supposedly minor corrections of $\mathbf{K}_{ef}$, to account for rapid variations of $\langle\phi\rangle$, is a difficult subject, which awaits further investigation.

The equations of conservation of mass equation are

$$\nabla.\langle\mathbf{q}\rangle = -\,s_{ef}\,\frac{\partial\langle\phi\rangle}{\partial t} \quad \text{(unsteady flow)} \quad ; \quad \nabla.\langle\mathbf{q}\rangle = 0 \quad \text{(steady flow)} \tag{3.5.2}$$

We shall consider isotropic formations, i.e. $K_{ef,ij} = K_{ef}\delta_{ij}$. Conversely, in the case of an anisotropic $\mathbf{K}_{ef}$, Eq. (3.5.1) can be rewritten in a Cartesian frame attached to the principal axes of $\mathbf{K}$.

Then, by a linear coordinates transformation (see Exercise 3.6.1) and by defining an equivalent conductivity, (3.5.1) is again expressed as in the isotropic case. It is common to adopt the horizontal conductivity as the equivalent one, and with $K_{ef,xx}=K_{ef,yy}=K_{efh}$ and $K_{ef,zz}=K_{efv}$, the only coordinate affected is the vertical one, which is transformed as follows: $z=z'(K_{efh}/K_{efv})^{1/2}$. This transformation is effective if boundaries and boundary conditions can be expressed with the aid of z' and K_{efh} solely, which we presume.

Elimination of $\langle \mathbf{q} \rangle$ from (3.5.1) and (3.5.2), in absence of trends of K_{ef} or s_{ef}, leads to

$$\nabla^2 \langle \phi \rangle = 0 \quad \text{(steady flow)} \; ; \quad s_{ef} \frac{\partial \langle \phi \rangle}{\partial t} = K_{ef} \; \nabla^2 \langle \phi \rangle \quad \text{(unsteady flow)} \qquad (3.5.3)$$

These partial differential equations are satisfied by $\langle \phi \rangle$ in the flow domain and they have to be supplemented by appropriate boundary and initial conditions. The latter have been analyzed in detail at the laboratory scale in Part 2. We shall adopt them at the local scale as well, assuming that they are deterministic. If this is not the case, random terms, additional to those related to the heterogeneous structure, will show up in the head residual and in its higher-order statistical moments. We shall return to the various types of boundary conditions in the sequel.

The Laplace and heat-conduction equations (3.5.3) are classical equations of the mathematical physics. Their properties and methods of solutions are the object of numerous books (e.g. the advanced compendium of Morse and Feshbach, 1953). These equations are identical to those describing flow in homogeneous media. Consequently, numerous analytical solutions, exact or approximate, pertinent to specific problems of water flow through porous formations have been developed and published in the past (see a list of relevant texts in Chap. 3.1). It is beyond the scope of this book to present a comprehensive review of the methodology developed in the past. Instead, we shall discuss a few specific and simple cases of exact analytical solutions for illustrative purposes. The derivation of these solutions is aimed at developing an understanding of a few typical flow patterns and at providing benchmark cases for validation of numerical solutions. Besides, we shall analyze in Chap. 3.6 a few classical approximate methods which simplify considerably the solution and are quite helpful in the numerical framework as well. The reader interested in the many existing solutions of various flow problems is advised to consult the recent monographs mentioned in Chap. 3.1.

Finally, to simplify notation, we shall delete the symbol <> and the subscript ef in this chapter and the following one, being understood that here, and only here, ϕ, $\mathbf{q}$, K and s stand for expected values and effective properties, respectively.

3.5.2 Illustration of exact solutions of a few classes of flow

The literature on groundwater flow distinguishes among a few categories of flow problems, the division being dictated mainly by the degree of complexity of the solution. We shall consider now a few such categories and derive simple examples of analytical solutions.

(i) Steady confined flow

These are flow problems in which the boundaries are fixed and given. Typical boundary conditions have been analyzed in Chap. 2.13 and are recalled here

$$\frac{\partial\phi}{\partial\nu} = 0 \quad \text{(impervious boundary)} \quad ; \quad \phi = \text{const} \quad \text{(free fluid)}$$

$$\phi = z + \text{const} \quad \text{(seepage face)} \tag{3.5.4}$$

The head ϕ is given, therefore, by the solution of Laplace equation (3.5.3) with boundary conditions (3.5.4). There is a very large body of mathematical literature dealing with the subject and various methods are available, e.g. separation of variables and use of series of orthogonal functions (Fourier, Fourier-Bessel) or formulation in terms of Green function. Major simplifications are possible for two-dimensional or axisymmetric flows, which we shall briefly examine.

Starting with flow in the x,y plane, it is possible to define the streamfunction $\Psi(x,y)$ (see, e.g. Batchelor, 1967) by the relationships

$$q_x = \frac{\partial\Psi}{\partial y} \quad ; \quad q_y = -\frac{\partial\Psi}{\partial x} \tag{3.5.5}$$

The existence of Ψ is ensured by the continuity equation (3.5.2), while Darcy's law (3.5.1) leads to $\nabla^2\Psi=0$. It is seen from (3.5.5) that $\mathbf{q}$ is normal to $\nabla\Psi$, i.e. lines Ψ=const are streamlines. Furthermore, the flux through any curve connecting two points A,B in the plane is given by

$$Q_{AB} = \int_{AB} \mathbf{q}.\boldsymbol{\nu}\, ds = \Psi(\mathbf{x}_B) - \Psi(\mathbf{x}_A) \tag{3.5.6}$$

where $\boldsymbol{\nu}$ is a unit vector normal to the curve.

These properties make the streamfunction a useful tool for depicting flows, particularly in conjunction with the specific discharge potential defined by $\Phi = -K\,\phi$, $\mathbf{q} = \nabla\Phi$. Lines Φ=const are normal to $\mathbf{q}$ and to streamlines and the network of lines of constant Φ and Ψ, usually drawn at constant and equal differences of their values, give a vivid picture of the flow pattern.

One of the most powerful methods for solving Laplace equation in two dimensions is by analytic functions of the complex variable $Z=x+iy$, where i is the imaginary unit. The basic property is that a complex potential can be defined by $f=\Phi+i\Psi$, and for Φ,Ψ harmonic and satisfying (3.5.1,5), f is an analytic function of Z. Conversely, any analytic function of Z may be regarded as the complex potential of a two-dimensional flow. Thus, rather than having to deal with functions of the two variables x,y, the problem is cast in terms of Z only. Another important result is that the complex specific discharge, defined by $w=q_x-iq_y$, is also analytic in Z with $w = df/dZ$. The transformation by which a domain in the plane Z is mapped onto the corresponding domain in the planes f or w by the analytic functions f(Z) or w(Z), is called a conformal mapping, which preserves angles and ratios between the length of infinitesimal intersecting segments. A rich collection of solutions of flows through porous media by analytic functions may be found, for instance, in the treatise by Polubarinova Kochina (1962).

To illustrate the approach we consider the flow towards a fully penetrating well in an aquifer of constant thickness (Fig. 3.5.1a), the head on the well perimeter $\phi=\phi_w$ being constant. The flow is two dimensional in the horizontal plane x,y and the solution is as follows

$$f = \Phi_w - \frac{Q}{2\pi D}\ell n\left(\frac{Z}{r_w}\right) \quad ; \quad w = \frac{df}{dZ} = -\frac{Q}{2\pi DZ} \tag{3.5.7}$$

where $\Phi_w=-K\phi_w$ and Q is the total well discharge. By using the polar coordinates r,θ with $Z=r\exp(i\theta)$, it is easy to see that (3.5.7) leads to

$$\phi-\phi_w = \frac{Q}{2\pi KD}\ell n\left(\frac{r}{r_w}\right) \quad (r\geq r_w) \quad ; \quad \Psi = -\frac{Q}{2\pi D}\theta \quad (0<\theta<2\pi) \tag{3.5.8}$$

If the head is given on a cylindrical boundary $r=R$, the discharge Q may be determined from (3.5.8) by the condition $\phi=\phi_R$ for $r=R$. The distribution of ϕ is depicted in Fig. 3.5.1a and the flow net is shown in Fig. 3.5.1b. This simple solution has applications to aquifer flow. Generally, any flow in an aquifer of constant thickness which is caused by insertion or withdrawal of fluid at constant discharge from a finite volume, has a logarithmic potential far from the source, while the velocity drops like $1/r$.

In the case of axisymmetric flow, referred to a spherical coordinate system in which z stands now for the symmetry axis, ϕ is a function of $r=|\mathbf{x}|$ and of θ, the angle between $\mathbf{x}$ and the z axis, and is independent of the azimuthal angle. A streamfunction, stemming from the continuity equation, can be defined as well, the following relationships being satisfied

$$q_r = \frac{\partial\Phi}{\partial r} = \frac{1}{r^2\sin\theta}\frac{\partial\Psi}{\partial\theta} \quad ; \quad q_\theta = \frac{1}{r}\frac{\partial\Phi}{\partial\theta} = -\frac{1}{r\sin\theta}\frac{\partial\Psi}{\partial r} \tag{3.5.9}$$

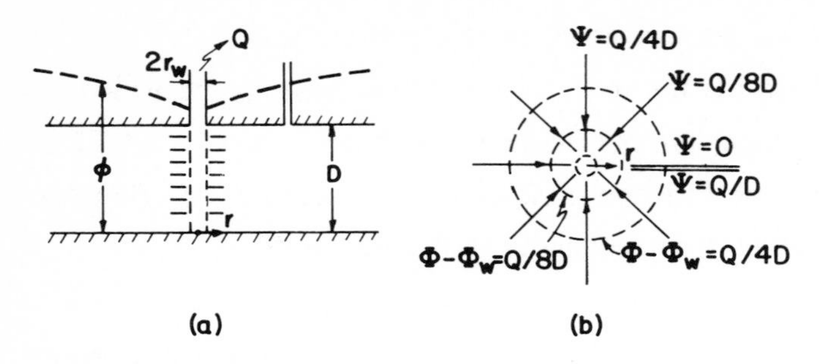

Fig. 3.5.1 Two-dimensional, steady, flow toward a fully penetrating well in a confined aquifer : (a) sketch of flow domain and (b) sketch of flow net in the potential plane.

The Stokes streamfunction Ψ (see, e.g. Batchelor, 1967) is related to the flux through a surface of revolution generated by a curve joining two points A,B by $Q_{AB} = 2\pi(\Psi_B - \Psi_A)$. The major simplification, as compared to the full three-dimensional case, is the reduction of the number of variables to two.

A simple example of axisymmetric flow (in fact, of spherical symmetry), is the one from a semi-spherical cavity at the impervious surface of a half-space (Fig. 3.5.2a). Although this example is of an idealized nature as compared with the previous one, it is quite instructive. The head and the streamfunction are now given by

$$\phi - \phi_w = - \frac{Q}{2\pi K} \left(\frac{1}{r} - \frac{1}{r_w} \right) \quad , \quad \Psi = - \frac{Q}{2\pi} \cos\theta \quad ; \quad q_r = \frac{Q}{2\pi r^2} \, , \, q_\theta = 0 \tag{3.5.10}$$

The flow net is depicted in Fig. 3.5.2b and again, if the external boundary is a half-sphere of radius R and of constant head ϕ_R, the discharge can be determined accordingly from (3.5.10). The far field behaviour of $\phi = 0(1/r)$ and $q_r = 0(1/r^2)$, corresponds to injection or withdrawal at constant discharge in a finite portion of the space near the origin.

We shall recall additional examples of confined flows in Chap. 3.6, which deals with some useful approximations.

(ii) Steady free-surface flow

We start with analyzing two-dimensional flow in the vertical plane, with x and y standing for horizontal and vertical axes, respectively. The boundary conditions prevailing on the free-surface (water table) have been derived in Sect. 2.13.4 and are reproduced here for convenience, as follows

$$q_y - q_x \frac{\partial \eta}{\partial x} = -R \quad (y = \eta) \quad ; \quad \phi = \eta - \psi_w \quad (y = \eta) \tag{3.5.11}$$

where $y = \eta(x)$ is the equation of the free-surface which separates the saturated and the unsaturated

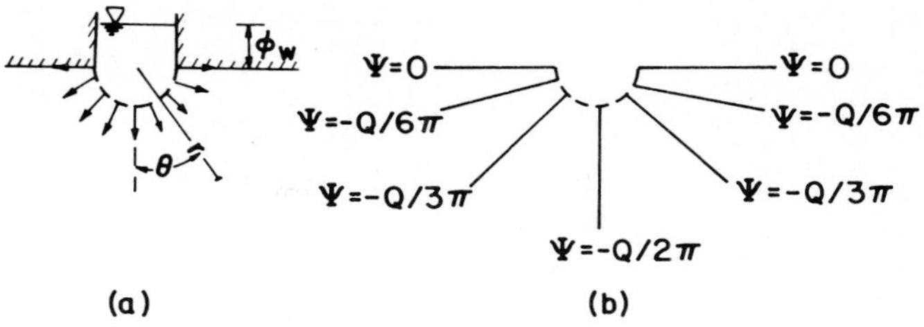

Fig. 3.5.2 Axisymmetric, steady, flow from a semi-spherical cavity, beneath an impervious boundary : (a) the flow domain and (b) streamsurfaces.

zones, the constant ψ_w is the air-entry value and R is the vertical recharge. It is emphasized that Eqs. (3.5.11) have been derived in Sect. 2.13.4 for homogeneous media and no analysis has been conducted so far to infer their validity for expected values of the flow variables in the case of heterogeneous media of random structure (more precisely, an additional term may show up from the averaging of $q_x \partial\eta/\partial x$). These subjects await further investigation, but it is reasonable to expect that these additional terms are quite small.

From a mathematical point of view, the solution of the Laplace equation with boundary conditions (3.5.11) is considerably more difficult than that of confined flow. Indeed, the boundary equation itself is an unknown of the problem and has to be determined together with $\phi(\mathbf{x})$. The technique which led to a few analytic solutions of the problem is the one of complex variables (for a few examples, see Polubarinova Kochina, 1962), particularly the hodograph method. To illustrate it, we shall rewrite first the two equations (3.5.11) in terms of $\mathbf{q}$. Differentiation of $\phi=\eta+\psi_w$ along the free-surface, i.e. $\partial\phi/\partial x+(\partial\phi/\partial y)(\partial\eta/\partial x) = \partial\eta/\partial x$, and subsequent substitution of $\partial\eta/\partial x = (\partial\phi/\partial x)/(1-\partial\phi/\partial y) = -Kq_x/(K+q_y)$ in the first equation, leads to

$$q_y^2 + q_x^2 + (K+R)q_y = -KR \quad \text{i.e.} \quad q_x^2 + \left(q_y + \frac{K+R}{2}\right)^2 = \left(\frac{K-R}{2}\right)^2 \tag{3.5.12}$$

It is seen that the unknown free-surface profile is mapped on a circle of radius (K-R)/2, centered at $q_x=0$, $-q_y=(K+R)/2$ in the complex velocity plane $w=q_x-iq_y$, which is called the hodograph plane. If the free-surface is of known shape in the complex potential plane as well, the mapping of w on f and a further integration of the equation $dZ=(1/w)df$ leads to the mapping of w or f onto the physical plane, i.e. to the solution of the problem. Although the number of close form solutions obtained this way is limited, such solutions are invaluable for comparison with numerical or approximate ones. To illustrate the procedure we shall consider a simple case (see Dagan, 1967d), namely free-surface flow of discharge Q over a semi-infinite, impervious, bottom toward a horizontal seepage face, in absence of recharge (Fig. 3.5.3a). This is a schematic case, but represents, sufficiently far from the exit, any steady horizontal flow. The flow domain OABO is mapped onto the

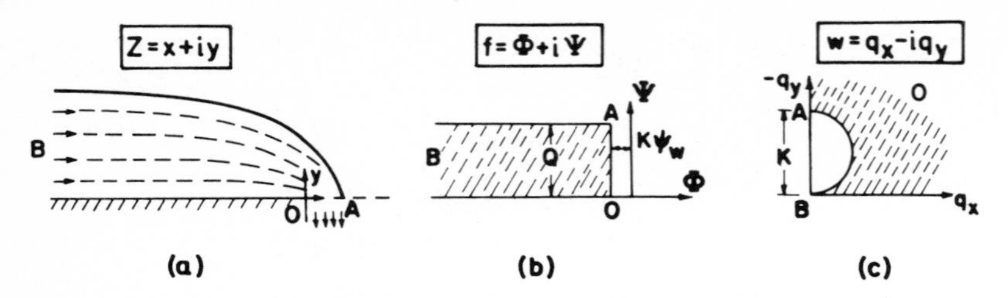

Fig. 3.5.3 Two-dimensional, free-surface, steady flow above an impervious step : (a) the physical plane, (b) the complex potential plane and (c) the complex specific discharge plane (the hodograph).

corresponding semi-infinite strip in the f plane (Fig. 3.5.3b). Indeed, the free-surface and the impervious bottom are streamlines, whereas the pressure is constant on OA. Finally, the hodograph is represented in Fig. 3.5.3c. By an inversion and a subsequent integration, the following simple relationships are obtained

$$w = -\frac{KQ}{f + 2K\psi_w} \quad ; \quad Z = -\frac{f(f + 2K\psi_w)}{2KQ} \tag{3.5.13}$$

Substitution of $f=\Phi+iQ$ in the expression of Z, separation of the real and imaginary parts and elimination of Φ leads to the simple equation of the free-surface profile

$$\eta^2 = \frac{2Q}{K}(x_A - x) \quad ; \quad x_A = \frac{Q}{2K} + \frac{K\psi_w^2}{2Q} \tag{3.5.14}$$

It is seen that the water table is of parabolic shape and far from the seepage face, i.e. for $x >> x_A$, $\eta \simeq \sqrt{2Q|x|/K}$. This is a general result for any free-surface flow over a horizontal bottom, in absence of recharge. The streamlines Ψ=const are confocal parabolas and the same is true for the equipotentials. Furthermore, far from the origin the equipotentials are almost vertical and $\phi \simeq \eta$, such that for a given η=H at x=-L, $Q \simeq KH^2/2L$.

Although no such simple solutions are available for more complicated geometries, it is quite instructive to depict the hodograph, even if its mapping onto the potential or physical plane is not feasible. To illustrate this point, the flow over a sloping impervious bottom resulting from constant recharge through soil, is depicted in Fig. 3.5.4a. This may be viewed as a model of underland flow in a watershed. The hodograph, obeying the various boundary conditions, is shown in Fig. 3.5.4b. Thus, the seepage face CE, on which ϕ=y (with neglect of ψ_w), is represented in the hodograph plane by the straight line $-q_y = -q_x \cot\alpha + K$ (see Chap. 2.13). The mapping of the two impervious boundaries AB and BC is obvious, whereas the water table has the circular shape (3.5.12). From this simple mapping we may learn about the magnitude of the specific discharge, which varies between q_y =R at A to $q_x = (K-R)\sin\alpha \cos\alpha$, $q_y = (K-R)\cos^2\alpha - K$ at E. At an intermediate point D the flow reverses itself, and its location in the w plane is found from the requirement that the normal velocity is zero.

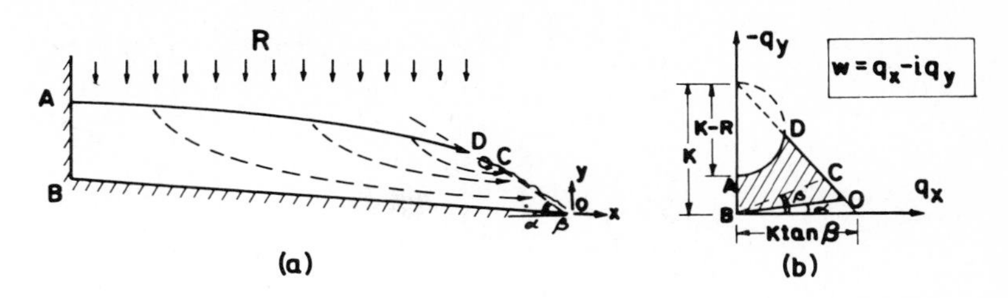

Fig. 3.5.4 Underland, steady, two-dimensional flow with a free-surface above a sloped bottom : (a) sketch of flow domain and (b) the hodograph.

More generally, the continuity requirements of mapping by analytical functions shows that the recharge R cannot exceed K and that the specific discharge on a water-table is always smaller than K, both results having a clear physical interpretation.

As mentioned already in the introduction, numerical methods have superseded the analytical procedures described here. Still, the qualitative analysis of the flow is facilitated with the aid of the tools just described above.

Unlike the case of two-dimensional flow, there are no known exact solutions of free-surface three-dimensional flow, although it is easy to generalize the boundary conditions (3.5.11) for this case (see Sect. 2.13.4)

$$q_z - q_x \frac{\partial \eta}{\partial x} - q_y \frac{\partial \eta}{\partial y} = -R \quad (z = \eta) \quad ; \quad \phi = \eta - \psi_w \quad (z = \eta) \tag{3.5.15}$$

where z is now a vertical coordinate and $z=\eta(x,y)$ is the equation of the free-surface. The first equation of (3.5.15) may be rewritten in the simple form $\mathbf{q}.\boldsymbol{\nu} = -R\,\nu_z$, or $\partial\phi/\partial\nu = (R/K)\nu_z$, where $\boldsymbol{\nu}$ is a unit vector normal to the free-surface. Additional forms of the free-surface boundary conditions may be obtained in terms of ϕ or $\mathbf{q}$ solely by differentiating the second equation of (3.5.13) along the free-surface, and this is left as an exercise.

Approximate solutions of three-dimensional flows will be discussed in the sequel, in Chap. 3.6.

(iii) Unsteady confined flow.

Eq. (3.5.3) has to be solved for given and fixed boundaries, and with appropriate boundary conditions, e.g. (3.5.4), and initial conditions. The flow equation may be rewritten as

$$\frac{\partial \phi}{\partial t} = \kappa \, \nabla^2 \phi \quad ; \quad \kappa = \frac{K}{s} \tag{3.5.16}$$

and it is identical to the equation of heat conduction in solids, with ϕ standing for the temperature and κ for the heat diffusivity. This linear parabolic partial differential equation has been investigated thoroughly in the literature and it has been solved by various methods, like separation of variables, Laplace transform and Green function. A comprehensive survey and a compendium of solu-

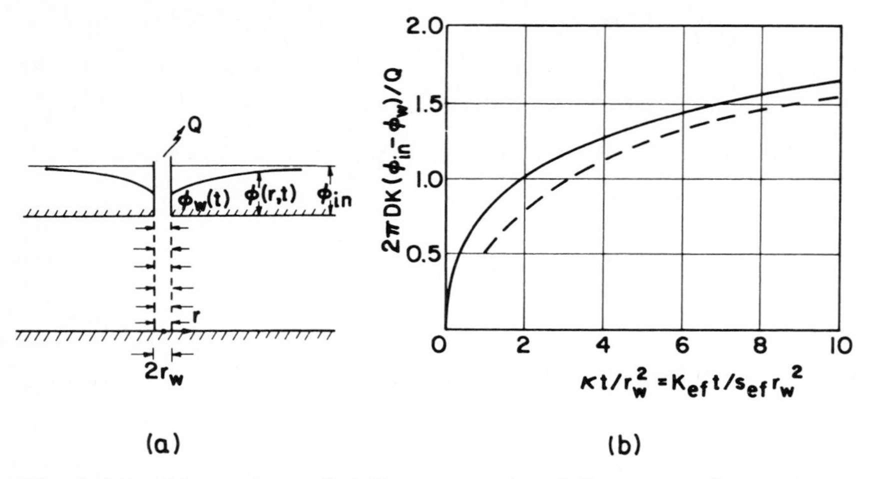

Fig. 3.5.5 Unsteady, radial flow, toward a fully penetrating well pumping at constant discharge, in a confined aquifer : (a) sketch of flow domain and (b) the variation of the head in the well with time (full line, exact solution Carslaw and Jaeger, 1959 ; dashed line, Theis solution Eq. 3.6.3).

tions for a variety of boundary and initial conditions may be found in the treatise by Carslaw and Jaeger (1959). Before illustrating the techniques, it is worthwhile to estimate the order of magnitude of κ by using the correlation (3.2.4), resulting in values of 10^1-10^5 m^2/day, for poorly to very pervious media, respectively.

The simple, but of wide application, example is the one of a fully penetrating well pumping at constant discharge Q at $t>0$ from an unbounded aquifer, initially at a constant head (Fig. 3.5.5a). The flow has radial symmetry and (3.5.16), as well as the boundary and initial conditions for $\phi(r,t)$, can be written as follows

$$\frac{\partial\phi}{\partial t} = \kappa\,\frac{1}{r}\,\frac{\partial}{\partial r}\left(r\,\frac{\partial\phi}{\partial r}\right) \qquad (r>r_w)$$

$$q_r = -K\,\frac{\partial\phi}{\partial r} = -\frac{Q}{2\pi D r_w} \quad (r=r_w) \quad , \quad \phi=\phi_{in} \quad (t=0) \tag{3.5.17}$$

where $r=(x^2+y^2)^{1/2}$.

The solution is provided in Carslaw and Jaeger (1959, Sect. 13.5) in the form of an integral resulting from the inversion of the Laplace transform of ϕ. The asymptotic expansion of ϕ for large dimensionless time is given by

$$\phi - \phi_{in} = -\frac{Q}{4\pi KD}\left[\ell n\,\frac{4\kappa t}{r^2} - E + O\left(\frac{r_w^{\,2}}{\kappa t}\right)\right] \qquad (\kappa t/r_w^{\,2}>>1) \tag{3.5.18}$$

where E = 0.5772... is the Euler number. The head on the well perimeter, as a function of time, is shown in Fig. 3.5.5b. In general, (3.5.18) represents the head behaviour sufficiently far from the origin for constant pumping from a finite region near the origin in a confined aquifer of thickness D. The logarithmic steady state solution (3.5.7) is embedded in (3.5.18). We shall return to the solution of (3.5.17) in Sect. 3.5.3.

(iv) Unsteady free-surface flow.

The head satisfies Eq. (3.5.16) in the flow domain, which has the free-surface $z=\eta(x,y,t)$ as an upper boundary. The boundary conditions are given by (2.13.23,4) and are reproduced here

$$n_{ef}\frac{\partial\eta}{\partial t} = q_z - q_x\frac{\partial\eta}{\partial x} - q_y\frac{\partial\eta}{\partial y} + R \quad ; \quad \phi = \eta - \psi_w \qquad (z = \eta) \tag{3.5.19}$$

where n_{ef}, the effective porosity for free-surface motion, is defined in Chap. 2.13. It is reminded that (3.5.19) is an approximation, even in the case of homogeneous media, valid only for slow drainage of the unsaturated zone above the free-surface (see Chap. 2.13).

Under a time-change of the head caused by the boundary conditions, fluid is released from the formation due to elastic storage and to the drop of the free-surface. The ratio between the two volumes is of the order $s_{ef}D/n_{ef}$, where D is the average formation thickness. An estimate of this ratio for thicknesses D up to 100 m, and with s based on (3.2.4), indicates that except for thick aquifers of very low conductivity, $s_{ef}D/n_{ef} \ll 1$. It is, therefore, customary to neglect the storativity, and to assume that ϕ satisfies the Laplace equation, rather than (3.5.16), in the flow domain.

Still, the mathematical problem posed by the nonlinear boundary conditions (3.5.19) on an unknown surface, which is part of the solution, has defied the derivation of exact solutions of two- or three-dimensional flows of interest. Approximate methods and solutions are discussed further on in the next chapter.

(v) Flow of two fluids separated by a sharp interface

We shall discuss briefly a somewhat different class of flows, which is beyond the central scope of this book, namely the flow of two liquids of different densities and viscosities, separated by a mixing zone of small thickness, as compared to other length scales of the formation. The conditions ensuring the existence of a thin transition zone are quite complex. Thus, in the case of two miscible fluids, e.g. fresh and salt waters in coastal aquifers, the heavier fluid has to underlie the lighter one, otherwise the configuration is unstable. Furthermore, the motion has to be such, that the mixing due to dispersion should not result in a thick mixing zone. Such conditions are attained approximately in the case of immobile salt water bodies over which fresh water is permanently flowing. Then, due to the smallness of the coefficient of lateral dispersion (see Chap. 2.10) and to the permanent flushing, the transition zone can be approximated by a sharp interface, similar to the free-surface. In the case of immiscible fluids, e.g. oil and water, the situation is closer to that of a free-surface, for which air and water play the role of the two fluids. Besides the problems discussed in Sect. 2.13.4, enhanced mixing may be caused by fingering, a phenomenon related to the displacement of a viscous fluid by one of lesser viscosity. Still, the approximation of a sharp interface may be quite accurate in particular cases, like the one mentioned above, and constitute the starting point for more refined approximations.

The equations of interface flow are quite easy to write down. Indeed, with subscripts 1 and 2 denoting the two fluids, the water heads are given by $\phi_1 = (p_1/\rho_1 g)+z$ and $\phi_2 = (p_2/\rho_2 g)+z$. Since elastic storage can be safely neglected, each potential satisfies the Laplace equation in its domain of definition. The new boundary conditions are the ones satisfied on the interface $z=\zeta(x,y,t)$, which separates the two fluids. Since it plays the role of a free-surface for each fluid, the first equation, of a kinematical nature, of (3.5.19) is easily generalized as follows

$$n_{ef}\frac{\partial\zeta}{\partial t} = q_{1z} - q_{1x}\frac{\partial\zeta}{\partial x} - q_{1y}\frac{\partial\zeta}{\partial y} = q_{2z} - q_{2x}\frac{\partial\zeta}{\partial x} - q_{2y}\frac{\partial\zeta}{\partial y} \qquad (z=\zeta) \qquad (3.5.20)$$

where n_{ef} stands now for the effective porosity with regard to the interface movement (it is close to the porosity n for miscible fluids) and $z=\zeta(x,y,t)$ is the equation of the interface.

The second relationship of (3.5.19) is replaced by the requirement of pressure continuity

$$p_1 - p_2 = p_c \quad \text{i.e.} \quad \rho_1 g(\phi_1-\zeta) = \rho_2 g(\phi_2-\zeta) + p_c \qquad (z=\zeta) \qquad (3.5.21)$$

where p_c is the capillary pressure, similar to the air-entry value, with $p_c=0$ for miscible fluids. Eqs. (3.5.20,21) can be rewritten in terms of the heads or specific discharges by employing Darcy's law in each fluid $\mathbf{q}_i = -(k/\mu_i)\,\nabla(p_i+\rho_i gz)$ $(i=1,2)$, with k the intrinsic effective permeability of the medium.

Determining the water heads in each domain, as well as the interface equation, with the aid of the three boundary conditions (3.5.20,19), is a formidable mathematical problem, which could be solved analytically only by approximate methods (for an extensive discussion, see Bear, 1972).

A particular, considerably simpler, case is the one of two-dimensional flow in the vertical plane, in which one of fluids, say fluid 2, is under rest, while the flow of fluid 1 is steady. Then $\phi_2\equiv$const and the interface boundary conditions (3.5.20,21) reduce to

$$q_{1z} - q_{1x}\frac{\partial\zeta}{\partial x} = 0 \quad ; \quad \phi_1 = -\frac{\rho_2-\rho_1}{\rho_1}\zeta + \text{const} \qquad \text{for } z=\zeta(x) \qquad (3.5.22)$$

Elimination of $\partial\zeta/\partial x$ leads to an unique equation in $\mathbf{q}$, namely

$$q_{1x}^2 + q_{1z}^2 - K'\,q_{1z} = 0 \qquad (z=\zeta) \qquad (3.5.22)$$

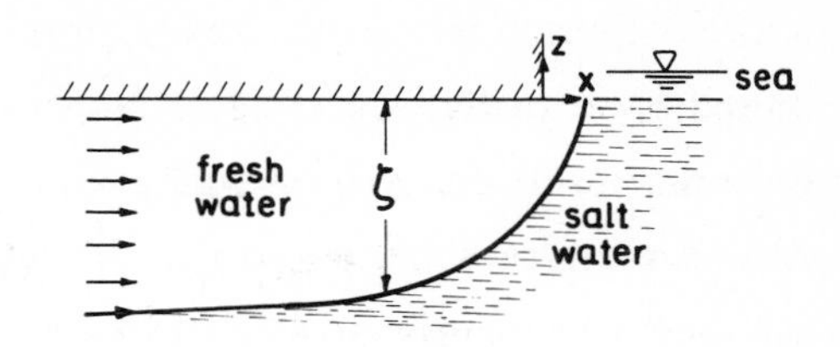

Fig. 3.5.6 Steady, two-dimensional, fresh water flow above a sharp interface and beneath an impervious upper boundary. The salt water is under rest.

where $K' = k(\rho_2-\rho_1)/\mu_1$. This equation is similar to (3.5.12) for a free-surface flow (in absence of recharge, i.e. R=0) and the interface is mapped in the hodograph plane on a circle of radius $K'/2$ centered at $-q_z=-K'/2$. Hence, with other boundaries fixed, steady interface flows can be solved by conformal mapping in a manner similar to that employed to solve free-surface flows. To illustrate the approach let us consider the case of fresh-water flow above a steady interface in a coastal aquifer, the coastline serving now as a seepage face. This flow is represented schematically in Fig. 3.5.6, which is a mirror image of 3.5.3a. For instance, the solution (3.5.15) for the interface profile becomes, with $\psi_w=0$,

$$\zeta^2 = \frac{2Q}{K'}(x_A - x) \quad ; \quad x_A = \frac{Q}{2K'} \tag{3.5.24}$$

It is seen that for a given discharge Q, the depth of the interface beneath the impervious top is much larger than the corresponding height of the free-surface above the bottom, by the ratio $K/K' = \rho_1/(\rho_2-\rho_1)$ and the same is true for the length of the outflow face x_A. This result justifies the approximate use of (3.5.24) for phreatic flow as well, since the water table is much flatter than the interface, and can be approximately replaced by a rigid boundary. We shall discuss approximate solutions of interface flows, for more complex examples, in the next section.

Exercises

3.5.1 Two-dimensional, steady, flow beneath a dam takes place in the vertical, lower half-plane $y<0$. The boundary conditions are : $\phi=H$ for $-\infty<x<-L/2$, $y=0$; $\partial\phi/\partial y=0$ for $-L/2<x<L/2$, $y=0$; $\phi=h$ for $L/2<x<\infty$, $y=0$, with $H\neq h$. Prove that the complex potential for this flow is given by $f= -Kh-(1/2\pi)[-iK(H-h)\,\ell n(Z^2-L^2/4)]$. Depict the flow domain in the potential plane. Compute the uplift on the dam. Derive the equations of streamlines and equipotentials and depict them for $|Z|>>L$. Show that by an appropriate interchange between Φ and Ψ the same solution applies to flow around a sheet-pile of length $L/2$.

3.5.2 Two-dimensional, steady, free-surface flow through a porous dam with vertical faces is caused by a difference of level between the upstream and downstream face. The boundary conditions are : ϕ=H for x=0, 0<y<H ; $\partial\phi/\partial y$=0 for 0<x<L, y=0 ; ϕ=h for x=L, 0<y<h. The upper boundary is the free-surface $y=\eta(x)$, with $\eta(0)$=H and $\eta(L)=\eta_L$ >h. The boundary x=L, $h<y<\eta_L$ is a seepage face. The air-entry value can be neglected.

Prove the well-known exact result for the discharge Q (per unit width), namely $Q = K(H^2-h^2)/2$. Hint : apply Green's formula for two harmonic functions α_1 and α_2, i.e. $\int(\alpha_1\ \partial\alpha_1/\partial n - \alpha_2\ \partial\alpha_1/\partial n)\ ds = 0$ on the domain contour, to the functions α_1=x and $\alpha_2=\Phi$. Extend the result to the axisymmetric flow toward a fully penetrating well of radius r_w in which the level is at z=h, which is pumping from a reservoir at r=R in which ϕ=H.

3.5.3 An artesian, fully penetrating well, of radius $r=r_w$, pumps steadily at constant head ϕ_w=h>D, from a confined aquifer of constant thickness D. At r=R the head is constant, i.e. ϕ=H>D. The aquifer is made up from two horizontal layers of constant thickness and of different effective conductivities, i.e. $0<z<D_1$, $K=K_1$ and $D_1<z<D_2$, $K=K_2$ with $D_1+D_2=D$. Compute the well discharge Q and depict the streamlines and equipotentials.

3.5.4 A confined aquifer of constant thickness D and of areal extent much larger than D, outflows in an estuary in which the head varies periodically due to tidal motions. Assume that the aquifer lies in the domain x<0, $-\infty<y<\infty$, 0<z<D and that steady flow with $q_x=Q/D$, $q_y=q_z=0$ takes place at $x\to-\infty$. The boundary condition at the outflow face x=0 is $\phi=H+a\sin(\omega t)$.

Derive the quasi-steady head $\phi(x,t)$ in the aquifer and the outflow discharge into the estuary. At what distance from x=0 is the head variation smaller than 0.01a ?

3.5.5 Depict in the hodograph plane the boundaries for flow above a steady interface and beneath a water-table of constant recharge.

3.6 SOLUTIONS OF THE MEAN FLOW EQUATIONS (APPROXIMATE METHODS)

In this chapter we shall examine a few approximate techniques which have been developed in the past in order to solve the equations satisfied by the average head. Although the availability of sophisticated codes has made these methods less indispensable than in the past, they are still quite useful, especially for treating free-surface or three-dimensional flows, which pose difficult numerical problems. They are also useful for deriving quick results and for acquiring a better understanding of the flow features. In the limited present context, we shall merely illustrate the various methods, rather than deriving them in complete generality.

3.6.1 The method of singularities

This method has a long and distinguished tradition in Fluid Mechanics (see, e.g. Van Dyke, 1964) and we shall illustrate it first by the simple example of a fully penetrating well pumping a confined aquifer (Fig. 3.5 and Eqs. 3.5.17). First, the problem is recast in the following dimensionless variables $\phi'=\phi KD/Q$, $r'=r/\sqrt{\kappa t}$, $r'_w=r_w/D$. Due to the linearity of Eqs. 3.5.17, we may write ϕ' in the general form $\phi'\equiv \mathrm{funct}(r',r'_w)$. Let us assume now that $r'_w<<1$, as it is generally the case, and expand ϕ' (see the similar problem of a thin or slender body in Van Dyke, 1964) as follows

$$\phi'(r',r'_w) = \phi'_0(r') + \delta_1(r'_w)\,\phi'_1(r') + \delta_2(r'_w)\,\phi'_2(r') + \dots \tag{3.6.1}$$

where the "gauge functions" δ_1, δ_2,... form an asymptotic sequence, i.e. $\lim \delta_{j+1}/\delta_j = 0$ $(j=1,2,\dots)$ for $(r'_w\to 0)$. These functions are determined by substituting (3.6.1) in Eqs. 3.5.17 and equating terms of the same order in the small parameter r'_w. The expression (3.6.1) is an asymptotic, small perturbation, expansion of ϕ'. Due to the linearity of the equation of flow (3.5.17), each ϕ'_j satisfies it. The boundary conditions satisfied by the leading order term ϕ_0 is obtained for $r_w=0$ in (3.5.17) as follows

$$\lim_{r\to 0}\left(-2\pi KrD\,\frac{\partial\phi_0}{\partial r}\right) = Q \quad \text{i.e.} \quad \lim_{r'\to 0}\left(r'\,\frac{\partial\phi'_0}{\partial r'}\right) = -\frac{1}{2\pi}\ ;$$
$$\phi_0 = \phi_{in} \qquad (t\to 0,\ \text{i.e. } r'\to\infty) \tag{3.6.2}$$

We shall derive ϕ_0 solely (for the higher order terms, see Dagan, 1967c). The equivalent problem in heat transfer is one of heat flow to a sink line of constant flux (Carslaw and Jaeger, 1965) and the solution of $\phi'=\phi'_0$ which satisfies the heat conduction equation and (3.6.2) is given by

$$\phi'_{in} - \phi' = \frac{1}{4\pi}\,\mathrm{Ei}\left(-\frac{r'^2}{4}\right) \quad \text{i.e.} \quad \phi_{in} - \phi = \frac{Q}{4\pi KD}\,\mathrm{Ei}\left(-\frac{r^2 s}{4Kt}\right) \tag{3.6.3}$$

where Ei is the exponential integral function (see, e.g. Abramowitz and Stegun, 1965). This is also known as Theis (1935) solution, who has derived it in the groundwater flow context. For $r'\to 0$, by using the expansion of Ei for small arguments, one gets $\phi'_{in}-\phi' \to (-1/2\pi)\,[\ell n(r'/2)+E+O(r'^2)]$, an approximation which is quite useful in applications (E is the Euler number). The approximate solution (3.6.3) is compared with the exact one for the extreme value $r=r_w$ in Fig. 3.5.5 and it is seen that it approaches it closely for $r_w^2 s/Kt<0.1$, and, furthermore, both have the same small r' expansion up to terms $O(r^2s/Kt)$. Hence, the Theis solution is applicable to most cases of interest, which explains its widespread use.

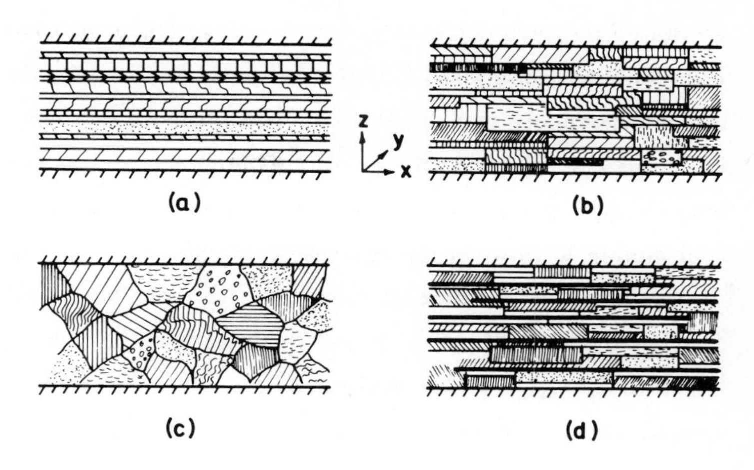

Fig. 3.7.1 Schematic representation of a few types of heterogeneous structures at the local scale: (a) stratified formation with horizontal continuous bedding, (b) anisotropic formation with horizontal logconductivity correlation scale larger than the vertical one, (c) isotropic three-dimensional structure and (d) thin impervious lenses separating heterogeneous layers of the matrix; the lenses horizontal extent is much larger than their vertical separation distance and than the horizontal correlation scale of the matrix.

assessment of the impact of parameter uncertainty is a relatively simple matter, which is summarized in Sect. 3.7.7 along the lines of Sect. 3.2.5.

The preceding three sections have dealt with the derivation of the expected values of the flow variables and we shall assume that in any given problem this task has been completed. This implies that the effective properties are known, either by analysis of a set of measured point values of Y and s, or by some tests which determine their space average. Here, we pursue the derivation of the higher statistical moments, and mainly the second-order ones, although a complete characterization of ϕ requires the evaluation of all moments.

Following the discussion of Chap. 3.2, we assume that K is lognormal and stationary, completely characterized by its mean m_Y and by its covariance $C_Y(\mathbf{r})$. The types of heterogeneities, at the local scale, we shall consider are illustrated by Figs. 3.7.1, and they are briefly discussed first. Thus, the first idealized case is of a perfectly stratified aquifer (Fig. 3.7.1a), which may be regarded as a limiting case of some sedimentary, lacustrine, geological structures. Its simple covariance is given by (3.2.11) and it may be useful for describing the correlation between ϕ values at points at not too large horizontal distances. It has the great theoretical advantage of lending itself to simple solutions of flow problems. As already mentioned in Sect. 3.2.2, a more realistic model is the one of Fig. 3.7.1b (see also Fig. 3.2.2), of stratification which is not continuous in the horizontal direction and is encapsuled by the anisotropic covariances of Sect. 3.2.2. We shall assume C_Y to be "scaled" (3.2.23) and axisymmetric, with isotropy in the horizontal plane. A third case, of lesser occurrence, is the isotropic one (Fig. 3.7.1c), for which C_Y is a function of $r=|\mathbf{r}|$ and one scale I_Y. Finally, the formation of the type depicted in Fig. 3.7.1d, is characterized by the existence of impervious lenses of small thickness and large horizontal extent. As a result, the vertically averaged flow between such lenses is essentialy two-dimensional. The existence of such an effect has been suggested by the results of tracer experiments (see Part 4), and the appropriate tool for analyzing the flow field in a layer is the isotropic two-dimensional $C_Y(r_h)$. Besides, the two-dimensional results are of wide applicability to flows at the regional scale (Part 5).

Again, along the lines of Chap. 3.2 it is reminded that the main scales of concern are the Y correlation scales (Sect. 3.2.1), and those representing the flow domain. The latter are the formation thickness $D \gg I_{Yv}$, and a horizontal scale of the order of a few D. Finally, the measurement or computational scale ℓ lies between the two.

With these definitions in mind, our aim is to derive the head second-order moments $C_{Y\phi}(\mathbf{x},\mathbf{x}')$ and $C_\phi(\mathbf{x},\mathbf{x}')$ and the associated specific discharge covariances $q_{ij}=\langle q'_i(\mathbf{x})\, q'_j(\mathbf{x}')\rangle$, where q' is the residual.

From a practical standpoint, one is merely interested in computing the moments of the space averages $\bar{\phi}$ or $\bar{\mathbf{q}}$, since these are generally the ones accessible by measurement, or of concern in predictions. The salient question is whether their variances are sufficiently small to warrant their neglect, i.e. to regard $\bar{\phi}$ or $\bar{\mathbf{q}}$ as macroscopic, deterministic, variables, in line with the traditional approach to groundwater flow. This question is also addressed in the sequel.

3.7.2 Steady, uniform in the average, flow in unbounded formations. First-order approximation

We start, like in Chap. 3.4 on effective conductivity, with the simplest case, of a steady flow of constant average head gradient $\nabla\langle\phi\rangle=-\mathbf{J}$ in an unbounded domain Ω. In the following sections we shall relax some of these restrictions, to examine the influence of boundaries and of large σ_Y^2 upon the head covariances. Hence, the aim is now to derive the first-order moments $C_{Y\phi}^{(1)}(\mathbf{x}_1,\mathbf{x}_2)$ and $C_\phi^{(1)}(\mathbf{x}_1,\mathbf{x}_2)$, as defined in Sect. 3.3.2(ii). The starting point maybe either the solution $\phi^{(1)}$ (3.4.11), expressed with the aid of the Green function, or the differential equations (3.3.12,13). The first has the advantage of incorporating the effects of boundaries, but the Green function loses its simplicity for bounded domains. Alternatively, we may seek the FT (Fourier transforms) of the covariances, which are easier to evaluate (this line has been pursued in the works of Gelhar, Gutjahr and their coworkers, e.g. Bakr et al, 1978, Gutjahr et al, 1978, Gelhar and Axness, 1983 and Gelhar, 1986). We shall use mainly the differential equations approach, along the lines of Dagan (1982a, 1985a), but also derive in each case the Fourier transforms, similarly to the developments of Sect. 3.2.4 for C_Y. It is reminded that due to the linear dependence of $\phi^{(1)}$ upon Y', the first order solution is normal, and completely characterized by its second-order moments.

The starting point is provided by Eqs. (3.3.12) and (3.3.13), which may be rewritten, after assuming stationarity, as follows

$$\nabla^2 C^{(1)}_{Y\phi}(\mathbf{r}) = -\mathbf{J}.\nabla C_Y(\mathbf{r}) \;\; ; \;\; \nabla^4 C^{(1)}_{\phi}(\mathbf{r}) = -\sum_j \sum_\ell J_j J_\ell \frac{\partial^2 C_Y(\mathbf{r})}{\partial r_j \partial r_\ell} \;\; ; \;\; \mathbf{r} = \mathbf{x}_1 - \mathbf{x}_2 \tag{3.7.1}$$

To simplify notations, we shall suppress in this section the upper index, being understood that all moments are of first-order in σ_Y^2. The covariances corresponding to the cases of Fig. 3.7.1 are derived, subsequently, by integrating in (3.7.1).

In the case the stratified formations, C_Y is a function of z solely, and the same is true for ϕ. Hence, covariances are different from zero only for flow normal to layering, in the vertical direction. As we have seen in the preceding sections such flows can be caused by recharge or by other three-dimensional effects. Generally, the vertical gradient is small compared to the horizontal one, and is localized in part of the formation depth, in variance with the assumption that it is applied to an unbounded domain. It is still of theoretical interest to examine this case, which has received attention in the literature, due to its simplicity. Hence, Eq. (3.7.1) and its solution for $C_{Y\phi}(r_z)$ becomes

$$\frac{d^2 C_{Y\phi}}{dr_z^{\,2}} = -J_z \frac{dC_Y}{dr_z} \quad \text{i.e.} \quad C_{Y\phi} = -J_z \int_0^{r_z} C_Y(z)\, dz \tag{3.7.2}$$

The two constants of integration have been determined by regarding (3.7.2) as the solution for a bounded domain whose extent tends to infinity. To illustrate the result, let us adopt the exponential C_Y (3.2.14), leading in (3.7.2) to

$$C_{Y\phi} = -J_z I_{Yv} \sigma_Y^2 [1-2H(r_z)] [1 - \exp(-|r_z'|)] \;\; ; \;\; r_z' = \frac{r_z}{I_{Yv}} \tag{3.7.3}$$

where H is the Heaviside step function. $C_{Y\phi}$ (3.7.3) is represented graphically in Fig. 3.7.3 for $r_z>0$. To understand the nature of the solution, as well as that of the following ones, we have represented in Fig. 3.7.2a a model of a stratified formation in which blocks of different conductivities are inserted in series (this is the model adopted by Freeze, 1975). The flow is driven by a given head difference between the ends and the specific discharge is constant in any realization, by the continuity equation. In the first-order, small σ_Y^2, approximation, the head residual at x_2 is the sum of the residuals for a flow in which each block is implanted in a homogeneous matrix of conductivity equal to the effective one (Fig. 3.7.2b). Thus, in order to evaluate $C_{Y\phi}(x_1,x_2)$, we have to solve the flow problem for a block implanted at x_1, to compute ϕ' at x_2, and to average the product $Y'(x_1)\,\phi'(x_2)$ for various realizations making up the ensemble. The head distribution for such a con-

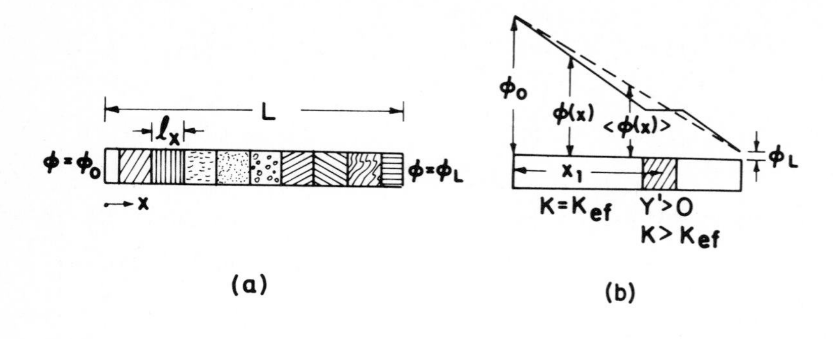

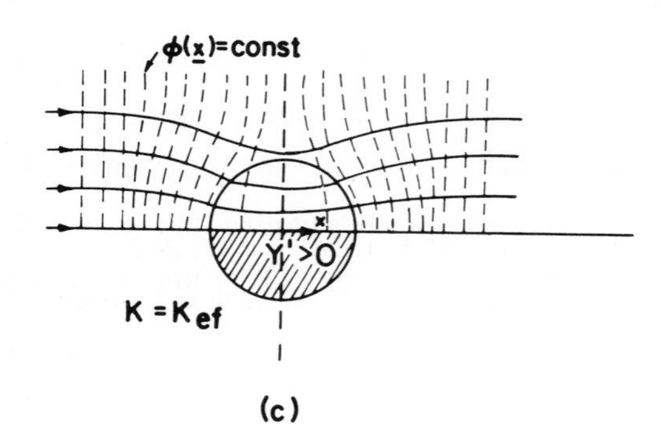

Fig. 3.7.2 Explanatory sketch for the behavior of the logconductivity-head cross-covariance $C_{Y\phi}$: (a) one-dimensional flow in a stratified formation made up from blocks set in series, (b) the head distribution for a block of $K>K_{ef}$ inserted in a matrix with $K=K_{ef}$; the head residual is $\phi'=\phi-\langle\phi\rangle$, and (c) streamlines and lines of constant head (represented only in the upper half-space) for a spherical or cylindrical inclusion of $K>K_{ef}$ in a matrix with $K=K_{ef}$. The flow at infinity is uniform. The head distribution on the x axis is similar to (b).

figuration is depicted schematically in Fig. 3.7.2b, for $Y'(x_1)>0$. Its shape is a straightforward consequence of Darcy's law, the head gradient in a more conductive block of $Y'>0$, being smaller than the average. The examination of Fig. 3.7.2b leads to a few conclusions, explaining the result in Eq. 3.7.3 : (i) it is seen that $C_{Y\phi}$ is indeed antisymmetrical, being negative for $r=x_1-x_2>0$; (ii) at the limit of an unbounded formation, i.e. $L/\ell_Y\to\infty$, $C_{Y\phi}$ tends to a constant value and (iii) this is not true for the actual case of finite dimensions, but for $r_z<<L$ and for points not too close to the boundary, the result is nevertheless valid.

With this picture in mind we proceed now with the computation of the head covariance C_ϕ, by integrating the one-dimensional version of (3.7.1), i.e. $d^4C_\phi/dr_z^4 = -J_z^2\, d^2C_Y/dr_z^2$. However, we face immediately a difficulty in trying to determine the four constants of integration by conditions of symmetry, smoothness at the origin and vanishing of C_ϕ at large r_z. We have discussed this type of difficulty at length in Chap. 1.7, in which we have shown that a function like $\int C_Y dx$, which results from the integration of an RSF of *finite integral scale*, has a variance which tends to infinity

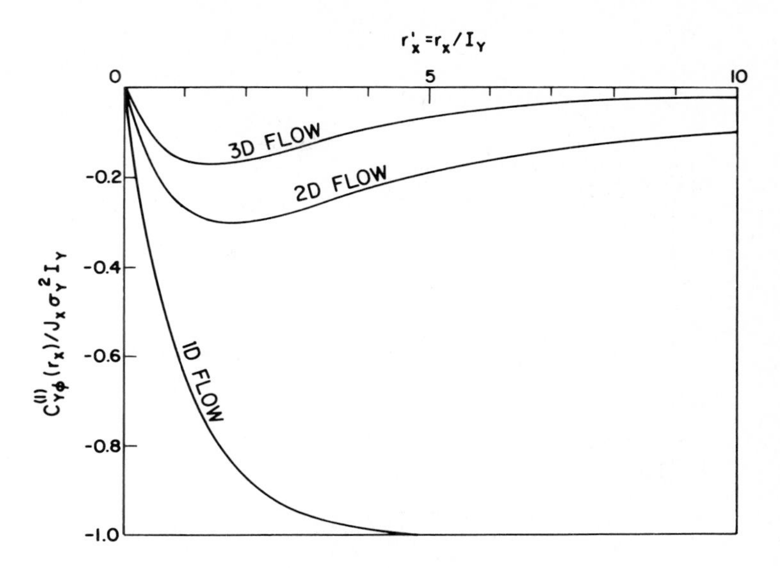

Fig. 3.7.3 The dimensionless, first-order, cross-covariance $C^{(1)}_{Y\phi}(\mathbf{x}_1,\mathbf{x}_2)$ as a function of r_x = $\mathbf{J}.(\mathbf{x}_1-\mathbf{x}_2)/J$. Uniform average flow in an unbounded formation and exponential C_Y. One-dimensional (Eq. 3.7.3), two-dimensional (Eq. 3.7.8) and three-dimensional (Eq. 3.7.8), flows, respectively.

for $r_z \to \infty$. This result can be easily understood by examining Fig. 3.7.2b and remembering that σ_Y^2 is the outcome of summation of ϕ'^2 for all realizations. Since this result is unacceptable from a physical point of view, there are a few ways to circumvent it : (i) to solve the problem for the actual case of a bounded domain, like in Fig. 3.7.2a (Freeze, 1975). This approach has the disadvantage that it complicates the solution, making it nonstationary, by its dependence on the distance from the boundary. Furthermore, the variance at a fixed point will increase monotonously with the ratio L/I_Y; (ii) to assume somewhat artificially that C_Y is a "hole" covariance, of zero integral scale. We do not pursue this line for reasons given in Chap. 3.2 and (iii) finally, to derive the head variogram, which is stationary for an RSF of stationary increments (Chap. 1.7). This approach, employed by Gutjahr and Gelhar (1981) for one-dimensional flow, has the definite advantage that it leads to a stationary moment, which encapsules the correlation structure of ϕ for two points which are not two close to the boundary. Along these lines, we obtain

$$\frac{d^4\Gamma_\phi}{dr_z^4} = J_z^2 \frac{d^2C_Y}{dr_z^2} \quad \text{i.e.} \; \Gamma_\phi = J_z^2 \sigma_Y^2 \int_0^{r_z} dz' \int_0^{z'} \rho_Y(z'')dz'' \tag{3.7.4}$$

To illustrate the result, Γ_ϕ (3.7.4) is calculated for the exponential C_Y (3.2.14), to obtain

$$\Gamma_\phi = J_z^2 I_{Yv}^2 \sigma_Y^2 [|r_z'| + \exp(-|r_z'|) - 1] \; ; \quad r_z' = r_z/I_{Yv} \tag{3.7.5}$$

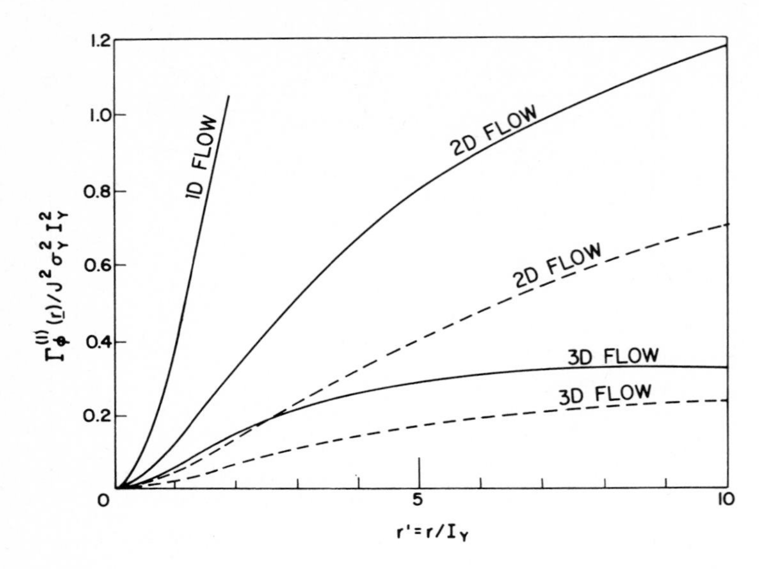

Fig. 3.7.4 Same as 3.7.3 for the longitudinal $\Gamma_{\phi L}$ (full line) and transverse $\Gamma_{\phi T}$ (dashed line) head variograms. One-dimensional flow (Eq. 3.7.5), two-dimensional (Eq. 3.7.11) and three-dimensional (Eq. 3.7.12).

which is represented graphically in Fig. 3.7.4. For large intervals, the variogram grows linearly with the distance, a result which can be interpreted along the lines of the previous discussion about $C_{Y\phi}$.

As we have already mentioned in Chap. 3.2, the flow in a stratified formation is rather of theoretical interest, and we turn now to the more realistic two- or three-dimensional structures depicted in Fig. 3.7.1.

First, we shall derive the head covariances for isotropic, two- or three-dimensional structures, i.e. for $C_Y(r)$, $r=|\mathbf{r}|$. Starting with $C_{Y\phi}$, Eq. (3.7.1) can be satisfied by using the auxiliary function P, as follows

$$\nabla^2 P(r) = -C_Y(r) \quad ; \quad C_{Y\phi}(\mathbf{r}) = \mathbf{J}.\nabla P = \frac{\mathbf{J}.\mathbf{r}}{r}\frac{dP}{dr} \tag{3.7.6}$$

Hence, one has to solve an ordinary differential equation for P, like in the one-dimensional flow. The solution is obtained by writing the ∇^2 operator in polar (2D) or spherical coordinates (3D), respectively, and by subsequent integration. Thus

$$\frac{1}{r}\frac{d}{dr}\left(r\frac{dP}{dr}\right) = -\sigma_Y^2\rho_Y \quad \text{i.e.} \quad \frac{dP}{dr} = -\frac{\sigma_Y^2}{r}\int_0^r r''\rho_Y(r'')\,dr'' \qquad (2D)$$

$$\frac{1}{r^2}\frac{d}{dr}\left(r^2\frac{dP}{dr}\right) = -\sigma_Y^2\rho_Y \quad \text{i.e.} \quad \frac{dP}{dr} = -\frac{\sigma_Y^2}{r^2}\int_0^r r''^2\rho_Y(r'')\,dr'' \qquad (3D) \tag{3.7.7}$$

It is seen that $C_{Y\phi}$ (3.7.6) have the general form dictated by the linear dependence on **J**, which can be derived from dimensional considerations and is antisymmetrical, like (3.7.3). Furthermore, $C_{Y\phi}$ is zero for two points normal to the mean gradient **J**, and the correlation is maximal for **r** parallel to **J**. To illustrate the results we substitute in (3.7.7) and (3.7.6) the exponential covariance (3.2.14), to obtain

$$C_{Y\phi} = I_Y \sigma_Y^2 \frac{\mathbf{J}.\mathbf{r}'}{r'^2} [(1+r')\exp(-r') - 1] \quad ; \quad (\text{2D}, \; r = \sqrt{r_x^2 + r_y^2})$$

$$\mathbf{r}' = \mathbf{r}/I_Y \quad ; \quad r' = |\mathbf{r}'| \tag{3.7.8}$$

$$C_{Y\phi} = I_Y \sigma_Y^2 \frac{\mathbf{J}.\mathbf{r}'}{r'^3} [\exp(-r')\,(r'^2+2r'+2) - 2] \quad ; \quad (\text{3D}, \; r = \sqrt{r_x^2 + r_y^2 + r_z^2})$$

These covariances are represented in Fig. 3.7.3, together with (3.7.3), for **J** parallel to **r**. The behaviour of $C_{Y\phi}$ in two- and three-dimensional flows, can be analyzed with the aid of models similar to that of Fig. 3.7.2, this time a circular or spherical block with $Y'>0$ being embedded in a homogeneous matrix (Fig. 3.7.2c). Obviously, the ϕ' field is now two- or three-dimensional, and antisymmetrical with respect to a plane normal to **J**. Unlike the stratified formation, $C_{Y\phi}$ decays at infinity, which can be explained by the fact that the disturbance caused by the block is more localized in two dimensions than in one, and even more so for three dimensions. Putting it in other words, streamlines may circumvent blocks of low conductivity, whereas in one-dimensional flow normal to stratification, any fluid particle has to pass through each layer. These results should serve as a warning against straightforward extrapolation from lower to higher dimensions.

We consider now the head auto-covariances for isotropic formations. Similarly to (3.7.6), we represent them with the aid of an auxiliary function Q in (3.7.1) as follows

$$\nabla^4 Q(r) = -C_Y(r) \quad ; \quad C_\phi(\mathbf{r}) = \sum_j \sum_\ell J_j J_\ell \frac{\partial^2 Q}{\partial r_j \partial r_\ell} \tag{3.7.9}$$

By differentiating Q, we may rewrite C_ϕ, or Γ_ϕ, in the invariant form (see Chap. 1.7, Eq. 1.7.15)

$$C_\phi(\mathbf{r}) = C_{\phi T}(r) + \frac{(\mathbf{J}.\mathbf{r})^2}{J^2 r^2} [\, C_{\phi L}(r) - C_{\phi T}(r) \,] \quad ; \quad C_{\phi T} = J^2 \frac{1}{r} \frac{dQ}{dr} \quad ; \quad C_{\phi L} = J^2 \frac{d^2 Q}{dr^2} \tag{3.7.10}$$

where $C_{\phi T}$ and $C_{\phi L}$ are the transverse (**r** normal to **J**) and longitudinal (**r** parallel to **J**) covariances, respectively. Since Q is function of r, Eq. (3.7.9) reduces to an ordinary differential equation for Q, by the repetitive application of the operator ∇^2, precisely like in (3.7.7). This equation can be solved

This expression will be explored at a large extent in Part 4, dealing with transport at the local scale. At any rate, it is seen that the specific discharge covariances are finite in all cases, since they result from the differentiation of C_ϕ. To illustrate the application of (3.7.26) we shall give here the final expressions of specific discharge variances for isotropic formations of exponential C_Y (3.2.14), namely

$$\langle q_1'^2 \rangle = \frac{3}{8} \sigma_Y^2 K_G^2 J_1^2 \;;\; \langle q_2'^2 \rangle = \frac{1}{8} \sigma_Y^2 K_G^2 J_1^2 \quad (2D)$$

(3.7.27)

$$\langle q_1'^2 \rangle = \frac{8}{15} \sigma_Y^2 K_G^2 J_1^2 \quad ; \quad \langle q_2'^2 \rangle = \langle q_3'^2 \rangle = \frac{1}{15} \sigma_Y^2 K_G^2 J_1^2 \quad (3D)$$

It is seen that there is a reduction by about half of the longitudinal variance $\langle q_1'^2 \rangle$ in two- and three-dimensional flows as compared to the one pertaining to the stratified formation. Furthermore, although the formation is isotropic, there is a pronounced anisotropy of the specific discharge covariance, with a much weaker variability of the velocity component normal to the mean flow than the one parallel to it. In the stratified case, of course, the transverse component is identically zero. Hence, the stratified medium provides bounds of the variances for the more complex heterogeneous structures of Fig. 3.7.1b,c,d , bounds equal to zero and to (3.7.25).

It is quite easy to calculate the linear integral scale of the specific discharge, in the first-order approximation, by integrating in (3.7.26) in the x_1 direction between zero and infinity. The simple result is that similarly to the stratified formation, we get for the longitudinal component $I_q = I_Y$ for both two- and three-dimensional flows. This is derived from the first term in (3.7.26), the other ones not contributing to I_q. In contrast, for the lateral components, $I_{q_2} = I_{q_3} = 0$, i.e. $q_{22}^{(1)}(x_1,0,0)$ is a "hole" covariance (this result can be interpreted with the aid of the flow model of Fig. 3.7.2c). We shall discuss again these findings in Part 4.

3.7.6 The effect of space averaging upon specific discharge variance

As we have already mentioned in Sect. 3.2.4, in most applications one is interested in the space averages of the flow variables. This need may be related to measurements, which are carried with the aid of devices which average over a finite volume. If this volume has a length scale smaller than I_Y, it provides the point value of the variable of interest. Then, due to spatial variability the measurements are subjected to uncertainty, which is encapsuled in the statistical moments discussed above. If the averaging volume is large at the I_Y scale, a reduction of variance occurs, discussed at length in Sect. 3.2.4 for Y itself. It is worthwhile to emphasize that determining the extent and the nature of the averaging volume associated with a particular measuring device is not a simple matter (see, for instance, a discussion in Cushman, 1984). The simplest, and most common, measuring device is a well, and we shall limit the discussion here to its averaging effect over a vertical segment (see Sect. 3.2.4).

An additional type of space averaging is related to computational needs, namely to the partition of the formation into elements over which the properties and the flow variables are averaged. Again, the smoothing effect of space averaging depends on the ratio between the magnitude of the elements, generally of a three-dimensional nature, and the heterogeneity scale.

We start the analysis of the effect of space averaging with the case of wells, represented as vertical segments of length ℓ, for the types of formations of Fig. 3.7.1. The simplest case is of a stratified formation (Fig. 3.7.1a), with the well axis normal to the bedding. In the case of horizontal flow, the head is deterministic and constant over the vertical, and is measured by the well with certainty, irrespective of its length. In the case of flow with a vertical component, the head residual varies irregularly in the formation. However, the well itself alters the nature of the flow, since the water head is constant in it. The whole issue is not of a practical interest, however, due to the smallness of the average vertical head gradient J_z which is encountered in applications. To assess the head difference prevailing at the extremities of a segment of length ℓ, we employ the variogram (3.7.20) represented in Fig. 3.7.5, by using the definition $\langle[\phi'(z+\ell)-\phi'(z)]^2\rangle = 2\,\Gamma_\phi(\ell)$. Furthermore, if $\ell >> \ell_Y$, where $\ell_Y = 2\,I_Y$ is the average layer thickness, Γ_ϕ (3.7.20) may be approximated by its linear part, i.e. $\Gamma_\phi(\ell) \simeq J_z^2 \ell_Y \ell\,[\exp(\sigma_Y^2)-1-\sigma_Y^2]/\sigma_Y^2$. A quick check shows that due to the smallness of J_z, say of order 0.001, the average fluctuation of the head difference is of the order of a few millimeters for wells whose length is in the range of meters. Hence, the well head may be assumed to represent the average one at its center. Although a similar analysis for the types of formations of Figs. 3.7.1 is more complex, the results are bound to be similar, since Γ_ϕ is smaller than (3.7.20), for same ℓ and J. The matter is somewhat different in the last type of formation, of Fig. 3.7.1d, for horizontal flow. Indeed, in each layer, the flow may be assumed to be two-dimensional in the plane, with $\overline{Y}$ averaged over an average thickness of say I_Y, and with covariance given by Eq. (3.2.40). If there is no correlation between the conductivities of two adjacent layers, the mean square of the head difference between them is equal to $2\sigma_\phi^2$, and since from a theoretical standpoint the latter may be quite large in two-dimensional flow, the same is true for the interlayer head drop. However, if we take into account the fact that the continuity of the thin impervious partitions may cease at a distance $L >> I_Y$, then by (3.7.11) $2\sigma_\phi^2 \simeq 4\Gamma_\phi(L) \simeq 2J^2 I_Y^2 \sigma_Y^2 \ln(L/I_Y)$, for $\sigma_Y^2 < 1$. Due to the logarithmic dependence on L, the magnitude of $2\sigma_\phi^2$ is still moderate, of the order of centimeters at most. If an observation well pierces N such layers and if it is assumed that the well head is the average of those prevailing in different layers, then its variance is approximately equal to σ_ϕ^2/N. Thus, the variance is further reduced, and it may be assumed to represent the average across the layers.

Unlike the head, the specific discharge is seldom measured directly for natural flows. The standard method is to carry out a pumping test (see Chap. 3,5), creating a flow which is nonuniform, rapidly varying, in the average. The difficulties encountered in the analysis of flow in a heterogeneous formation under these conditions have been already discussed in Sect. 3.4.5. A distinguished case, lending itself to a simple and exact solution, is the one of a fully penetrating well in a stratified formation (see Fig. 3.5.1). Since for a given I_{Yv} and σ_Y^2, an upper bound for the specific discharge variance is achieved in this case, its analysis is of interest for other types of formations as well. Thus, with the configuration of Fig. 3.5.1, and with the well axis perpendicular to the bedding, the head gradient is deterministic and given by $\partial\phi/\partial r = -Q_w/2\pi K_A \ell r$, where $K_{ef}=\langle K\rangle=K_A$ in this case, Q_w is the well discharge and ℓ is its length. By Darcy's law the mean radial component of the specific discharge is given by $\langle q_r\rangle = -Q_w/2\pi\ell r$, whereas the residual is $q_r'=-K'Q_w/2\pi\ell r$. For K lognormal, the covariance has, therefore, the expression $q_{rr}(z,z')/\langle q_r\rangle^2 = \exp[C_Y(z,z')]-1$, exactly like in the case of uniform flow (3.7.25). The variance of the space averaged specific discharge $\bar{q}_r = (1/\ell)\int q_r(z)dz$ can be calculated by the same procedure as Γ_ϕ (3.7.18). For the linear Y covariance the result is, therefore, given by

$$\sigma_{\bar{q}}^2/(\langle q\rangle^2) = \frac{2}{\ell^2}\int_0^r (r-r'')\{\exp[C_Y(r'')]-r^2\}\, dr'' \simeq 2\,\frac{\ell_Y}{\ell}\,\frac{\exp(\sigma_Y^2)-1-\sigma_Y^2}{\sigma_Y^2} \qquad \text{for } \frac{\ell}{\ell_Y}>>1 \quad (3.7.28)$$

To grasp the dependence of the ratio between the well length ℓ and the layers correlation scale $\ell_Y=2I_Y$ upon σ_Y^2, we have represented it in Fig. 3.7.8 for the value $\sigma_{\bar{q}}^2/\langle q\rangle^2=0.01$, i.e. for a value for which $\bar{q}$ is practically deterministic. The striking feature of Fig. 3.7.8 is the rapid growth of the required well length with σ_Y^2. At any rate, for $\sigma_Y^2 \leq 1$ the well length must be of the order of a hundred correlation scales. This length is many times achieved in practice by pumping wells, but in some tests it may be smaller. On the other hand, in the case of discontinuous stratification (Fig. 3.7.1d), the variance is somewhat reduced, but not too much if the well radius is small compared to I_{Yh}.

3.7.7 The effect of parameters estimation errors

We have discussed in Sect. 3.2.5 the effect of the estimation errors of the parameters $\boldsymbol{\theta}$ characterizing the statistical moments of the random space function Y upon any functional W of Y. It was shown that estimation errors manifest in an additional term showing up in the variance of W. These results apply directly to any of the flow quantities like the head ϕ or the specific discharge $\mathbf{q}$, whose variances have to be supplemented by terms like (3.2.44).

An additional source of uncertainty is, however, associated with other flow parameters, which were assumed to be constant, like the components of the head gradient $\mathbf{J}$, the effective porosity n or the discharge of a well Q_w. These parameters are generally determined by measurements and are affected by errors and they can be estimated only. For simplicity of notation we shall denote such

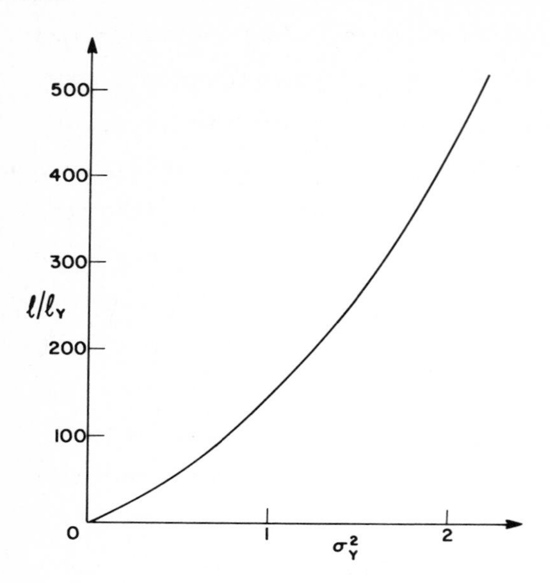

Fig. 3.7.8 The ratio between the length ℓ of a fully penetrating well and the logconductivity correlation scale ℓ_Y required in order to render $\sigma_{\bar{q}}^2/\langle q\rangle^2=0.01$. Steady flow through a stratified formation of type depicted in Fig. 3.7.1a.

parameters by the vector $\boldsymbol{\theta}$ as well, by just adding them to the vector of moments of Y. Statistical analysis of data can provide us with estimates $\tilde{\boldsymbol{\theta}}$ of $\boldsymbol{\theta}$, as well as with the variance-covariance matrix of the estimates. Thus, in the case of a sufficiently large number of independent measurements of a parameter the estimate is equal to the arithmetic mean, whereas the variance of the estimate is equal to the variance of the sample divided by the number of measurements. Obviously, as the number of measurements increases or their quality improves, the estimate approaches the "true" value.

The impact of the errors of estimation of various parameters upon the statistical moments of the flow variables can be evaluated along the lines of Sect. 3.2.5. In particular Eq. (3.2.42), which states that the estimate of a flow variable depending on $\boldsymbol{\theta}$ is equal to its expected value in which $\tilde{\boldsymbol{\theta}}$ is substituted in the result, applies. Similarly, the variance is given by Eq. (3.2.44), i.e. the variance in which $\tilde{\boldsymbol{\theta}}$ are substituted, has to be supplemented by the variances of the estimate. To illustrate this point we shall rewrite (3.2.44), but this time for the covariance of a generic flow variable $W(\mathbf{x}|\boldsymbol{\theta})$, in which the dependence upon $\mathbf{x}$ results from that of W on the random space $Y(\mathbf{x})$. Thus,

$$C_W(\mathbf{x}_1,\mathbf{x}_2) = \langle[W(\mathbf{x}_1)-\langle W(\mathbf{x}_1)\rangle][W(\mathbf{x}_2)-\langle W(\mathbf{x}_2)]\rangle =$$

$$= \langle[W(\mathbf{x}_1-\tilde{W}(\mathbf{x}_1)\rangle][W(\mathbf{x}_2)-\tilde{W}(\mathbf{x}_2)]\rangle + \langle[\tilde{W}(\mathbf{x}_1)-\langle W(\mathbf{x}_1)\rangle][\tilde{W}(\mathbf{x}_2)-\langle W(\mathbf{x}_2)\rangle]\rangle =$$

$$= C_W(\mathbf{x}_1,\mathbf{x}_2|\tilde{\boldsymbol{\theta}}) + \langle[\tilde{W}(\mathbf{x}_1)-\langle W(\mathbf{x}_1)\rangle][\tilde{W}(\mathbf{x}_2)-\langle W(\mathbf{x}_2)\rangle]\rangle \tag{3.7.30}$$

where it is recalled that $\tilde{W}(\mathbf{x}) = \langle W(\mathbf{x}|\tilde{\boldsymbol{\theta}})\rangle$ is the estimate of W, obtained by substituting the estimate $\tilde{\boldsymbol{\theta}}$ for $\boldsymbol{\theta}$ in the conditional mean of W for fixed $\boldsymbol{\theta}$.

This relationship simplifies considerably if we adopt a first-order Taylor expansion in the residuals of $\boldsymbol{\theta}$, valid for small coefficients of variation of the estimates. Then, similarly to (3.2.45), the last term of (3.7.30) becomes

$$\langle[\tilde{W}(\mathbf{x}_1)-\langle W(\mathbf{x}_1)\rangle][\tilde{W}(\mathbf{x}_2)-\langle W(\mathbf{x}_2)\rangle]\rangle \simeq \sum_i\sum_j \frac{\partial\tilde{W}(\mathbf{x}_1)}{\partial\tilde{\theta}_i}\,\frac{\partial\tilde{W}(\mathbf{x}_2)}{\partial\tilde{\theta}_j}\,\Theta_{ij} \qquad (3.7.31)$$

rendering it in terms of the sensitivity coefficients $\partial\tilde{W}/\partial\tilde{\theta}_i$ and of the estimates variance-covariance matrix Θ_{ij}.

We shall evaluate effectively the impact of these errors upon the head field in regional flows in Chap. 5.5. The main point to remember, however, is that the second-order moments of the flow variables, and particularly the variance, determined in the preceding chapters for fixed parameters, have to be supplemented by additional terms related to the parameters errors. Furthermore, space averaging does not reduce the uncertainty stemming from these terms, which are small if the estimates are accurate.

Exercises

In the first three exercises the following assumptions are adopted : the average flow is uniform and a first-order approximation in σ_Y^2 is employed.

3.7.1 Derive the cross-covariance $C_{Y\phi}$ and the variogram Γ_ϕ for flow in an unbounded domain and for the semi-spherical C_Y (Eq. 3.2.16). Compare the results for one-, two- and three-dimensional flows with those of Fig. 3.7.3.

3.7.2 Derive the expressions of the variances of the components of the specific discharge for flow in an unbounded domain and anisotropic, exponential, axisymmetric C_Y (3.2.10). Hint: follow the procedure leading to the head variances (3.7.16).

3.7.3 Derive $C_{Y\phi}$, Γ_ϕ and σ_ϕ^2 for one-dimensional flow through a stratified formation (Fig. 3.7.2a) of finite length L. Consider two cases: (i) deterministic and constant specific discharge q and random J and (ii) deterministic and constant head drop $\phi_0-\phi_L$ at the boundaries. Illustrate for linear C_Y (3.2.20). Compare Γ_ϕ with the one pertaining to an unbounded formation ($L\to\infty$).

3.7.4 Under same conditions as in exercise 3.7.3, examine the effect of space averaging ϕ, upon the reduction of its variance. The ratio $\sigma_{\bar{\phi}}^2/\sigma_\phi^2$ depends on x/ℓ_Y, ℓ/ℓ_Y and L/ℓ_Y. Discuss the impact of these parameters for their values much larger than unity.

PART 4. SOLUTE TRANSPORT AT THE LOCAL (FORMATION) SCALE

4.1 INTRODUCTION

In this part we shall study the transport of solutes by water in saturated flow through porous formations. Unlike the case of transport under laboratory conditions, discussed in Chap. 2.10, we are concerned here with motion and spread of solutes at the large scale and under the natural conditions prevailing in aquifers. This subject has important applications to pollution processes, salt water intrusion, recharge of miscible fluids, waste disposal, etc. Its experimental investigation under field conditions is hampered by a few major difficulties, which are not encountered in the laboratory. First, the monitoring of solute distribution in space and time requires drilling a large number of observation wells and frequent measurements of the solute concentration. These are costly operations which cannot be justified in most standard cases. Furthermore, the transport under natural flow is a slow process and the monitoring of the motion of a solute body or a plume, to obtain a comprehensive picture of its development, may require many years of continuous measurements over a large area. Besides the large expenditure, this extended period may not be available to the practitioners. In view of these difficulties one should not wonder that comprehensive field studies (see next section) have been carried out only recently.

The role of the theory of transport is to help understanding the complex phenomena occurring in nature and to provide the tools needed for predicting the concentration pattern for given flow conditions. In view of the difficulties mentioned above, one cannot overestimate the service rendered by theory toward the long range assessment of solute spread.

The traditional approach to modeling transport in natural formations was to assume that the dispersion equation (e.g. Eq. 2.12.11) for an inert solute, derived on the basis of laboratory experiments, i.e.

$$\frac{\partial C}{\partial t} + \sum_{j=1}^{3} V_j \frac{\partial C}{\partial x_j} = \sum_{j=1}^{3} \sum_{\ell=1}^{3} \frac{\partial}{\partial x_j} (D_{j\ell} \frac{\partial C}{\partial X_\ell}) \tag{4.1.1}$$

holds also at large scale. This equation can be solved in principle to determine $C(\mathbf{x},t)$, the concentration as function of the space coordinates and time, once the fluid velocity $\mathbf{V}$ and the dispersion coefficients $D_{j\ell}$ are known. A few sophisticated computer codes, which can accommodate quite general flows and boundary conditions, have been developed toward this aim (e.g. Konikow and Bredehoeft, 1978). In spite of this progress, the numerical solution of the transport equation in three-dimensional complex flows poses serious difficulties and the subject is still open to investigations.

Field investigations, though scarce and incomplete, show in a consistent manner that the values of the dispersion coefficients derived under laboratory conditions (see Sect. 2.10) do not apply to large scale transport. The latter have been found to be larger, by a few orders of magnitude, than random motions. In the case of heterogeneous aquifers this is no more the case, since the spatial variations of the velocity, caused by that of formation properties, cannot be modeled as deterministic, for the reasons given in Part 3. This is the main difference between the treatment of the subject here and in a few other existing texts on transport. The situation is similar to that encountered in turbulent flows, in which diffusion is governed by the large turbulent eddies rather than by molecular diffusion, and we shall indeed rely heavily here on the developments of the literature on turbulent transport. There are some differences, however, the main one being that in turbulent flows the timewise fluctuations of the velocity field play a key role in the spreading process, whereas the time changes are slow and deterministic for flow through porous formations. In contrast with this relative simplicity, the presence of large heterogeneity scales compared with that of the solute body, encountered in some groundwater applications, make the problem of ergodicity and uncertainty more complex.

We shall describe now a few basic approaches to the solution of the transport problem represented by Eqs. (4.3.1-4.3.4).

4.3.2 The statistics of fluid particles displacements (the Lagrangian framework)

The Lagrangian representation of the velocity field is a classical one in Fluid Mechanics (see, e.g. Batchelor, 1967). Let $\mathbf{V}_t(t,t_0,\mathbf{a})$ be defined as the velocity of a fluid particle at time t which was at time t_0 at the point of coordinate $\mathbf{x}=\mathbf{a}$. The Lagrangian velocity is the natural representation in kinematics of material points. It is defined as the speed perceived by an observer which follows a particle in its motion, and this was the representation adopted by Taylor (1921) in developing his theory of diffusion by continuous motions. In the present context $\mathbf{V}_t$ is an RSF (random space function) depending indirectly on the space coordinates throughout $\mathbf{a}$ and time, i.e., in each realization of the formation and transport, $\mathbf{V}_t$ has a different functional dependence upon $\mathbf{a}$ and t. Considering now a fixed $\mathbf{a}$, the statistical structure of $\mathbf{V}_t$ is defined by the joint probability density function of $\mathbf{V}_t$ at different times t_1, t_2,... and the associated moments. Thus, $\langle\mathbf{V}_t\rangle$ is the expected value and $v_{t,j\ell}$ is the Lagrangian autocovariance tensor, defined as follows

$$v_{t,j\ell}(t_1,t_2,\mathbf{a}) = \langle[v_{t,j}(t_2,t_0,\mathbf{a})\ v_{t,\ell}(t_1,t_0,\mathbf{a})]\rangle \qquad (j,\ell=1,2,3) \qquad (4.3.6)$$

where $\mathbf{v}_t(t,t_0,\mathbf{a})=\mathbf{V}_t(t,t_0,\mathbf{a})-\langle\mathbf{V}_t(t,t_0,\mathbf{a})\rangle$ is the velocity fluctuation (residual). The statistical moments of the particle total displacement $\mathbf{X}_t$, as defined by (4.3.1), are therefore, related to those of the velocity by

$$\frac{d\langle \mathbf{X}_t \rangle}{dt} = \langle \mathbf{V}_t(t,t_0,\mathbf{a}) \rangle \qquad \text{with } \langle \mathbf{X}_t \rangle = \mathbf{a} \text{ for } t=t_0$$

$$\frac{\partial^2 X_{t,j\ell}}{\partial t_1\, \partial t_2} = v_{t,j\ell}(t_1,t_2,\mathbf{a}) \qquad (j,\ell=1,2,3) \tag{4.3.7}$$

where $\mathbf{X}'_t=\mathbf{X}_t-\langle \mathbf{X}_t \rangle$ is the residual and $X_{t,j\ell}=\langle X'_{t,j}X'_{t,\ell}\rangle$ is the displacements autocovariance. The simplest p.d.f. of $\mathbf{X}_t$ is $f(\mathbf{X}_t;t,\mathbf{a},t_0)$ which can be defined as follows : $f\, d\mathbf{X}$ is the probability that the particle originating from $\mathbf{x}=\mathbf{a}$ at $t=t_0$ is in the volume element $d\mathbf{x}=d\mathbf{X}$ at time t. This particle position p.d.f. has been employed by Taylor (1921), Batchelor (1949) and many other studies as a starting point for investigating transport. The first two moments defined with the aid of f are, therefore, found from (4.3.7) as follows

$$\langle \mathbf{X}_t \rangle = \int \mathbf{X}_t\, f(\mathbf{X}_t)\, d\mathbf{X}_t = \int_{t_0}^{t} \langle \mathbf{V}_t(t',t_0,\mathbf{a}) \rangle\, dt'$$

$$X_{t,j\ell}(t,\mathbf{a}) = \int X'_{t,j}X'_{t,\ell}\, f(\mathbf{X}_t)\, d\mathbf{X}_t = \int_{t_0}^{t}\int_{t_0}^{t} v_{t,j\ell}(t',t'',t_0,\mathbf{a})\, dt'\, dt'' \qquad (j,\ell=1,2,3 \text{ ; fixed } \mathbf{a},t_0) \tag{4.3.8}$$

and higher-order moments of $\mathbf{X}_t$ can be related in a similar manner to those of $\mathbf{V}_t$. Hence, the derivation of the particles displacement statistics in terms of that of the velocity field is carried out in principle this way. At this point it is worthwhile to separate the velocity field into two components, $\mathbf{V}_t=\mathbf{V}+\mathbf{v}_d$, where $\mathbf{V}$ is convective velocity of the fluid, which is related to the Darcian specific discharge, and $\mathbf{v}_d$ is a "brownian motion" component, related to pore-scale dispersion or other dispersive mechanism characterized by a small correlation scale. Following classical statistical concepts (see, e.g., Risken, 1984), $\mathbf{v}_d$ is assumed to have the following properties : (i) it is of zero mean, (ii) it is uncorrelated to $\mathbf{V}$ and (iii) $v_{d,j\ell}$ is a "white noise" covariance, i.e. $v_{d,j}$ and $v_{d,\ell}$ are correlated only for a short time compared to any other time scale of the process, or alternatively they are the limit of random walk process of a small, but finite, step. Hence, we have the decomposition $v_{t,j\ell}=v_{j\ell}+v_{d,j\ell}$, and by (4.3.8) $\mathbf{X}_t=\mathbf{X}+\mathbf{X}_d$, $\langle \mathbf{X}_t \rangle = \langle \mathbf{X} \rangle = \int \langle \mathbf{V} \rangle dt'$ and $X_{t,j\ell} = X_{j\ell}+X_{d,j\ell}$. The covariance $v_{d,j\ell}$ has the white noise (see Chap. 1.4) expression $v_{d,j\ell}(t',t'') = 2\, D_{d,j\ell}(t')\, \delta(t'-t'')$, where δ stands for the Dirac distribution and $D_{d,j\ell}$ is the tensor of pore-scale dispersion coefficients. Substituting this correlation into the last equation of (4.3.8) after differentiating once with respect to time, we obtain

$$\frac{dX_{d,j\ell}}{dt} = 2\int_{t_0}^{t} v_{d,j\ell}(t,t')\,dt' = 4\int_{t_0}^{t} D_{d,j\ell}(t')\,\delta(t-t')\,dt' = 2\,D_{d,j\ell}(t) \qquad (j,\ell=1,2,3) \tag{4.3.9}$$

By the central limit theorem, the "brownian motion" particle displacement $\mathbf{X}_d = \int \mathbf{v}_d\,dt'$ is Gaussian, since it is the result of summation of an infinite number of uncorrelated infinitesimal displacements $\mathbf{v}_d\,dt$. Hence, we may write for the three-dimensional space (see Chap. 1.2)

$$f(\mathbf{X}_d) = \frac{1}{(2\pi)^{3/2}|\boldsymbol{X}_d|^{1/2}} \exp[-\frac{1}{2}\sum_{j=1}^{3}\sum_{\ell=1}^{3} X_{d,j}X_{d,\ell}X_{d,j\ell}^{-1}\,] \tag{4.3.10}$$

where $X_{d,j\ell}^{-1}$ is the inverse of the matrix $\boldsymbol{X}_d$ and $|\boldsymbol{X}_d|$ is its determinant. Hence, the p.d.f. of $\mathbf{X}_d$ (4.3.10) is completely characterized by the dispersion tensor $D_{d,j\ell}$, and any statistical moment can be derived in terms of it. It can be shown indeed (see exercise 4.3.1) that the concentration field associated with $\mathbf{X}_d$ satisfies the diffusion equation.

We consider now the convective component $\mathbf{V}$ of the Lagrangian velocity field, and restrict the following developments to Lagrangian stationarity, for which the theory leads to its most useful results. Thus, it is assumed that the Lagrangian velocity autocovariance may be written as follows : $v_{j\ell}(t',t'',t_0,\mathbf{a}) = v_{j\ell}(|t'-t''|,\mathbf{a})$. Putting it into words, it is assumed that the correlation between the velocities of the fluid particle at times t' and t'' depends only on the time lag $t'-t''$. Substituting this relationship in (4.3.8) and employing Cauchy algorithm (see Chap. 1.4) yields for the displacements autocovariance

$$X_{j\ell}(t) = \int_0^t\int_0^t v_{j\ell}(t'-t'')dt'dt'' = 2\int_0^t (t-t')\,v_{j\ell}(t')\,dt'$$

$$\text{i.e.} \tag{4.3.11}$$

$$\frac{d^2X_{j\ell}}{dt^2} = 2\,v_{j\ell}(t) \qquad (j,\ell=1,2,3)$$

and we have taken, without loss of generality $t_0=0$, with $v_{j\ell}(t) = \langle v_j(0)\,v_\ell(t)\rangle$. Furthermore, for the sake of simplicity of notation, we have suppressed the dependence on $\mathbf{a}$. Next, we shall assume that the particle velocity residuals v_j and v_ℓ cease to be correlated after a finite time $T_{j\ell}$, which in turbulent motions is related to the eddies size. We shall show later that for flow through heterogeneous formations it is connected to the conductivity correlation scale. Furthermore, with displacements residuals given by $\mathbf{X}' = \int_0^t \mathbf{v}(t')dt'$, by invoking the central limit theorem, $\mathbf{X}'$ can be regarded at large travel times $t \gg T_{j\ell}$ as the sum of a large number of uncorrelated displacements, which tends to normality. At the same limit (4.3.11) yields

$$X_{j\ell}(t) \rightarrow 2t\int_0^\infty v_{j\ell}(t')\,dt' + O(T_{j\ell}) \qquad \text{for } t \gg T_{j\ell} \tag{4.3.12}$$

Eq. (4.3.12) can be rewritten, after defining the Lagrangian integral scale (macroscale) by $T_{j\ell} = (1/\sigma_{v,j\ell})\int_0^\infty v_{j\ell}(t')dt'$, as follows : $X_{j\ell}(t) \rightarrow 2\sigma_{v,j\ell} T_{j\ell} t$. Here, the symbol $\sigma_{v,j\ell} = v_{j\ell}(0)$ represents the velocity variance matrix. Finally, for $t \gg T_{j\ell}$, the multivariate normal p.d.f. of $\mathbf{X}$ can be written, similarly to (4.3.10), as follows

$$f(\mathbf{X}) = \frac{1}{(2\pi)^{3/2}|\boldsymbol{X}|^{1/2}} \exp[-\frac{1}{2}\sum_{j=1}^{3}\sum_{\ell=1}^{3}(X_j-\langle X_j\rangle)(X_\ell-\langle X_\ell\rangle\, X_{j\ell}^{-1}\,] \tag{4.3.13}$$

In contrast, for small travel time $t \ll T_{j\ell}$, the Taylor expansion of $v_{j\ell}(t)$ near t=0, i.e. $v_{j\ell}(t) = \sigma_{v,j\ell} + (t^2/2)(\partial^2 v_{j\ell}/\partial t^2) + ...$ yields, after substitution in (4.3.9)

$$X_{j\ell}(t) \rightarrow \sigma_{v,j\ell} t^2 + O(t^4) \tag{4.3.14}$$

but at this limit $\mathbf{X}$ is no more Gaussian, unless $\mathbf{v}$ is so. Summarizing these important results of Taylor's theory of diffusion by continuous motions, we have found that the displacement vector $\mathbf{X}_t = \mathbf{X} + \mathbf{X}_d$, of a solute particle which originate at $\mathbf{x}=\mathbf{a}$ for t=0, tends to a multivariate normal p.d.f. for large travel time if the Lagrangian velocity is stationary. It is given by (4.3.13) in which $X_{j\ell}$ is replaced by $X_{t,j\ell} = X_{j\ell} + X_{d,j\ell}$. Furthermore, with $D_{t,j\ell} = (1/2)(dX_{t,j\ell}/dt) = D_{j\ell} + D_{d,j\ell}$, $D_{t,j\ell}$ is constant and from (4.3.12) it follows that $D_{j\ell} = \sigma_{v,j\ell}\, T_{j\ell}$. In contrast, for small travel time, by (4.3.14) $D_{t,j\ell} = \sigma_v^2 t + D_{d,j\ell}$. Hence, the tensor of the apparent dispersion coefficients $D_{t,j\ell}$ is the sum of the pore-scale dispersion $D_{d,j\ell}$ and that of dispersion by the convective random motions $D_{j\ell}$. The latter grows linearly with time at small t and tends to a constant limit for travel time much larger than the Lagrangian integral scale.

In spite of its elegance and generality, the Lagrangian framework is of limited usefulness in relating transport to the heterogeneous structure, because the fluid convective velocity is expressed as a rule in the Eulerian form, and this is the topic of the following Section.

4.3.3 The statistics of particles displacements (the Eulerian framework)

The water macroscopic velocity in the porous formation is determined from Darcy's law by the basic relationship

$$\mathbf{V}(\mathbf{x},t) = \mathbf{q}/n \quad ; \quad \mathbf{q} = -K\,\nabla\phi \tag{4.3.15}$$

where, as usual, $\mathbf{q}$ is the specific discharge, K is the hydraulic conductivity, n is the effective porosity and ϕ is the water head. Generally, these are functions of the space coordinate $\mathbf{x}$, fixed with respect to the formation boundary, and of the time t. Hence, $\mathbf{V}$ is here the Eulerian velocity field, the one which can be determined either by computations or by measurements, with the aid of the other variables of (4.3.15) or by the less common field measurements of $\mathbf{V}$ with the aid of tracers. In the case of a heterogeneous porous formation of random distribution of its hydraulic properties, both $\mathbf{q}$ and n are random space function, whose properties have been analyzed in Part 3. At this point we shall denote by $\mathbf{U}(\mathbf{x},t)=\langle\mathbf{V}(\mathbf{x},t)\rangle$ the expected value (ensemble average) of the pore velocity and by $\mathbf{u}(\mathbf{x},t) = \mathbf{V}-\mathbf{U}$ its fluctuation (residual). We shall restrict here the further developments to the case of statistically homogeneous fields in space. The statistical structure of the velocity field is determined by the joint p.d.f. of its values at different points $\mathbf{x}_1,\mathbf{x}_2,...$ or in turn by the associated various moments. As shown in Part 3, a central role is played by the autocovariance tensor $u_{j\ell}(\mathbf{r}) = \langle u_j(\mathbf{x}_1)\, u_\ell(\mathbf{x}_2)\rangle$, with $\mathbf{r}=\mathbf{x}_1-\mathbf{x}_2$ and $j,\ell=1,2,3$.

The basic differential equation (4.3.1), relating the displacement $\mathbf{X}_t$ to the velocity field, becomes now

$$\frac{d\mathbf{X}_t}{dt} = \mathbf{U}(\mathbf{X}_t,t) + \mathbf{u}(\mathbf{X}_t,t) + \mathbf{v}_d \quad \text{with initial condition : } \mathbf{X}_t=\mathbf{a} \quad \text{for } t=t_0 \tag{4.3.16}$$

where $\mathbf{v}_d$ is the "brownian motion" velocity associated with pore-scale dispersion and $\mathbf{x}=\mathbf{a}$ is the initial coordinate of the particle. Unlike (4.3.1) which is an explicit ordinary differential equation for $\mathbf{X}_t$, once the Lagrangian velocity field $\mathbf{V}(t,\mathbf{a})= \mathbf{U}(\mathbf{X}_t,t) + \mathbf{u}(\mathbf{X}_t,t)$ is given, the stochastic Eq. (4.3.16) is more complex due to the appearance of the dependent variable in its right-hand-side. Consequently, it is generally not possible to derive in an explicit manner the relationships between the statistical moments of the displacement $\mathbf{X}_t$ and those of the Eulerian velocity field. To arrive at such relationships, which is our main aim, some approximations, of an analytical or numerical nature, are needed. The numerical solution of (4.3.16), which is straightforward, but cumbersome and time consuming, will be discussed in Chap. 4.4. We shall also investigate later some approximations pursued in the literature, which lend themselves to simple formulations.

Staying within a general frame here, it seems that the most promising avenue for investigating the statistical structure of $\mathbf{X}_t$ is by the Fourier transform (FT) methodology. Thus, the FT of the velocity residual and its inverse are defined as usual (see Chap. 1.6) by

$$\hat{\mathbf{u}}(\mathbf{k},t) = \frac{1}{(2\pi)^{m/2}} \int \mathbf{u}(\mathbf{x},t)\, e^{i\mathbf{k}\cdot\mathbf{x}} d\mathbf{x} \quad ; \quad \mathbf{u}(\mathbf{x},t) = \frac{1}{(2\pi)^{m/2}} \int \hat{\mathbf{u}}(\mathbf{k},t)\, e^{-i\mathbf{k}\cdot\mathbf{x}} d\mathbf{k} \tag{4.3.17}$$

Since $\mathbf{u}$ is real, it follows that $\hat{\mathbf{u}}\exp(-i\mathbf{k}\cdot\mathbf{x}) = \hat{\mathbf{u}}^*\exp(i\mathbf{k}\cdot\mathbf{x})$, with $\hat{\mathbf{u}}^*$ the complex conjugate of $\hat{\mathbf{u}}$. Here $\mathbf{k}$ is the wave-number vector of components (k_1,k_2,k_3) or (k_x,k_y,k_z), integration is from $-\infty$ to $+\infty$, and m is the $\mathbf{x}$ and $\mathbf{k}$ spaces dimensionality. It is reminded here that a rigorous formulation of (4.3.17) employs Stieltjes integrals, but the same results can be formally obtained by incorporating in the class of functions and their transforms the Dirac distribution and its derivatives (see Chap. 1.6). To further simplify computations we separate the total displacement into two components $\mathbf{X}_t=\mathbf{X}+\mathbf{X}_d$, like in the previous section. We assume, furthermore, that $\mathbf{U}$ is constant and $\mathbf{X}_d$ does not depend on $\mathbf{u}$, i.e. the pore-scale dispersion coefficients are taken as constant, an approximation which is justified in view of their limited impact upon large scale transport. Then, we write formally $\mathbf{X}=\mathbf{a}+\mathbf{U}(t-t_0)+\mathbf{X}'$ and due to the statistical independence between $\mathbf{u}$ and $\mathbf{v}_d$ discussed in the previous section, we have from (4.3.16)

$$\frac{d\mathbf{X}'}{dt} = \mathbf{u}(\mathbf{X}_t,t) \; ; \quad \frac{d\mathbf{X}_d}{dt} = \mathbf{v}_d \tag{4.3.18}$$

with $d\mathbf{X}'$ and $d\mathbf{X}_d$ independent infinitesimal displacements.

The crux of the matter is the stochastic differential equation for $\mathbf{X}'$ (4.3.18), in which the right-hand-side depends on the total displacement $\mathbf{X}_t$. It can be rewritten in terms of the FT of $\mathbf{u}$(4.3.17) after substitution in (4.3.18), as follows

$$\frac{d\mathbf{X}'}{dt} = \frac{1}{(2\pi)^{m/2}} \int \hat{\mathbf{u}}(\mathbf{k},t) \exp(-i\mathbf{k}\cdot\mathbf{X}_t) d\mathbf{k} \tag{4.3.19}$$

The first basic question is whether $\mathbf{X}'$ is a fluctuation, i.e. $\langle\mathbf{X}'\rangle=0$, as implied by our notation. This would presume that the Lagrangian velocity fluctuation $\mathbf{v}(t,\mathbf{a})$ is equal to $\mathbf{u}[\mathbf{X}_t(t,\mathbf{a}),t]$, which is definitely true for $t\to 0$, i.e. for the regime in which (4.3.14) is valid and for which $\mathbf{X}_t \simeq \mathbf{a}$. On the other hand, for large t and for a velocity field $\mathbf{u}$ of finite integral scale, it is reasonable to assume that $\hat{\mathbf{u}}$ and $\exp(-i\mathbf{k}.\mathbf{X}_t)$ become uncorrelated (see Sect. 4.8.1), which leads to $d\langle\mathbf{X}'\rangle/dt=0$ and $\langle\mathbf{X}'\rangle=\mathbf{U}_\infty$=const. If $\mathbf{U}_\infty$ is different from $\mathbf{U}$, a drift between the average Eulerian and Lagrangian velocities develops. We shall neglect here the presence of such a possible drift, which is presumably small. At any rate, the equation (4.3.19) and the developments based on it are going to be employed here mainly for deriving approximate, linearized, solutions which presume that the velocity coefficients of variations are small, in which case the drift is a higher-order effect. Its impact is a matter of further investigations in which various nonlinear effects have to be accounted for (see Sect. 4.8.1).

From (4.3.19) we can obtain the time derivatives of the various statistical moments of $\mathbf{X}$, and in particular of the autocovariance $X_{j\ell}$, in terms of the FT of the Eulerian velocity field. Thus, taking the ensemble average of the product $dX'_j/dt_1 . dX'_\ell/dt_2$, yields

$$\frac{\partial^2 X_{j\ell}(t_1,t_2,\mathbf{a},t_0)}{\partial t_1\, \partial t_2} = \langle u_j(\mathbf{X}_{t1},t_1)\, u_\ell(\mathbf{X}_{t2},t_2)\rangle =$$

$$\frac{1}{(2\pi)^m}\iint \langle \hat{u}_j(\mathbf{k}',t_1)\, \hat{u}^*_\ell(\mathbf{k}'',t_2)\, \exp[i\mathbf{k}''\cdot\mathbf{X}_{t2} - \mathbf{k}'\cdot\mathbf{X}_{t1}]\rangle\, d\mathbf{k}' d\mathbf{k}''$$

$$(j,\ell=1,...,m\ ;\ m=1,2,\text{ or }3) \tag{4.3.20}$$

where $\mathbf{X}_{t1} = \mathbf{X}_t(t_1,\mathbf{a},t_0)$, $\mathbf{X}_{t2} = \mathbf{X}_t(t_2,\mathbf{a},t_0)$ and m is the number of space dimensions.

A particular, but important, case is the one of statistical Lagrangian stationarity, discussed in the previous section. It is not a simple matter to establish the general conditions to be satisfied by **U** and **u** in order to ensure Lagrangian stationarity. They are bound to prevail, for instance, if **U** is constant and if **u** is a stationary, time independent, solenoidal function of **x**. At any rate, assuming that Lagrangian stationarity prevails, we obtain from (4.3.11) and (4.3.20)

$$\frac{d^2 X_{j\ell}(t,\mathbf{a})}{dt^2} = 2\,\langle u_j(0,0)\, u_\ell(\mathbf{X}_t,t)\rangle =$$

$$\frac{2}{(2\pi)^m}\iint \langle \hat{u}_j(\mathbf{k}',0)\, \hat{u}^*_\ell(\mathbf{k}'',t)\, \exp(i\mathbf{k}''\cdot\mathbf{X}_t)\rangle d\mathbf{k}' d\mathbf{k}'' \quad (j,\ell=1,...,m\ ;\ m=1,2,\text{ or }3) \tag{4.3.21}$$

where, without loss of generality, $t_0=0$ (for arbitrary t_0 we have to replace t by $t-t_0$).

The above basic equations relate the particle-displacement covariance to the Fourier components of the Eulerian velocity field. However, due to the statistical dependence of $\mathbf{X}_t$ upon **u**, $X_{j\ell}$ cannot be generally expressed in terms of the spectrum of **u**, and we face the same difficulty as encountered before, in Eq. (4.3.18). However, (4.3.19) and especially (4.3.21) lend themselves to a few useful approximations (see, e.g. Lundgren & Pointin 1975), which will be explored in the sequel. A simplification which is already adopted here is the one stemming from the normality of $\mathbf{X}_d$, as expressed by Eqs. (4.3.9) and (4.3.10). Towards this aim we first observe that for a Gaussian process of zero mean we have the following relationship, which is easily derived with the aid of the characteristic function for multivariate normal distributions (Chap. 1.2),

$$\langle \exp(i\mathbf{k}.\mathbf{X}_d)\rangle = \exp\Big(-\frac{1}{2}\sum_{j=1}^{m}\sum_{\ell=1}^{m} X_{d,j\ell}\, k_j k_\ell\Big) \tag{4.3.22}$$

Hence, by substituting in (4.3.21) $\mathbf{X}_t=\langle\mathbf{X}\rangle+\mathbf{X}'+\mathbf{X}_d$, by assuming that the "Brownian motion" component of $\mathbf{X}_d$, of zero mean and of covariance $X_{d,j\ell}$, is independent of $\mathbf{u}$, and by using (4.3.22), we obtain

$$\frac{d^2X_{j\ell}}{dt^2}=\frac{2}{(2\pi)^m}\iint \langle\hat{u}_j(\mathbf{k}',0)\,\hat{u}^*_\ell(\mathbf{k}'',t)\,\exp(i\mathbf{k}''\cdot\mathbf{X}')\rangle\exp[i\mathbf{k}''\cdot\langle\mathbf{X}\rangle-\frac{1}{2}\sum_{p=1}^{m}\sum_{q=1}^{m}X_{d,pq}k''_p k''_q]\,d\mathbf{k}'\,d\mathbf{k}''$$

$$(j,\ell=1,...,m\ ;\ \text{and m is the space dimension}) \qquad (4.3.23)$$

replacing (4.3.21).

Eq. (4.3.23) is the starting point for a few important applications of transport in porous formations to be developed in the sequel. This will be shown to be possible when the exponential term in the expression between the averaging signs in (4.3.23) can be separated from the FT of the velocity components. Higher-order moments of $\mathbf{X}'$ can be evaluated in terms of the FT of $\mathbf{u}$ in a similar manner, but such computations are not pursued here. It is reminded here that in the case in which $\mathbf{X}(t,\mathbf{a},t_0)$ is Gaussian, the two moments $\langle\mathbf{X}\rangle$ and $X_{t,j\ell}$ determine its p.d.f. in a complete manner.

4.3.4 The concentration expected value

The concentration field has been defined with the aid of the particles displacements by the basic equation (4.3.2) and by its generalizations (4.3.3), (4.3.4) and (4.3.5). ΔC and C are random space functions, since $\mathbf{X}_t$, on which they depend, are random. In this section we shall derive the relationships between the expected values of C and the p.d.f. $f(\mathbf{X}_t)$. Thus, averaging (4.3.2) yields, by the definition of the Dirac distribution,

$$\langle\Delta C(\mathbf{x},t,\mathbf{a},t_0)\rangle=\int\frac{\Delta M}{n}\,\delta(\mathbf{x}-\mathbf{X}_t)\,f(\mathbf{X}_t;t,\mathbf{a},t_0)\,d\mathbf{X}_t=\frac{\Delta M}{n}\,f(\mathbf{x};t,\mathbf{a},t_0) \qquad (4.3.24)$$

This fundamental result can be put into words as follows : the expected value of the concentration field ΔC is proportional to the probability distribution function of the particles displacements which originate at $\mathbf{x}=\mathbf{a}$ for $t=t_0$, in which the displacement $\mathbf{X}_t$ *is replaced* by the space coordinate $\mathbf{x}$.

This relationship can be understood intuitively if we consider a volume element $\Delta\mathbf{x}$ and define the concentration as the mass of solute in this volume, divided by the fluid volume in it. Let N be the total number of realizations and let P be the number of realizations in which the particle trajectory terminates in $\Delta\mathbf{x}$ at time t. The expected value of the concentration is by definition equal to

$\lim_{N\to\infty}$ P ΔM/(N n Δ**x**) (this definition is employed conveniently in numerical simulations of transport, see Chap. 4.4). But, by the definition of the probability density function, f Δ**x** = $\lim_{N\to\infty}$ P/N, and elimination of P between the two expressions leads to (4.3.24).

In obtaining (4.3.24) it was assumed that ΔM, the mass, and n, the effective porosity, are fixed. If the latter is also regarded as a random space function, the result in (4.3.24) should be regarded as conditioned on a realization of n.

The concentrations related to a finite solute cloud (4.3.3) or a plume (4.3.4) are given by similar formulae, due to the commutativity of the averaging and integration operations. Hence, we have

$$\langle C(\mathbf{x},t,t_0)\rangle = \int_{V_0} \frac{n}{n_0} C_0(\mathbf{a})\, f(\mathbf{x};t,\mathbf{a},t_0)\, d\mathbf{a} \;\; ; \;\; \langle C(\mathbf{x},t)\rangle = \int_{V_0}\int_0^t \frac{n}{n_0} \dot{M}(t_0,\mathbf{a})\, f(\mathbf{x};t,\mathbf{a},t_0)\, dt_0\, d\mathbf{a} \quad (4.3.25)$$

respectively, and similarly for the space average (4.3.5).

To arrive at the partial differential equation satisfied by f, it is reminded first that the particles displacements depend on the random velocities $\mathbf{v}_d$ associated with the pore-scale dispersion and on the fluid convective motion represented by **V**=**U**+**u**, as shown in the previous sections. In a first stage we consider a given realization of the convective field **V**, i.e. the conditional p.d.f. f($\mathbf{X}_d$|**V**) and the associated concentration. In other words, the convective velocity field is regarded as deterministic, while the only random component is due to the "Brownian motion". In this case f(**x**,t|**u**) can be shown to satisfy a classical Focker-Planck equation (see, for instance, Risken, 1984)

$$\frac{\partial f}{\partial t} + \sum_{j=1}^{m} \frac{\partial}{\partial x_j}(V_j f) = \sum_{j=1}^{m}\sum_{\ell=1}^{m} D_{d,j\ell} \frac{\partial^2 f}{\partial x_j \partial x_\ell} \quad (4.3.26)$$

where **V** satisfies the continuity equation and the tensor $\mathbf{D}_d$ is constant or time dependent. The case in which $\mathbf{D}_d$ is spatially varying is a delicate one, since the passage from the kinematic equation (4.3.1) to (4.3.26) depends on the type of integral, Itô or Stratonovich, one chooses for the "Brownian motion" related term (for a succinct discussion of this classical issue see Bodo et al, 1987). Since anyway the impact of $\mathbf{D}_d$ upon transport is small compared to that of heterogeneity, we shall be satisfied with taking it as constant herein.

This leads for $\langle\Delta C(\mathbf{x},t|\mathbf{V})\rangle$ (4.3.24) or $\langle C(\mathbf{x},t|\mathbf{V})\rangle$ (4.3.25), conditioned on **u**, and for constant n, to

$$\frac{\partial\langle C\rangle}{\partial t} + \sum_{j=1}^{m} \frac{\partial}{\partial x_j}(V_j \langle C\rangle) = \sum_{j=1}^{m}\sum_{\ell=1}^{m} [D_{d,j\ell} \frac{\partial^2\langle C\rangle}{\partial x_\ell \partial x_j}] \quad (4.3.27)$$

Then, it can be shown by using (4.3.40), (4.3.41) and (4.3.42) that

$$\mathrm{VAR}[\overline{X}_{jk}(t)] = \bar{\sigma}^{2}_{j\ell} + \bar{\sigma}_{jj}\bar{\sigma}_{\ell\ell} + \frac{1}{V_0^2}\int [Z^2_{jk}(t,\mathbf{a},\mathbf{b}) + Z_{jj}(t,\mathbf{a},\mathbf{b})Z_{kk}(t,\mathbf{a},\mathbf{b})]\, d\mathbf{a}\, d\mathbf{b}\ -$$

$$\frac{2}{V_0^2}\int\!\!\int\!\!\int [Z_{jk}(t,\mathbf{a},\mathbf{b})Z_{jk}(t,\mathbf{b},\mathbf{c}) + Z_{jj}(t,\mathbf{a},\mathbf{b})Z_{kk}(t,\mathbf{b},\mathbf{c})]\, d\mathbf{a}\, d\mathbf{b}\, d\mathbf{c} \qquad (j.k=1,2,3) \qquad (4.3.43)$$

where integrations are over V_0. Again, it is easy to examine the behavior of (4.3.43) at the two extreme limits discussed above. For a small solute body $\bar{\sigma}_{j\ell} \simeq X_{j\ell}$ and $\mathrm{VAR}[\overline{X}_{jk}(t)] \simeq 0$, whereas for $A_0 >> I^2$ the variance cancels again because Z_{jk} are correlated transversally over I. Hence, it is reasonable to assume that the variance is limited and that $\overline{X}_{t,j\ell}$ (4.3.40) is bound to be ergodic.

Another point of interest is that both (4.3.41) and (4.3.42) depend on the velocity variogram rather than on its covariance. This follows from the definition (4.3.40) in which the differences between the displacement of two particles, rather than the displacements, do appear. It follows that $\overline{X}_{t,j\ell}$ can be defined even if the velocity field is of stationary increments rather than stationary in the ordinary sense. However, the variance of the displacement of the center of mass $\bar{\sigma}_{j\ell}$ is undefined in this case.

If $\mathbf{X}$ is Gaussian, which is always the case for a sufficiently large t if the macroscales are finite, the same is true for the expected value of ⟨C⟩, as shown by the above relationships. In particular the third-order spatial moment of ⟨C⟩ is zero and this is also obvious from Eq. (4.3.29), satisfied by ⟨C⟩. In such a case some space averages of the concentration, e.g. the mass flux through a plane, can be evaluated by using the solution $\langle C(\mathbf{x},t|\overline{X}_j, \overline{X}_{t,j\ell})\rangle$ (4.3.29), *conditioned* on the values of $\overline{\mathbf{X}}$ and $\overline{X}_{t,j\ell}$. The unconditional moments of C are subsequently obtained by employing the p.d.f. of $\overline{\mathbf{X}}$ and $\overline{X}_{t,j\ell}$ and the basic relationships of conditional probability (see Sect. 4.9.2).

In the general case, however, the calculation of higher-order moments require the knowledge of the multiple correlation of particles displacement, e.g. the one pertaining to three particles at different initial locations. The computation of such moments is not pursued here.

Exercises

4.3.1 Prove that the probability density function of a multivariate normal vector, e.g. (4.3.10) and (4.3.13), satisfies the partial differential equation (4.3.28). Hint : employ the Fourier transform of a Gaussian distribution, i.e. the characteristic function of Chap. 1.2, and the relationships between the FT of derivatives and the FT of the function.

4.3.2 The concentration field in a plume has been defined in (4.3.4) with the aid of the rate of injection of the solute mass $\dot{M}$. Relate, in a simple manner, this entity to the initial concentration in a plane normal to the average flow vector.

4.3.3 Write down the various moments of the displacement vector for small travel time compared to the macroscales, in line with (4.3.14), in terms of the Eulerian velocity field.

4.3.4 By using the two particles displacements covariance tensor $Z_{j\ell}$ (4.3.30), derive the expected value of the rate of change of the distance between two particles which originate at different points at same t. Analyze the extreme cases of close and distant particles.

4.3.5 Derive the concentration variance at the center of a solute body, whose initial shape is a sphere of radius R_0. The Eulerian velocity field is stationary and $\langle C \rangle$ satisfies Eq. (4.3.30) with constant dispersion coefficients. Neglect pore-scale dispersion and assume that initially the concentration C_0 is constant. Analyze the dependence of the variance upon dimensionless R_0 and t.

4.3.6 Define in a convenient manner first- and second-order spatial moments to characterize a plume in which solute is injected at constant rate. Express the expected value of these moments and the variance of the first with the aid of the particle displacement statistical moments.

4.4 A FEW NUMERICAL SIMULATIONS OF SOLUTE TRANSPORT IN HETEROGENEOUS FORMATIONS

As we have already mentioned in Sect. 4.1, a few numerical codes have been developed in order to solve the transport equation (4.1.1) within the traditional framework, namely for homogeneous formations or for ones of deterministic, slowly varying in space, properties. Numerical simulations of transport in heterogeneous aquifers in which the hydraulic conductivity changes in an irregular manner in space, with highly contrasting values, are in their infancy, due to the difficult numerical problems they have to overcome. Generally, such simulations in formations of random structures involve three consecutive stages, after the space domain has been discretized : (i) first, realizations of the K and n spatial fields have to be generated. For K lognormal and of a given stationary structure, this can be achieved, for instance, by using a multivariate normal generator. For a large number of nodes, computer time can be saved by using approximate techniques, like turning bands (see, e.g., Journel and Huijbregts, 1978). As we have mentioned already in Part 3, a valid representation of the random structure requires that the mesh size should be significantly smaller than the properties integral scales, which may result in a very large number of elements ; (ii) second, the flow equations have to be solved, with appropriate boundary and initial conditions, in order to derive the head field. This already poses serious numerical problems if K, for instance, changes abruptly between neighboring nodes. Next, the velocity field has to be determined by numerical differentiation of the head and by the use of Darcy's law. This an error-prone procedure; a technique which alleviates some of the difficulties has been suggested recently by Frind and Matanga (1985), which formulate the flow problem in terms of the head and the streamfunction. So far, this technique is limited to two-dimensional flows in incompressible formations, and (iii) once the velocity field has been derived numerically, the transport problem is solved either by a numerical scheme applied directly to the transport equation (4.1.1) to determine $C(\mathbf{x},t)$ in the given realization, or by a particle tracking procedure. The latter is in the spirit of the preceding Chapter, i.e. the solute body is represented by a large number of indivisible particles which are moved along the velocity vector, a random component associated with the pore-scale dispersion effect being added at each time step. To obtain meaningful statistical results, the above sequence has to be repeated a large number of times, to generate an ensemble of realizations. From this brief description it should be clear that numerical modeling of transport in heterogeneous formations is a formidable task. No wonder that so far, the few attempts in this area, have been limited to steady flow through heterogeneous structures of a two-dimensional nature. In this respect, the pioneering study of Smith and Schwartz (1980) should be mentioned first. They have used a Monte-Carlo technique and in each realization the formation was represented by a collection of square blocks of different conductivities, which were generated from a lognormal population. The blocks were either of uncorrelated $Y=\ell nK$, or slightly correlated by using a nearest-neighbor autoregressive relation. The formation was of a rectangular shape, with opposite boundaries of either of constant head or of no flow, achieving a uniform one-dimensional flow parallel to the impervious boundaries, in the average. The selected values of σ_Y^2 were relatively large, namely 1.3 and 2.8, whereas n was taken of much smaller variability and very little was revealed about the details of the numerical scheme and its validation. As we shall show later for another simulation, a vivid picture of the flow pattern is achieved by depicting the streamtubes. For instance, the fact that the latter do not intersect, is a good omen for the numerical simulation. No such picture or similar ones were provided by the authors, the accuracy of the velocity distribution within blocks being taken for granted. We shall not dwell here upon the results of the simulations of Smith and Schwartz (1980) not only because of these limitations, but also because of the relative small numbers of blocks between the impervious boundaries, namely ten, resulting in a width of the flow zone of the order of eight conductivity correlation scales. This is probably much less than the number encountered in actual formations at local scale in which, due to stratification, the vertical correlation scale may be much smaller than the formation depth. Thus, the simulations are rather representative of regional flows, to be discussed in Part 5. Smith and Schwartz (1980, 1981) used the particle tracking technique and their results are not represented in terms of concentration, but of arrival times distribution or spatial moments. Furthermore, they refrained from ensemble averaging results and rather concentrated on analyzing the spatial moments in a few realizations. A few of their results are of general interest: the large spread due to the heterogeneous structure as compared to the negligible role of pore-scale dispersion; the lack of normality of the

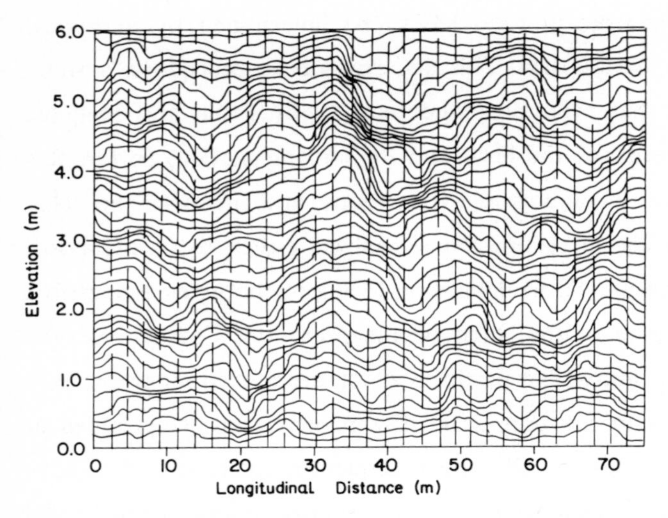

Fig. 4.4.1 Representation of a few streamlines, based on a numerical simulation of two-dimensional flow in a heterogeneous formation (reproduced from Frind et al, 1987).

particle spatial distribution in many realizations; large variability of spatial moments and time variation of the macrodispersion coefficient, which were defined with the aid of the time derivative of the second spatial moment of the particles cloud.

A more detailed representation in an example of two-dimensional flow has been achieved by Frind et al (1987), who have used the dual approach of Frind and Matanga (1985) mentioned above. They have modeled flow and transport in a few realizations only, which is understandable in view of the required numerical effort. To underscore the degree of detail needed in order to achieve an accurate representation of the flow pattern, it is worthwhile to recall that in their simulation of a vertical cross-section of a formation of 4 meters depth and 100 meters length, the number of nodes was 10^6, which was later reduced to 2.83×10^5. The conductivity statistical structure was the one pertaining to the Borden site (Mackay et al, 1986) experiment, mentioned in Chap. 4.2, with $\sigma_Y^2=0.38$, $I_v=0.12$ m and $I_h=2.8$ m, but in two dimensions, rather than for the actual three-dimensional heterogeneity. To illustrate some of the results we have reproduced in Fig. 4.4.1 the streamline pattern in one of the realizations of the flow (the upper and lower boundaries are impervious). This picture is suggestive in demonstrating the effect of spatial heterogeneity and the considerable numerical effort required in order to achieve an accurate description of the flow field.

Summarizing this brief discussion, it can be said that numerical simulations of transport in heterogeneous formations of three-dimensional structure are still not available and at best, they shall serve in the near future as numerical experiments, rather than predictive tools. In the case of two-dimensional heterogeneity, pertaining to regional flows, numerical modeling is in the reach, but serious computational problems have to be overcome before making it a routine tool. In view of this state of affairs, approximate methods of solving transport problems are extremely useful, and they form the subject of the remaining chapters of this part.

4.5 TRANSPORT THROUGH STRATIFIED FORMATIONS

4.5.1 Introduction

In this and in the following sections we are going to investigate transport in heterogeneous porous formations at the local scale, by combining the general methodology of Chap. 4.3 with the results pertaining to flow of Part 3. It should be reminded here that the local scale heterogeneity is the one related to the spatial variability of hydraulic conductivity and effective porosity throughout the thickness of the formation and over a similar scale in the horizontal plane. The assumption of stationarity implies that the vertical correlation scales of Y and n are much smaller than the thickness, and this is bound to be true due to layering which is often present in sedimentary formations.

In Part 3 we have selected a few idealized representations of heterogeneous structures, shown schematically in Figs. 3.7.1, and we have derived subsequently the statistical moments of the flow variables pertaining to each case. This is precisely the line followed in the remaining sections of this part, namely to investigate transport through the same type of formations.

The first, and simplest, case is the one of a perfectly stratified formation, depicted in Fig. 3.7.1a. As we have already mentioned in Chap. 3.7, this is an idealization which resembles the layering present in many natural formations, except that the continuity of such layers over large distances is not warranted. Thus, theoretical results for which continuity is an essential prerequisite may be unrealistic. Nevertheless, the investigation of transport through stratified formations is of interest for two main reasons. First, this is the only case in which simple and exact results for the concentration statistical moments can be achieved. Hence, it constitutes a convenient tool for grasping more complex cases, and it can also serve as a benchmark for numerical simulations. Second, layers continuity may prevail for transport over short travel distances, e.g. for injection of solute by a well and measurement of concentration in another nearby one, and the results of the theory might be applicable in such cases.

Due to its simplicity, transport through stratified formations, regarded as random or deterministic, has been studied quite extensively and relatively early (Mercado, 1967, Marle et al, 1967, Gelhar et al, 1979) and its investigation is still pursued (e.g. Matheron and de Marsily, 1980, Güven et al, 1984, 1986). The theoretical basis of these studies has been laid by Taylor (1953) and Aris (1956), who have considered solute transport in laminar, parallel, flow in tubes (see Sect. 2.10.2).

In this chapter we shall examine a few of the results obtained in the past by following, however, the approach outlined in Chap. 4.3.

4.5.2 Steady flow parallel to the bedding

(i) Velocity distribution and particles displacements moments.

The coordinate system employed in Sect. 3.7 will be adopted here as well, namely $x_3=z$ is a vertical coordinate normal to the bedding, and $x_1=x$ and $x_2=y$ are horizontal coordinates (see Fig. 4.5.1a for a stratified formation made up from distinct layers). With B, the formation thickness, and with $\partial\Phi/\partial x_1=-J$, a constant head gradient applied parallel to the bedding, the flow is horizontal and the velocity vector $\mathbf{V}(V_1,0,0)$ is given exactly (see Chap. 3.7) by

$$\langle V_1\rangle = U = \frac{JK_A}{n} \quad ; \quad V_1'(x_3) = u(x_3) = \frac{UK'(x_3)}{K_A} \tag{4.5.1}$$

where $K_A=\langle K(x_3)\rangle$ is the hydraulic conductivity arithmetic mean and $K'(x_3)$ is the residual. To simplify matters, we have assumed that the effective porosity is constant, in view of its much smaller variability than that of K (see Chap. 3.2); the impact of its estimation error will be examined in Chap. 4.9. With K assumed to be a stationary random function of $z=x_3$, the velocity covariances $u_{j\ell}$ are equal to zero, except for $u_{11}(x_3,x_3')=u_{11}(r_3)$, $r_3=|x_3-x_3'|$, and we have the following simple relationships

$$u_{11}(r_3) = U^2\frac{C_K(r_3)}{K_A^2} \quad ; \quad \hat{u}_{11}(k_3) = U^2\frac{\hat{C}(k_3)}{K_A^2} \tag{4.5.2}$$

where the Fourier transforms are with respect to the variable r_3. Since we can regard the transport, without loss of generality, as two-dimensional in the vertical x_1,x_3 plane, we shall make use of the following FT relationships (see Chap. 1.6) $\hat{u}(k_1,k_3)=\sqrt{2\pi}\,\hat{u}(k_3)\delta(k_1)$ and $\hat{u}_{11}(k_3,k_3')=\langle\hat{u}(k_3)\hat{u}(k_3')\rangle=\sqrt{2\pi}\,\hat{u}_{11}(k_3)\delta(k_3-k_3')$, the latter resulting from the stationarity of u. Next, the pore-scale dispersion is represented by the following covariances in the vertical plane: $X_{d,11}=2D_L t=2\alpha_L Ut$ and $X_{d,33}=2D_T t=2\alpha_T Ut$, where we have taken the dispersion coefficients as constant, and α_L and α_T are the longitudinal and transverse pore-scale dispersivities, respectively. We are now in a position to derive the first two moments of the displacements covariances, in terms of the spectrum, by substituting the preceding relationships into the general Eq. (4.3.23) for m=2. Thus

$$\langle X_1\rangle = U\,t \quad ; \quad \langle X_3\rangle = X'_3 = 0$$

$$\frac{d^2X_{11}}{dt^2} = \frac{1}{\pi}\iiiint <\hat{u}(k_3')\,\hat{u}(k_3'')\,\exp(ik_1'X_1'>.$$

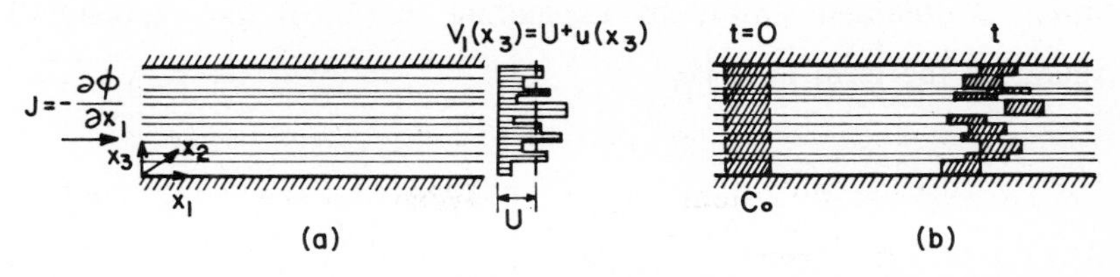

Fig. 4.5.1 Definition sketch for flow parallel to bedding in a stratified formation made up from distinct layers: (a) velocity distribution and (b) the motion of a solute body in the first regime.

$$\exp[ik''_1\langle X_1\rangle-\frac{1}{2}X_{d,11}\,k''^2_1-\frac{1}{2}X_{d,33}\,k''^2_3]\;\delta(k'_1)\;\delta(k''_1)\;dk'_1\;dk''_1\;dk'_3\;dk''_3$$

i.e

$$\frac{d^2X_{11}}{dt^2}=\sqrt{\frac{2}{\pi}}\int_{-\infty}^{\infty}\hat{u}_{11}(k)\exp(-D_T k^2t)\,dk=\sqrt{\frac{2}{\pi}}\frac{U^2}{K_A^2}\int_{-\infty}^{\infty}\hat{C}(k)\exp(-D_T k^2t)\,dk$$

$$X_{t,11}=X_{11}+2D_L t\;;\;X_{t,33}=2D_T t \tag{4.5.3}$$

These formulae are exact and the separation of the velocity spectrum $\hat{u}_{11}$ and the convective displacement X'_1 in the general equation (4.3.23) for X_{11}, is possible since the velocity depends only on x_3 while $\mathbf{X}'$ has its only component in the x_1 direction. Unfortunately, such a separation is not possible for a general two- or three-dimensional structure (see Chaps. 4.6, 4.7).

Summarizing these results, Eqs. (4.5.3) render the first two moments of the particles displacements in terms of the average velocity U, the conductivity autocovariance C_K and the pore scale dispersion coefficients D_L and D_T. We shall analyze now the dependence of the convective "macrodispersion" coefficient $D_{11}=(1/2)dX_{11}/dt$ upon time, to characterize the various stages of the transport process. D_{11} is obtained by an integration over time in (4.5.3), with $D_{11}(0)=0$, leading to the exact expression

$$D_{11}(t)=\frac{1}{2}\frac{dX_{11}}{dt}=\sqrt{\frac{1}{2\pi}}\frac{1}{D_T}\int_{-\infty}^{\infty}\hat{u}_{11}(k)\frac{[1-\exp(-D_T k^2t)]}{k^2}\,dk \tag{4.5.4}$$

(ii) The three regimes of solute transport.

We assume that K has a vertical integral scale I and that I<<B. Under these conditions, we define two dimensionless times, namely $t'=tD_T/I^2$ and $t''=tD_T/B^2=t'(I^2/B^2)$.

The first regime we consider is for $t'<<1$, which can be characterized in words as a time from the beginning of the solute body motion, which is small compared to the time needed for lateral pore-scale dispersion to ensure vertical mixing over the distance I. The expression of D_{11} at leading order is obtained simply by expanding in (4.5.4) the exponential function in a power series and retaining the term in t, i.e.

$$D_{11}(t) \to \sqrt{\frac{1}{2\pi}}\, t \int_{-\infty}^{\infty} \hat{u}_{11}(k)\, dk = u_{11}(0)\, t = \frac{\sigma_K^2}{K_A^2} U^2 t \quad (t' << 1) \tag{4.5.5}$$

Hence, we have arrived at the fundamental result that D_{11} grows linearly with time and is independent of D_T in this early period. This is precisely the early Taylor time of Eq. (4.3.12) and of Exercise 4.3.3. It is easy to arrive at the same result from a simple physical model. Indeed, in the absence of transversal mixing any solute particle moves only horizontally with a velocity $V_{t,1}=U+u(z)+u_d$ and its displacement is simply $X_{t,1}=V_{t,1}\, t$. Since $X'_1=u\, t=U(K'/K_A)t$, we immediately get $X_{11}=U^2(\sigma_K^2/K_A{}^2)t^2$ which leads by differentiation to (4.5.5). It is easy to depict the concentration field in this regime. Indeed, C satisfies the one dimensional dispersion equation

$$\frac{\partial C}{\partial t} + [U+u(z)]\, \frac{\partial C}{\partial x} = D_L \frac{\partial^2 C}{\partial x^2} \tag{4..5.6}$$

of solution

$$C(x,z,t)=\frac{1}{2\sqrt{\pi D_L t}} \exp\{-\frac{[x-Ut-u(z)t-a_1]^2}{4D_L t}\}\, \delta(z-a_3) \quad \text{for} \quad C(x,z,0)=\delta(x-a_1)\, \delta(z-a_3)\, \delta(t) \tag{4.5.7}$$

Furthermore, if the effect of longitudinal pore-scale dispersion is neglected ($D_L=0$), slugs of solute are translated in the horizontal direction by the velocity U+u. Thus, the shape of the solute body can be easily depicted, as shown in Fig. 4.5.1a. This approximation, with some additional requirements of small σ_K^2 has been employed by Mercado (1967) in order to model transport through heterogeneous formations.

Eq. (4.5.7) can be used as the starting point for computing the various statistical moments of C, once those of K are given. However, the expected value of the second spatial moment of a solute body is related (see Eq. 4.3.41) to X_{11}, which has been already evaluated above in Eq. (4.5.3). As for $\langle C \rangle$, its computation is simplified if K is normal; then, $\langle C \rangle$ satisfies (4.3.29) with D_{11} given by (4.5.5), and its solution is similar to (4.5.7) in which u=0 and $D_{d,L}$ is replaced by $D_{t,11}=D_{11}+D_{d,L}$.

The second asymptotic regime is for t'>>1 and t"<<1, i.e. for a time which is large compared to the one needed in order to ensure vertical mixing across I, but is small compared to the one required for mixing across the entire formation thickness. In this regime, the presence of the aquifer lower and upper boundaries are not felt, except for their immediate neighborhood, and B is immaterial. The appropriate limit in (4.5.4) is for $t \to \infty$ and for u_{11} pertaining to an infinite domain. Since the main contribution in the integrand is in the neighborhood of k=0, we get at leading order

$$D_{11}(t) \rightarrow \sqrt{\frac{1}{2\pi}} \frac{\hat{u}_{11}(0)}{D_{d,T}} \int_{-\infty}^{\infty} \frac{1-\exp(-D_{d,T}k^2t)}{k^2} dk \qquad (4.5.8)$$

By the definition of the FT, $\hat{u}_{11}(0) = (1/\sqrt{2\pi})\int u_{11}(r_3)dr_3 = \sqrt{2/\pi}(U^2\sigma_K^2/K_A^2)\,I$. Substituting it in (4.5.8), and performing the integration yield

$$D_{11}(t) = \frac{2}{\sqrt{\pi}} \frac{\sigma_K^2}{K_A^2} \frac{U^2I^2}{D_{d,T}} \left[\frac{tD_{d,T}}{I^2}\right]^{1/2} \qquad (4.5.9)$$

which is the result obtained by Matheron and de Marsily (1980). At this limit $\langle C\rangle$ is Gaussian, i.e. it satisfies the transport Eq. (4.3.29). However, the transport is not Fickian, since D_{11} grows like $t^{1/2}$ rather than tending to a constant value. This apparent contradiction with the asymptotic Taylor's result (4.3.12) stems from the peculiar nature of a stratified formation, in which conductivities and velocities are correlated in the horizontal plane over an infinite distance, violating the basic requirement of Taylor's theory which leads to (4.3.12). Hence, the vertical mixing effect of $D_{d,T}$ is to offset only partially the tendency of linear growth with time caused by stratification. Although this mechanism may seem to explain the large values of longitudinal macrodispersivities measured in field tests (see Chap. 4.2) and its change with the travel distance L=Ut, one should realize that perfect stratification does not generally persist over the very large distances needed to warrant the use of (4.5.9).

To further illustrate the two above regimes, an exponential covariance for K has been adopted, i.e. $C_K = \sigma_K^2 \exp(-r_3/I)$ and $\hat{u}_{11}(k) = \sqrt{2/\pi}\, U^2\sigma_K^2\, I/K_A^2(1+k^2I^2)$. Substitution in (4.5.4) yields, after integration over k, the close form result

$$\frac{D_{11}}{UI} = \frac{UI}{D_{d,T}} \frac{\sigma_K^2}{K_A^2} f(t') \quad \text{with } f(t') = \{2\sqrt{\frac{t'}{\pi}} - 1 + [1-\mathrm{erf}(\sqrt{t'})]\exp(t')\} \quad ; \quad t' = tD_{d,T}/I^2$$

and (4.5.10)

$$f \rightarrow t'[1 - \frac{4}{3}\sqrt{\frac{t'}{\pi}} + O(t')] \quad t' \rightarrow 0 \quad ; \quad f \rightarrow 2\sqrt{\frac{t'}{\pi}} - 1 + O(\frac{1}{\sqrt{t'}}) \quad t' \rightarrow \infty$$

which has been represented in Fig. 4.5.2. It is seen that in this particular case the small t' regime is valid for say $L/I \leq 0.25(I/\alpha_{d,T})$ whereas the large t' one is an accurate approximation for say

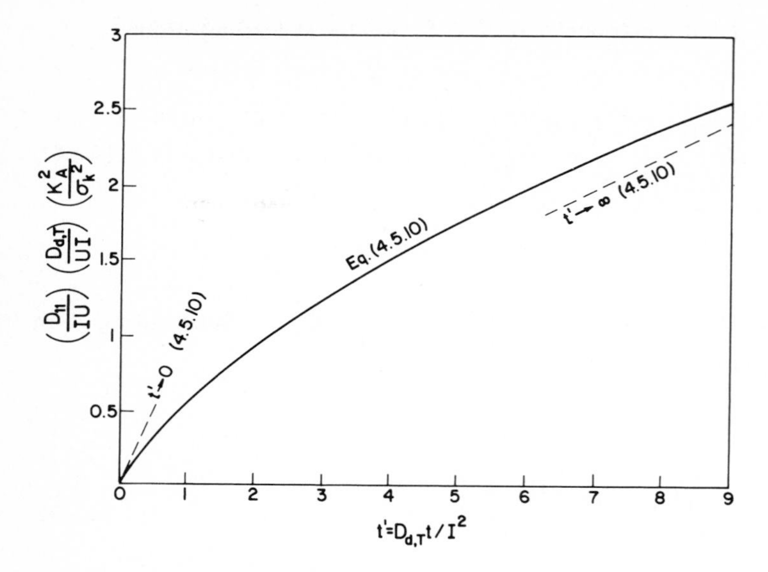

Fig. 4.5.2 The dependence of the "macrodispersivity" D_{11} (4.5.10) upon dimensionless time.

$L/I>18(I/\alpha_{d,T})$. Assuming, for instance, $I/\alpha_{d,T}\simeq 100$ and $I\simeq 0.1$ m leads to $L<2.5$ m and $L>180$ m, respectively, provided that perfect stratification occurs at such a large scale. This result has been obtained in the fundamental study of Matheron and de Marsily (1980), who have also derived $D_{11}(t)$ for a Gaussian (3.2.23) conductivity covariance and also a "hole" covariance of zero integral scale. The "macrodispersion" D_{11} is represented as a function of time in their Fig. 6. One of the striking results is that D_{11} for the exponential and Gaussian covariances are indistinguishable, for equal integral scales. This is indicative of the lack of sensitivity of D_{11} to the precise shapes of different C_K curves, provided they have the same integral scales.

The last regime we consider is for $t''>>1$, i.e. for the case in which lateral pore-scale dispersion causes mixing over the entire thickness of the formation. This is the regime investigated by Taylor (1953) and Aris (1956) for laminar flow, of deterministic convective velocity distribution (see Sect. 2.10.2) and applied to flow through heterogeneous formation by Marle et al (1967), Gelhar et al (1979), Güven et al (1984,1986). The main result of Taylor-Aris analysis is that the dispersive regime is Fickian at the above limit, and D_{11} tends to a constant value, of the order $U^2B^2/D_{d,T}$, i.e. to a macrodispersivity $\alpha_{ef,L}=D_{11}/U=O(B^2/\alpha_{d,T})$. This value may be extremely large due to the smallness of the transverse pore-scale dispersivity (see Chap. 2.10), and this was regarded as a possible explanation of large dispersivity values encountered in the field. To attain this regime, however, stratification has to be present over travel distances of hundreds or thousands of aquifer thicknesses, and such an occurrence is very improbable, if not impossible. For this reason we shall not dwell here upon the results related to the third regime. Still, it is interesting to show how can one obtain quick estimates of Taylor-Aris dispersion coefficient by employing an idealized flow model suggested by Güguven and Molz (1986). The latter assume a periodic and deterministic velocity distribution, i.e. $u=U\cos(N\pi r_3/B)$, where N is an integer. This is generalized to a random distribution by introducing a phase, i.e. $u=U\cos[(N\pi r_3/B)+\beta]$, where β is a random variable of uniform distribution in the interval 0-π. In words, the velocity profile is fixed except for a random translation and it is assumed to exist over the entire space, i.e. the profile is reflected across the boundaries. The pertinent covariance is easily obtained in the form suggested by Güven and Molz (1986), namely $u_{11}(r_3)=(1/2)U^2\cos(N\pi r_3/B)$. This can be used in the general equation (4.5.4) in order to compute D_{11}, as pointed out by Sposito and Barry (1987). Indeed, $\hat{u}(k)=(1/2)\sqrt{2\pi}\,U^2\,[\delta(k+N\pi/B)+\delta(k-N\pi/B)]$ (see Chap. 1.6) and substitution in (4.5.4) yields the simple close form result

$$\frac{D_{11}(t)}{UB}=\frac{1}{2\pi^2N^2}\frac{UB}{D_{d,T}}[1-\exp(D_{d,T}N^2\pi^2t/B^2)] \tag{4.5.11}$$

Thus, in the case of a two-layered formation, i.e. N=1, the asymptotic limit is $D_{11}/UB\to(1/2\pi^2)(UB/D_{d,T})$, of the type mentioned before. It is pointed out that the periodic velocity covariance is of a zero integral scale, i.e. it has the character of a hole covariance, which is typical for a periodic stationary process. Consequently, the second regime mentioned above is not present in such an idealized case.

Further generalizations of transport for three-dimensional periodic media of a deterministic structure have been investigated by Gupta and Bhattacharya (1986).

(iii) The spatial moments of the concentration expected value.

For the sake of illustration, we shall limit the discussion to the first regime of transport, i.e. we may assume an infinite vertical extent and a conductivity covariance of finite integral scale.

In order to evaluate the spatial moments of a solute body in the first regime, we may use directly the close form solution for the concentration (4.5.7). Leaving this as exercises, we shall rely, for the sake of illustration, on the approach of Sect. (4.3.6), of a more general nature. To evaluate the second-order spatial moment, we need to derive first $Z_{j\ell}$, the covariance of displacements of two particles (Sect. 4.3.6). In the present case the only component different from zero is Z_{11}, whose expression (4.3.31) is particularized to a stratified formation by the same substitutions which led to X_{11} (4.5.3). The immediate result (see also Dagan, 1987a) is

$$\frac{dZ_{11}(t,\mathbf{a},\mathbf{b})}{dt}=\sqrt{\frac{2}{\pi}}\frac{1}{D_{d,T}}\int_{-\infty}^{\infty}\hat{u}(k)\,\frac{1-\exp(-k^2D_{d,T}t)}{k^2}\cos[k(b_3-a_3)]dk \tag{4.5.12}$$

with no dependence, in the present case, on the horizontal coordinates of the initial particles positions. Obviously, for $b_3=a_3$ Eq. (4.5.12) degenerates into (4.5.3). We are now in a position to derive the expected value of the "moment of inertia" $\overline{X}_{11}$ (4.3.41) of a solute body inserted over a volume

V_0 (the expected value of the abscissa of the center of gravity is obviously equal to Ut). Since the horizontal coordinates are immaterial, V_0 is taken as a pulse along the vertical axis on a segment of length L_0. Hence, $\bar{\sigma}_{11}$ (4.3.39), the variance of the horizontal displacement of the center of gravity, reduces now to

$$\bar{\sigma}_{11}(t) = \frac{2}{L_0^2} \int_0^{L_0} (L_0-a)\, Z_{11}(t,a)\, da \tag{4.5.13}$$

where the integration variable replaces b_3-a_3 in (4.5.12) and Cauchy algorithm has been used to reduce the double integration to one. Starting again with the first regime, i.e. $t' \ll 1$, we get from (4.5.12), after expanding the exponential, the simple result $Z_{11}=X_{11}C_K(b_3-a_3)/\sigma_K^2$, which after substitution in (4.5.13) yields $\bar{\sigma}_{11}=X_{11}\sigma_{\bar{K}}^2/\sigma_K^2$. It is reminded (Sect. 3.2.4) that $\sigma_{\bar{K}}^2$ is the variance of the conductivity space average $\bar{K}=(1/L_0)\int K(z)dz$. Hence, we arrive at the following expression for the expected value of the second-order spatial moment (4.3.41) of the solute body

$$\langle \bar{X}_{11}(t)\rangle = \bar{X}_{11}(0) + X_{11}(t)\,[1-\sigma_{\bar{K}}^2/\sigma_K^2] + 2D_{d,L}\, t \tag{4.5.14}$$

It is seen, therefore, that $\langle \bar{X}_{11}\rangle$ grows quadratically with time, precisely like X_{11} (4.5.5). In particular, if $L_0 \gg I$, i.e. if the initial vertical extent of the solute body is much larger than the conductivity integral scale, we have from Sect. (3.2.4) $\sigma_{\bar{K}}^2/\sigma_K^2 \to 2I/L_0$ and in (4.5.14) the square bracket, which is equal to $1-2I/L_0$ tends to unity. In the opposite case of $L_0 \ll I$, the ratio in the square bracket tends to zero and the solute body does not undergo any longitudinal spread, except for the effect of pore scale dispersion. This is easily understandable if we refer to Fig. 4.5.1b: the solute body in any realization is now one of the horizontal slabs, which are translating at velocity U+u.

As we have already mentioned, if K is normal, $\langle C\rangle$ is Gaussian and dX_{11}/dt (4.5.14) characterizes the rate of spread of the solute body. It was shown in Chap. (3.2) that K is generally of a lognormal distribution, the negative values of the conductivity present in the normal distribution being thus excluded (still, for $\sigma_Y^2 \ll 1$ the two distribution are very close, and $\langle C\rangle$ is approximately Gaussian even in the first regime). Since the solution (4.5.14) is exact, we are in a position to investigate the validity of a small perturbation approximation in σ_Y^2 upon transport. Indeed, returning to D_{11} (4.5.5) and by using the relationships between the statistical moments of K and $Y=\ell n\, K$ (Chap. 3.4), we get

$$D_{11}(t) = [\exp(\sigma_Y^2)-1]U^2 t \tag{4.5.15}$$

If U is known, as it happens for instance in the case of a tracer test in which U has been measured, a first order approximation for small σ_Y^2 is valid if $\exp(\sigma_Y^2)-1 \simeq \sigma_Y^2$. Thus, for $\sigma_Y^2<0.1$ the error is less than 0.05. If, however, the only available information is based on head and conductivity measurements, U is derived indirectly from $U=K_{ef}J/n$. In the case of a stratified formation and flow parallel to bedding $K_{ef}=K_A$ (Part. 3), and consequently (4.5.5) leads to

$$D_{11}(t) = \frac{\sigma_K^2 J^2}{n} t = \exp(\sigma_Y^2)[\exp(\sigma_Y^2)-1] \frac{K_G J^2}{n} t \tag{4.5.16}$$

and the same degree of accuracy is now achieved only for $\sigma_Y^2<0.03$. Thus, it is concluded that for this type of flow the first-order approximation is of a quite limited validity. Furthermore, an inconsistent approximation in which the effective conductivity is kept in its nonlinear dependence upon σ_Y^2 is beneficial.

In the case of a lognormal distribution $\langle C\rangle$ is no more Gaussian and it does not satisfy Eq. (4.3.29). Higher-order spatial moments are then needed in order to characterize $\langle C\rangle$, e.g. the third-order one, which is related to skewness. Thus, from the simple definition of $X_1'=UK'/K_A\, t= U[\exp(Y'-\sigma_Y^2/2)-1]t$, we immediately have

$$\langle[X_1'(t)]^3\rangle = U^3\,[\exp(3\sigma_Y^2) - 3\exp(\sigma_Y^2) +2]\, t^3 \tag{4.5.17}$$

which is of the order $3\sigma_Y^4$ for $\sigma_Y^2<1$. The lateral pore-scale dispersion, effective in the second regime, would tend to render $\langle C\rangle$ Gaussian. This is, however, a slow process and the second regime is anyway not attainable in most applications due to the requirements of extensive stratification mentioned above.

(iv) Concentration variance.

The general approach to the computation of the concentration variance in a formation of random heterogeneity has been described in Sect. 4.3.5. As we have shown, the variance depends on a few factors, like the extent of V_0, the initial solute body, of V, the averaging volume surrounding the point $\mathbf{x}$ and, of course, on time. Due to this multitude of factors it is quite difficult to analyze in a general manner the various possibilities. Instead, we shall consider two extreme cases which are easy to grasp. The first one pertains to the first, $t'<<1$, regime in which the solute is convected horizontally (Fig. 4.5.1b), the initial concentration C_0 is constant and the point value of the concentration is considered. In this case the derivation of Sect. 4.3.5 applies, namely C has the bimodal p.d.f. of Fig. 4.3.2b and the variance is given by Eq. (4.3.35). The coefficient of variation $\sigma_C/\langle C\rangle=C_0/\langle C\rangle-1$ may attain quite large values, which can be easily evaluated after determining $\langle C\rangle$ as shown before. We shall retake this discussion in the next section, dealing with a few possible field applications. At the other extreme, we consider the gross features of the solute body, namely its spatial moments, which are also random variables, analyzed in Sect. (4.3.6). Thus, starting with the motion of the center of obtained in (4.3.23) by taking $\mathbf{X}'=0$ in the exponent (the implications are discussed in Chap. 4.8). Eq. (4.3.23) simplifies considerably and becomes

$$\frac{d^2X_{j\ell}}{dt^2} = \frac{2}{(2\pi)^{3/2}} \int \hat{u}_{j\ell}(\mathbf{k}) \left[\exp(i\mathbf{k}\cdot\mathbf{U}t - \sum_{p=1}^{3}\sum_{q=1}^{3} D_{d,rs} k_r k_s t)\right] d\mathbf{k} \quad (j,\ell=1,2,3) \tag{4.6.2}$$

Unlike (4.3.23), Eq. (4.6.2) renders $X_{j\ell}$ as an explicit function of the velocity spectrum $\hat{u}_{j\ell}$. Before connecting the latter with the properties of the heterogeneous structure, we shall discuss a few implications of (4.6.2) (see Dagan, 1987). First, by inverting $\hat{u}_{j\ell}$ in (4.6.2) we obtain

$$\frac{d^2X_{j\ell}}{dt^2} = \frac{2}{(2\pi)^{3/2}} \iint \hat{u}_{jl}(\mathbf{x}') \exp[-i\mathbf{k}\cdot(i\mathbf{x}'-\mathbf{U}t) - \sum_{q=1}^{3}\sum_{p=1}^{3} D_{d,pq} k_p k_q t]\, d\mathbf{k}\, d\mathbf{x}' \tag{4.6.3}$$

which can be integrated exactly over $\mathbf{k}$ to yield the alternative formulation

$$\frac{d^2X_{j\ell}}{dt^2} = \frac{1}{4\pi^{3/2} |\mathbf{D}_d|^{1/2} t^{1/2}} \int u_{jl}(\mathbf{x}') \exp\left[-\sum_{p=1}^{3}\sum_{q=1}^{3} \frac{(x'_p - U_p t)(x'_q - U_q t) D^{-1}_{d,pq}}{4t}\right] d\mathbf{x}' \tag{4.6.4}$$

where $|\mathbf{D}_d|$ is the determinant and $D^{-1}_{d,pq}$ is the inverse of the matrix $\mathbf{D}_d$. In a coordinate system in which x_1 is parallel to $\mathbf{U}$, this matrix is diagonal and of components $D_{d,L}$, $D_{d,T}$ and $D_{d,T}$, so that $|D_d| = D_{d,L} D_{d,T}^2$ and $D^{-1}_{d,pq}$ is also diagonal, of components $1/D_{d,L}$, $1/D_{d,T}$ and $1/D_{d,T}$, respectively. Taking the components of the pore-scale dispersion as constant, and related to the mean velocity U, is consistent with the other approximations.

Eq. (4.6.4) has a simple physical interpretation: it could be obtained from the outset by assuming that fluid particles are convected by the velocity field $\mathbf{U}+\mathbf{u}$ and diffuse independently at the constant rate $\mathbf{D}_d$, neglecting therefore transfer between streamlines due to pore-scale dispersion, which leads to higher-order effects in σ_Y^2 (this scheme has constituted the starting point in Dagan, 1982b).

Results similar to (4.6.2) have been obtained by using other asymptotic expansions, with the equation satisfied by concentration (4.1.1) as the starting point, by Chu and Sposito (1980) and by Winter et al (1984)

Returning to (4.6.2), we shall substitute in it the expression of the first-order small perturbation approximation of the velocity spectrum. Recalling the first-order approximations of the specific discharge (3.4.7) and head (3.3.9) we have

$$\langle \mathbf{q} \rangle = K_G \mathbf{J} \quad ; \quad \mathbf{q}' = K_G \ (\mathbf{J}\ Y' - \nabla\phi) \quad ; \quad \nabla^2\phi = \mathbf{J}.\nabla Y' \qquad \text{i.e.}$$

$$\mathbf{U} = K_G \mathbf{J}/n \quad ; \quad \hat{\phi} = -\frac{i\mathbf{k}\cdot\mathbf{J}}{k^2}\hat{Y}' \quad ; \quad \hat{u}_j = \frac{K_G}{n}(J_j \hat{Y}' + i\, k_j \hat{\phi}) = U_j \frac{1 - ik_j}{k^2}\hat{Y}'$$

which leads to the final expression for the spectrum

$$\hat{u}_{j\ell}(\mathbf{k}) = \sum_{p=1}^{3}\sum_{q=1}^{3} U_p U_q (\delta_{pj} - \frac{k_j k_q}{k^2})(\delta_{q\ell} - \frac{k_\ell k_q}{k^2}) \hat{C}_Y(\mathbf{k}) + O(\sigma_Y^4) \tag{4.6.5}$$

Substitution of (4.6.5) into (4.6.2) yields

$$\frac{dX^2_{j\ell}}{dt^2} = \frac{2}{(2\pi)^{3/2}}) \sum_{p=1}^{3}\sum_{q=1}^{3} \frac{K_G^2 J_p J_q}{n^2} \int (\delta_{pj} - \frac{k_j k_p}{k^2})(\delta_{q\ell} - \frac{k_\ell k_q}{k^2}).$$

$$\hat{C}_Y(\mathbf{k}) \exp(i\mathbf{k}\cdot\mathbf{U}t - \sum_{r=1}^{3}\sum_{s=1}^{3} D_{d,rs} k_r k_s t)\, d\mathbf{k} \qquad (j,\ell=1,2,3) \tag{4.6.6}$$

This fundamental equation renders the second derivative of the particles displacement covariance at first-order in σ_Y^2 as function of the average head gradient $-\mathbf{J}$, of the heterogeneous structure represented by C_Y and of pore-scale dispersion. It is emphasized that in (4.6.6) $\mathbf{U}$ has been replaced by $K_G \mathbf{J}/n$, resulting from the first term of the expansion of $\langle \mathbf{q} \rangle = K_{eff} \mathbf{J}$, where the effective conductivity tensor can be written as $K_{eff} = K_G(1+\gamma)$, with $\gamma = 0(\sigma_Y^2)$ (Sect. 3.4.2). We shall touch this point again in Sect. 4.6.5.

By the same token, we may use the alternative formulation (4.6.4), which does not employ Fourier transforms, and replace in it the velocity covariance by its first-order expression (3.7.26).

Finally, the further approximation of high Peclet number is obtained from (4.6.5) by taking $D_{d,j\ell}=0$. This leads to

$$\frac{dX^2_{j\ell}}{dt^2} = \frac{2}{(2\pi)^{3/2}} \sum_{p=1}^{3}\sum_{q=1}^{3} \frac{K_G^2 J_p J_q}{n^2} \int (\delta_{pj} - \frac{k_j k_p}{k^2})(\delta_{q\ell} - \frac{k_\ell k_q}{k^2}).$$

$$\hat{C}_Y(\mathbf{k}) \exp(i\mathbf{k}\cdot\mathbf{U}t)\, d\mathbf{k} \qquad (j,\ell=1,2,3) \tag{4.6.7}$$

or in (4.6.4) to

$$\frac{d^2X_{j\ell}}{dt^2} = 2\, u_{j\ell}(Ut) \qquad (j,\ell=1,2,3) \tag{4.6.8}$$

Eq. (4.6.8) yields by straightforward integration

$$X_{j\ell}(t) = 2 \int_0^t (t-\tau)\, u_{j\ell}(U\tau)\, d\tau \tag{4.6.9}$$

which was the starting point in Dagan (1982b, 1984). It has a simple physical interpretation: the actual particle displacement $\mathbf{X}_t$ in the fundamental kinematical equation $d\mathbf{X}/dt=\mathbf{u}(\mathbf{X}_t)$ (4.3.18), is replaced by the average trajectory, i.e. straight lines of equation $\langle\mathbf{X}\rangle$=Ut. Intuitively speaking, this approximation is bound to apply if the velocities spanned along the average streamline do not differ too much from those encountered along the actual trajectory.

Once $\langle\mathbf{X}\rangle$ and $X_{j\ell}$ $(j,\ell=1,2,3)$ are determined in terms of the average head gradient and of the formation properties, the expected value of the trajectory of the center of gravity of the solute body (4.3.37, 4.3.39) and of its second-order moment (4.3.41) become known, provided that its lateral extent is large compared to the transverse correlation scales. Otherwise, one needs to evaluate also the two-particle displacements correlation (4.3.31), which have expressions similar to (4.6.2). Indeed, starting with (4.3.31) we obtain, for instance, for small σ_Y^2 and 1/Pe

$$\frac{d^2Z_{j\ell}}{dt^2} = \frac{2}{(2\pi)^{(3/2)}} \sum_{p=1}^{3}\sum_{q=1}^{3} \frac{K_G^2 J_p J_q}{n^2} \int (\delta_{pj} - \frac{k_j k_p}{k^2})\, (\delta_{q\ell} - \frac{k_\ell k_q}{k^2}).$$

$$\hat{C}_Y(\mathbf{k})\, \exp[i\mathbf{k}\cdot(U t+\mathbf{b}-\mathbf{a})]\, d\mathbf{k} \qquad (j,\ell=1,2,3) \tag{4.6.10}$$

which reduces to (4.6.2) for $\mathbf{a}=\mathbf{b}$.

An important point is that the *first-order approximation* of the particle displacement is multivariate normal at *any time*, and not only for large t. Indeed, by (3.4.7) **u** is a linear functional of Y′ and, furthermore, **X′** (4.3.18) is linear in **u**. Therefore, the expected value of the concentration $\langle C(\mathbf{x},t)\rangle$ is Gaussian and satisfies the transport equation (4.3.27) with $d\langle\mathbf{X}\rangle/dt=$ U, $D_{t,j\ell}=D_{j\ell}+D_{d,j\ell}$ and $D_{j\ell}=(1/2)\, dX_{j\ell}/dt$. Hence, the basic Eqs. (4.6.5, 4.6.6) above can be employed as the starting point for the computation of the "macrodispersivity" $D_{j\ell}$ as well (see next section).

We proceed now with the effective computation of $X_{j\ell}$ and of $D_{j\ell}$ for a specific covariance C_Y.

4.6.3 Time dependent "macrodispersivity" (first-order approximation, high Pe)

To illustrate and apply the preceding general results, we adopt the exponential C_Y (3.2.14). Furthermore, to simplify the computations we consider the axisymmetric case, i.e. $I_{Y1}=I_{Y2}=I_{Yh}$ and $I_{Y3}=I_{Yv}$, the anisotropy ratio being defined as $e=I_{Yv}/I_{Yh}$. The same anisotropic covariance has been adopted in Sect. (3.4.2) to compute the effective conductivity and in Sect. (3.7.2) to evaluate various statistical moments of the head. The following expressions are recalled from Sect. (3.4.2)

$$C_Y(\mathbf{r}) = w[1-H(\mathbf{r})] + \sigma_Y^2 \exp(-r') \quad ; \quad r'=[(r_1/I_{Yh})^2+(r_2/I_{Yh})^2+(r_3/I_{Yv})^2]^{1/2}$$

$$\hat{C}_Y(\mathbf{k}) = I_{Yh}^2 I_{Yv}\, \hat{C}_Y(\mathbf{k}')$$

$$\hat{C}_Y(\mathbf{k}') = \frac{1}{(2\pi)^{3/2}} \int C_Y(\mathbf{r}') \exp(i\mathbf{k}'\cdot\mathbf{r}')\, d\mathbf{r}' = \sqrt{\frac{8}{\pi}}\, \sigma_Y^2 \frac{1}{(1+k'^2)^2}$$

$$k'_1=k_1 I_{Yh} \; ; \; k'_2=k_2 I_{Yh} \; ; \; k'_3=k_3 I_{Yv} \; ; \; k'=|\mathbf{k}'|$$

where the "nugget" effect w is immaterial for transport and is subsequently dropped out. To derive $X_{j\ell}$, $\hat{C}_Y(\mathbf{k})$ is substituted into (4.6.7) first, followed by five integrations : two with respect to the travel time t (with the initial conditions $X_{j\ell}=dX_{j\ell}/dt=0$ for t=0) and three with respect to the components of the wave-number vector **k**. As for the latter, it is convenient to switch to polar coordinates in the horizontal plane of the **k**′ space, i.e. $k'_1=K\cos\theta$, $k'_2=K\sin\theta$ and $dk'_1\, dk'_2= K\, dK\, d\theta$. After these steps, four integrations, with respect to t, k'_3 and θ can be carried out in a close form. Herewith the results for horizontal flow, i.e. the average head gradient **-J** parallel to the x_1 axis, in which case $X_{j\ell}$ is a diagonal tensor of principal axes parallel to the coordinates,

$$\frac{X'_{11}(t';e)}{\sigma_Y^2} = 2t' + 2(e^{-t'}-1) +$$

$$8e\int_0^\infty [J_0(Kt')-1]\left[\frac{1}{(1+K^2-e^2K^2)^2} - \frac{eK}{(1+K^2-e^2K^2)^2(1+K^2)^{1/2}} - \frac{eK}{2(1+K^2-e^2K^2)(1+K^2)^{3/2}}\right]dK\; -$$

$$-2e\int_0^\infty [J_0(Kt')-\frac{J_1(Kt')}{Kt'}-\frac{1}{2}]\left[\frac{e^3K^3(e^2K^2-5-5K^2)}{(e^2K^2-1-K^2)^3(1+K^2)^{3/2}} + \frac{1+K^2-5e^2K^2}{(1+K^2-e^2K^2)^3}\right]dK \tag{4.6.11}$$

$$\frac{X'_{22}(t';e)}{\sigma_Y^2} = -2e\int_0^\infty [\frac{J_1(Kt')}{t'}-\frac{K}{2}]\left[\frac{e^3K^2(e^2K^2-5K^2-5)}{(e^2K^2-1-K^2)^3(1+K^2)^{3/2}} + \frac{1+K^2-5e^2K^2}{K(1+K^2-e^2K^2)^3}\right] dK \tag{4.6.12}$$

$$\frac{X'_{33}(t';e)}{\sigma_Y^2} = -4e\int_0^\infty [J_0(Kt')-1]\left\{\frac{1}{(e^2K^2-1-K^2)^2}\left[\frac{1}{2}+\frac{2e^2K^2}{1+K^2-e^2K^2}+\frac{eK(e^2K^2+3+3K^2)}{2(e^2K^2-1-K^2)(1+K^2)^{1/2}}\right]\right\}dK$$

(4.6.13)

where J_0 and J_1 are the Bessel functions of zero and first order, respectively. These results are reproduced from Dagan (1988) and as expected for this small σ_Y^2 and large Pe approximation, the covariances $X'_{j\ell} = \sigma_Y^2$ funct(t′,e), where the dimensionless variables are $X'_{j\ell}=X_{j\ell}/I_{Yh}^2$ and $t'=tU/I_{Yh}$. The integrations in (4.6.10)-(4.6.13) have been carried out numerically and the results are given in Fig. (4.6.1) for the the dependence of the longitudinal covariance X_{11} upon t′, and in Figs. (4.6.2) and (4.6.3) for the lateral X_{22} and vertical X_{33} ones, respectively, for a few values of the anisotropy ratio $e \leq 1$. In particular, in the isotropic case e=1 the integration with respect to K in (4.6.10)-(4.6.13) can be carried out in a close form, the result being as follows

$$\frac{X'_{11}(t')}{\sigma_Y^2} = 2t' - 2\left[\frac{8}{3} - \frac{4}{t'} + \frac{8}{t'^3} - \frac{8}{t'^2}\left(1+\frac{1}{t'}\right)e^{-t'}\right] \quad ; \quad X'_{j\ell}=X_{j\ell}/I_{Yh}^2 \ , \ t'=tU/I_{Yh} \qquad (4.6.14)$$

$$\frac{X'_{22}}{\sigma_Y^2} = \frac{X'_{33}}{\sigma_Y^2} = 2\left[\frac{1}{3} - \frac{1}{t'} + \frac{4}{t'^3} - \left(\frac{4}{t'^3}+\frac{4}{t'^2}+\frac{1}{t'}\right)e^{-t'}\right] \qquad (4.6.15)$$

These relationships have been obtained by Dagan (1984) by using the formulation (4.6.9), (3.7.26), the covariance $C_{Y\phi}$ (3.7.8) and the variogram Γ_ϕ (3.7.12) directly. We shall discuss now these results by referring to Figs. 4.6.1-4.6.3.

The longitudinal covariance X_{11} (Fig. 4.6.1), in the direction of the mean flow and in the plane of isotropy of C_Y, grows monotonously with the the travel time (or equivalently, with the average travel distance $\langle X_1 \rangle = Ut$) of the solute body. At the limit $t' \to 0$ its behavior is the general one resulting from Taylor's theory (4.3.14). Thus, in the isotropic case one has from (4.6.3), as shown in Dagan (1984),

$$X_{11} \to u_{11}(0)\, t^2 = \frac{8}{15}\sigma_Y^2 U^2 t^2 \qquad (t'=tU/I_Y \ll 1,\ U=K_G J/n) \qquad (4.6.16)$$

whereas, for $t' \gg 1$, X_{11} grows linearly with time, again in agreement with the general Taylor's result (4.3.12) ; the large t′ behavior will be discussed separately in the sequel. This is in contrast with the case of stratified formations (Chap. 4.5), since in the present case the displacements of fluid particles separated by a time lag which is large compared to I_Y/U, become uncorrelated. One of the striking results revealed by Fig. 4.6.1 is that the anisotropy ratio e has a relatively small impact upon

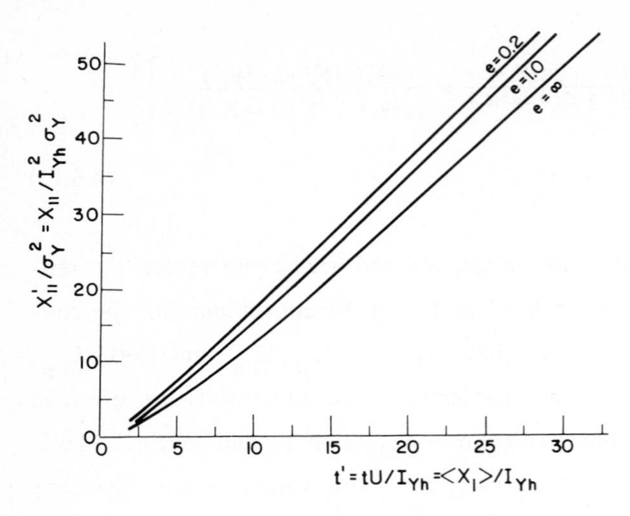

Fig. 4.6.1 The longitudinal displacements covariance X_{11} (4.6.11) as a function of the travel time t. Anisotropic exponential logconductivity covariance C_Y, with Cartesian coordinates in its principal directions, of integral scales $I_{Y1}=I_{Y2}=I_{Yh}$, $I_{Y3}=I_{Yv}$ and with $e=I_{Yv}/I_{Yh}$. Uniform average flow in the x_1 direction. First approximation in σ_Y^2, high Peclet number (reproduced from Dagan, 1988).

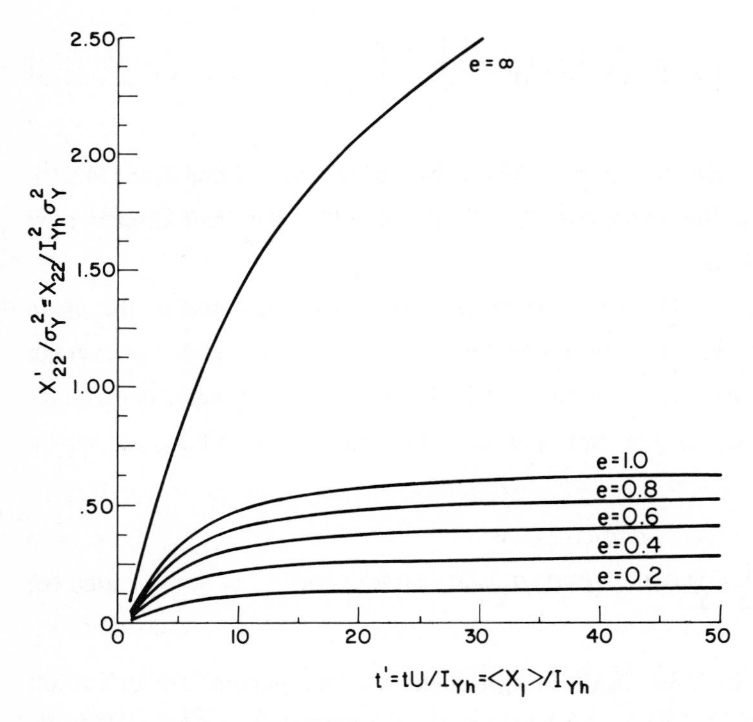

Fig. 4.6.2 The lateral displacements covariance X_{22} (4.6.12) as a function of the travel time t. Same conditions as in Fig. 4.6.1 (reproduced from Dagan, 1988).

(iii) The expected value of the concentration moments

The derivations of the last two sections permit one to evaluate the displacements expected values $\langle \mathbf{X} \rangle$ and covariances $X_{j\ell}$ or the associated effective dispersion coefficients $D_{j\ell}=(1/2)(dX_{j\ell}/dt)$ $(j,\ell=1,2,3)$. With these quantities the expected value of the first $\langle \overline{\mathbf{X}} \rangle$ and of the second spatial moment of the concentration $\langle \overline{X}_{t,j\ell} \rangle$ (4.3.41) can be computed, provided that the extent of the solute body is large enough to warrant the neglect of $\bar{\sigma}_{j\ell}$ (4.3.39), i.e. if ergodicity is obeyed for $\overline{\mathbf{X}}$, the displacement of the center of mass. It is relatively easy to examine the dependence of $\bar{\sigma}_{11}$ in the asymptotic stage upon the initial size of the solute body. Toward this aim, we have first to calculate the two particles displacements covariance Z_{11} (4.6.10), which for large t and large Pe has the simple expression, generalization of (4.6.9)

$$Z_{11}(t,\mathbf{a},\mathbf{b}) = 2t\int_0^\infty u_{11}(Ut+|a_1-b_1|,|a_2-b_2|,|a_3-b_3|)\, dt =$$

$$2tU^2\sigma_Y^2\int_0^\infty \rho_Y(Ut+|a_1-b_1|,|a_2-b_2|,|a_3-b_3|)\, dt \tag{4.6.25}$$

where the mean flow is in the x_1 direction. To get some quick estimates we shall employ the separated exponential C_Y (3.2.25), i.e. $\rho_Y=\exp(-r_1/I_{Y1} - r_2/I_{Y2} - r_3/I_{Y3})$. This leads immediately in (4.6.25) to the following close form expression

$$Z_{11}(t,\mathbf{a},\mathbf{b}) = 2\, t\, U\, \sigma_Y^2\, I_{Y1}\, \exp\left[-\frac{|a_1-b_1|}{I_{Y1}}-\frac{|a_2-b_2|}{I_{Y2}}-\frac{|a_3-b_3|}{I_{Y3}}\right] \tag{4.6.26}$$

which for $\mathbf{a}=\mathbf{b}$ degenerates into the previously obtained $X_{11}= 2\, t\, U\, \sigma_Y^2\, I_{Y1}$. To arrive at $\bar{\sigma}_{11}$ (4.6.10), we have to integrate Z_{11} over the initial volume V_0 of the solute body, which is taken for simplicity to be a rectangle of sides L_1,L_2 and L_3. Then, by using Cauchy algorithm, we get

$$\bar{\sigma}_{11} = \frac{1}{V_0^2}\int\!\!\int Z_{11}(t,\mathbf{a},\mathbf{b})\, d\mathbf{a}\, d\mathbf{b} = X_{11}\, f(I_{Y1}/L_1)\, f(I_{Y2}/L_2)\, f(I_{Y3}/L_3)$$

where $X_{11}=2tU\sigma_Y^2I_{Y1}$ and $f(I_Y/L) = \frac{1}{L}\int_0^L (L-\alpha)\exp(-\frac{\alpha}{I_Y})\, d\alpha =$

$$\frac{2I_Y}{L}[1-\frac{I_Y}{L}+\frac{I_Y}{L}\exp(-\frac{L}{I_Y})] \tag{4.6.27}$$

The function f embodies the effect of space averaging and is close to unity for $I_Y/L>4$ and tends to I_Y/L for $I_Y/L<1/20$. Since generally the vertical integral scale is much smaller than the horizontal ones, the main effect in reducing $\bar{\sigma}_{11}$ results from $f(I_{Y3}/L_3)$, and for a vertical extent of the solute body say larger then twenty integral scales, $\bar{\sigma}_{11}\simeq 0$. Hence, this may be used as a practical rule to check whether the ergodic requirement is obeyed by the trajectory of the center of mass and the "moment of inertia" of the solute body. In the latter case $\bar{\sigma}_{j\ell}$ can be dropped from $\langle \bar{X}_{t,j\ell}\rangle$ (4.3.41).

Exercises

4.6.1 By using Eq. (4.6.7) as the starting point and the relationships between the specific discharge covariances at first-order and the Y, ϕ covariances (Sect. 3.7.2), rewrite $X_{j\ell}$ and $D_{j\ell}$ in terms of C_Y, $C_{Y\phi}$ and C_ϕ.

4.6.2 A solute body of a spherical shape is inserted at t=0 in a formation of isotropic C_Y By using the close form results (4.6.14) and (4.6.15), represent the distortion of the solute body as function of time.

4.6.3 With the aid of the same Eqs. (4.6.14), (4..6.15) determine the rate of decay of lateral dispersivities with travel time, for large t.

4.6.4 Demonstrate that for $t\to\infty$ the only surviving term of $D_{j\ell} = (1/2)\,(dZ_{j\ell}/dt)$ (4.6.10), for two particles at initial separation **b-a**, is the longitudinal one.

4.6.5 A solute body of initial rectangular shape is inserted in an anisotropic formation. The relative dimensions of the body are $L_1/I_{Y1}=1$, $L_2/I_{Y2}=1$ and $L_3/I_{Y3}=10$. Compute the variance of the displacement of the center of mass and the expected value of the "moment of inertia" of the solute body, for large travel times.

4.6.4 A solute body of an initial spherical shape of radius $R= 5\,I_Y$ is inserted in an anisotropic formation. With neglect of pore-scale dispersion, determine the concentration variance at the center of mass as function of time.

4.7 TWO-DIMENSIONAL TRANSPORT AND COMPARISON WITH A FIELD EXPERIMENT

The transport at the local scale is essentially a three-dimensional process, since heterogeneous structures are such, while a stratified formation is a notable exception. However, as already emphasized in Part 3 and Chap. 4.5, it is quite improbable that ideal layering persists over large horizontal distances, and eventually spatial effects prevail. Furthermore, in some circumstances one is interested in predicting the concentration averaged along the vertical, rather than its point value. We have discussed in Sect. 3.2.4 the general effect of space averaging, and particularly its impact upon the spatial moments of the solute body (Sect. 4.3.6). Returning to $\overline{C}$, averaged along the vertical coordinate x_3, its expected value $\langle\overline{C}\rangle$ satisfies a two-dimensional transport equation in the horizontal plane, with $D_{t,j\ell}=D_{j\ell}+D_{d,j\ell}$ the longitudinal and lateral total dispersion tensor components for $j,\ell=1,2$. This, of course, is valid under the first-order perturbation approximation, for which transport is Gaussian at any time. The additional effect of space averaging is to reduce the concentration variance, along the lines of the previous section.

An essentially different situation is the one depicted in Fig. 3.7.1d, namely of a formation in which thin, impervious, layers of a large horizontal extent, prevent vertical transport. In such a hypothetical formation, transport is three-dimensional in the inter-layers conductive blocks, but the block average over the vertical, which we shall denote by C as well, is a two-dimensional process, of zero vertical transport. The simplified scheme to model this configuration, suggested in Dagan (1987), is to regard the flow as two-dimensional, but, however, with the three-dimensional logconductivity $Y(\mathbf{x})$ replaced by its space average $\overline{Y}$ over a vertical segment of length ℓ, the latter being equal to the average thickness of the inter-layers blocks. This is an approximate scheme, but more refined ones are precluded by the fact that accurate mappings of the impervious layers, are difficult, if not impossible, to achieve. As a matter of fact, we shall show in the sequel that the entire concept is motivated indirectly by the results of a transport experiment, rather than by direct detection of the structural elements which prevent vertical motion. Thus, the suggested scheme should be regarded as a conjecture, awaiting further field verification. At any rate, the theoretical results obtained here, are useful for modeling transport at the regional scale (Part 5), which is a two-dimensional process.

The effect of vertical averaging of Y has been analyzed in Sect. 3.2.4 and we shall use the results here. Thus, for the simplified, separated covariance C_Y, suggested there, the covariance of $\overline{Y}$ (3.2.40) can be written as follows

$$C_{\overline{Y}}(x_1,x_2) = \sigma^2_{\overline{Y}}\, C_{Yh}(x_1,x_2) \;\; ; \;\; \sigma^2_{\overline{Y}} = 2\left[\frac{\ell}{I_{Y3}}-1+\exp\left(-\frac{\ell}{I_{Y3}}\right)\right]\sigma^2_Y \tag{4.7.1}$$

In (4.7.1) the plane $x=x_1$, $y=x_2$ is the horizontal one, C_{Yh} is the logconductivity covariance in the same plane, and I_{Y3} is the vertical integral scale. The suggested scheme (Dagan, 1987) is therefore, to regard the heterogeneous formation as a random, stationary, two-dimensional one, of log-normal $\overline{Y}$, with $\langle\overline{Y}\rangle=m_Y$ and $C_{\overline{Y}}$ given by (4.7.1).

Along the lines of Chap. 4.6, we proceed now with the derivation of the particles displacements covariances $X_{j\ell}$ (j,ℓ=1,2) under the assumptions of a mean, steady, uniform flow and a first-order approximation in $\sigma^2_{\overline{Y}}$. The basic formulations are again given by Eqs. (4.6.2-4.6.4) for $X_{j\ell}$ and (4.6.5) for $\hat{u}_{j\ell}$, with j,ℓ=1,2 and C_Y replaced by (4.7.1) this time. For the sake of illustration we consider again an exponential, isotropic, covariance, i.e. $C_{Yh}=\sigma^2_{\overline{Y}}\exp[-(r_1^2+r_2^2)^{1/2}/I_{Yh}]$, and an average gradient -J in the x_1 direction, i.e. J(J,0,0). Furthermore, we adopt the large Pe approximation, which leads to (4.6.7), which can be integrated in a close form. The result, derived by Dagan (1982), is a particular case of (4.6.11) and (4.6.12) in which e→∞. Herein are the final results for the longitudinal and transversal covariances, respectively

$$\frac{X'_{11}(t')}{\sigma^2_{\overline{Y}}} = 2t' - 3\ell nt' + \frac{3}{2} - 3E + 3[Ei(-t') + \frac{e^{-t'}(1+t')-1}{t'^2}] \tag{4.7.2}$$

$$\frac{X'_{22}}{\sigma^2_{\overline{Y}}} = \ell nt' - \frac{3}{2} + E - Ei(-t') + 3[\frac{1-(1+t')e^{-t'}}{t'^2}] \tag{4.7.3}$$

where E=0.577... is the Euler number and Ei is the exponential integral. The dimensionless variables are $X'_{j\ell}=X_{j\ell}/I^2_{Yh}$ and $t'=tU/I_{Yh}$, and x_1,x_2 are, obviously, the principal directions of $X_{j\ell}$.

The dependence of X_{11} (4.7.2) upon the travel time, or equivalently upon the travel distance $\langle X_1\rangle/U$, is represented in Fig. 4.6.1 by the curve e=∞. It is seen that it in its dimensionless form it somewhat differs from the results for three-dimensional structures. At the small t' limit, X_{11} has Taylor's general form (4.6.3), namely $X_{11}\to(3/8)\sigma^2_{\overline{Y}}U^2t^2+...$ obtained from (4.7.2) for t'<<1, which is quite close to (4.6.16). A more detailed picture of the behavior at large t' is obtained for the apparent dispersivity (see Sect. 4.6.) $X_{11}/2tU$, which is represented as function of the travel time in Fig. 4.6.4. It is seen that the tendency to the asymptotic value $D_{11}=\sigma^2_{\overline{Y}}U\,I_{Yh}$, i.e. to the macrodispersivity or field scale asymptotic dispersivity $\alpha_{eff,L}=\sigma^2_{\overline{Y}}I_{Yh}$, is slower than for three-dimensional structures. This can be seen by examining (4.7.2), in which $X'_{11}/2t'\to 1-(3/2)\ell nt'+...$ for t'→∞. Due to the presence of the logarithmic term, X_{11} reaches 0.9 of its asymptotic value only for a travel distance larger than 60 horizontal integral scales. Hence, the neglect of the transient term is of lesser accuracy for two-dimensional structures than for three-dimensional ones, unless the travel time is very large.

The difference between transport in 2D and 3D heterogeneous structures is more dramatic for the transversal component X_{22} (4.7.3), which is represented in Fig. 4.6.2. Again, for small t', Eq. (4.7.3) leads to $X_{22}\to(1/8)\sigma^2_{\overline{Y}}U^2t^2+...$, which is not too different from (4.6.17). A marked contrast exists, however, at large t'. Indeed, in the 3D cases X_{22} (Fig. 4.6.2) tends to a constant value, whereas X_{22} (4.7.3) grows indefinitely with time (Fig. 4.6.2). The limit of (4.7.3) is

$$X_{22} \rightarrow \sigma_Y^2 I_{Yh}^2 \left[\ell n\, t' + 0.933 + O(1/t')\right] \qquad (t'>>1) \tag{4.7.4}$$

i.e. the growth is logarithmic, unlike the linear one predicted by Taylor's theory. The physical interpretation is that in a two-dimensional structure, the fluid particles can circumvent zones of low conductivity only laterally, causing them to depart from the average trajectory at a much larger extent than in the three-dimensional case, in which they can move also vertically. Still, the velocity covariance u_{22} is of zero integral scale, precluding the linear dependence upon t. If we evaluate from (4.7.4) the apparent $X_{22}/2Ut$ or the effective $\alpha_T = (1/2U)(dX_{22}/dt)$ dispersivity, we shall find that it grows, reaches a maximum, and then tends to zero like $(\ell n\, t')/t'$ or $1/t'$, respectively. Consequently, if the $t=\infty$ limit (4.6.18) of (4.6.2) is sought directly, as in the developments of Gelhar and Axness (1983) or Neuman et al (1987), the terms of (4.7.4) are lost. Furthermore, a constant macrodispersivity can be arrived at only by keeping the pore-scale dispersion term in (4.6.18). Thus, the result of Gelhar and Axness (1982) for large Pe number is $\alpha_T = (\sigma_Y^2\, \alpha_{d,L}/8)(1 + 3\alpha_{d,T}/\alpha_{d,L})$, which is very similar with the three-dimensional relationship (4.6.24). However, the travel time needed for X_{22} (4.7.4) and $X_{22}=U\sigma_Y^2\alpha_{d,L}\, t/4$ to become of same order is very large, depending on $Pe=I_{Yh}/\alpha_{d,L}$. Thus, $t'\simeq 260$ for Pe=10, whereas $t'=3600$ for Pe=100, or in words the solute body has to travel a distance equal to 3600 logconductivity horizontal integral scales for the pore-scale related term to become of the same order as (4.7.4). This result indicates that in any conceivable application, the neglect of the pre-asymptotic, transient, term of the transverse covariance or of the macrodispersivity is not warranted. Furthermore, for large Pe, the pore-scale dispersion additional term $D_{d,T}$ is negligible as well, and the lateral spread is governed by the convective effect embodied in (4.7.3) solely.

We turn now to the comparison between the measurements at the Borden-site field test and the present theoretical results. The extensive conductivity measurements carried out at the site (Sudicky, 1986) have been presented already in Sect. 3.2.1, whereas the findings about the motion of the solute body (Mackay et al, 1986 and Freyberg, 1986) have been described in Chap. 4.2. The shape of the solute body, for the inert chloride component to be considered here, is shown in Fig. 4.2.2a. It is reminded (Sect. 3.2.1) that Sudicky (1986) has found the logconductivity to be normal, of anisotropic, exponential, C_Y (3.2.2) with nugget w=0.1, variance of correlated Y $\sigma_Y^2=0.28$ and integral scales $I_{Yh}=2.8$ m and $I_{Yv}=0.12$ m, respectively. Furthermore, the velocity of the center of gravity $U=d\langle \overline{X}\rangle/dt$ was (after a short initial transient), horizontal, constant and equal to 0.091 m/day.

The analysis of the considerable amount of concentration point measurements (Freyberg, 1986) showed indeed that C varied in an irregular manner in space, as expected for a transport governed by heterogeneity effects. This is also reflected by the irregular contours displayed in Fig. 4.2.2c. To analyze the spread process, Freyberg (1986) has limited the calculations to evaluating the trajectory of the center of gravity $\overline{X}$ (4.3.38) and the second spatial moment $\overline{X}_{t,j\ell}$ (4.3.40), as functions of the travel time. As mentioned before, the trajectory was found to be rectilinear and the velocity to be constant, which from a theoretical standpoint is equivalent to $\bar{\sigma}_{j\ell}=0$. This is expectable, since the volume of injected solute was of a vertical extent much larger than I_{Yv}. The next finding of interest was that the vertical moment $\overline{X}_{33}(t)$ was approximately equal with its initial value $\overline{X}_{33}(0)$. Hence, practically speaking, the solute body did not spread in the x_3 direction (the coordinate system is selected as in the preceding Sections, i.e. x_1 horizontal and parallel to U, x_2 horizontal and x_3 vertical). This finding is again in agreement with the theoretical results presented here, since for $e=I_{Yv}/I_{Yh}=0.12/2.8=0.043$ and for $\sigma_Y^2=0.28$, X_{33} (Fig. 4.6.3) is exceedingly small, and the same is true for the pore-scale lateral dispersion. As for the other two moments, $\overline{X}_{11}$ and $\overline{X}_{22}$, it was found that they are indeed the principal components of the tensor tensor $\overline{X}_{j\ell}$. However, the lateral $\overline{X}_{22}$ grew with time at a rate much higher than the one predicted by the theory, for a three-dimensional heterogeneous structure of low e (Fig. 4.6.2). This finding has motivated the conjecture by Dagan (1987) that the formation should be modeled as one of a two-dimensional structure, due to the presence of impervious thin layers of large horizontal extent, along the lines of this section. However, since the analysis of cores (Sudicky, 1986) did not indicate clearly the existence of such layers, this should be regarded as a working hypothesis, no other convincing mechanism being suggested so far to explain the field findings. Thus, the results of the present section, and particularly Eqs.(4.7.2, 4.7.3) were applied in order to analyze the field findings. Since no measured ℓ, the interlayer distance was available, it was suggested (Dagan, 1987) to assume that $\ell \simeq I_{Yv}$, i.e. if impervious thin lenses are present, they should probably be at a distance of the order of the correlation scale. This choice led in (4.7.1) to $\sigma_{\overline{Y}}^2 \simeq 0.74\ \sigma_Y^2$. The last step toward comparing theory and field data is adopting the ergodic hypothesis for the spatial moments, i.e. the actual $\overline{X}_{j\ell}$ is equal to $\langle \overline{X}_{j\ell} \rangle$, which is definitely justifiable for the large size of the initial solute body. Under these conditions one can finally write in (4.3.41) $\overline{X}_{jj}(t) \simeq \overline{X}_{jj}(0) + X_{jj}(t)$, where the latter are given by Eqs. (4.7.2, 4.7.3) with $\sigma_{\overline{Y}}^2=0.74\ \sigma_Y^2$, the use of the first-order approximation being justified in the present case. Freyberg (1986) has adopted this scheme and has sought the values of σ_Y^2 and I_{Yh}, the only parameters left unknown, which would result in a best fit of measurements and theory. The result is represented in Fig. 4.2.3, and it was achieved for $\sigma_Y^2=0.24$ and $I_{Yh}=2.7$ m. The agreement between these values and the ones measured directly by core analysis should be considered as very good, if the numerous approximations involved in the process are taken into account (the outlier in Fig. 4.3.2b is apparently related to the encounter with a large scale structural change). Barry and Sposito (1988) discuss extensively the problems associated with the analysis of field data. Although many more careful field experiments are needed in order to arrive at definite conclusions about the predictive capability of the theory, the findings described here are definitely encouraging.

Exercises

4.7.1 Depict graphically the dependence of the lateral, apparent and effective, macrodispersivities upon travel time, based on (4.7.3).

4.7.2 Under the assumptions adopted in the present section, derive the dependence of the expected value of the "moment of inertia" of the concentration in the transverse plane of a plume on x_1 and t. The solute is injected continuously in an area in the plane $x_1=0$ and the mean flow is parallel to x_1.

4.8 EFFECTS OF NONLINEARITY AND UNSTEADINESS.

4.8.1 Effect of nonlinearity in σ_Y^2 upon transport

In the preceding two chapters we have examined in detail the solution of the transport problem at first-order in σ_Y^2, and the salient question is how accurate is the approximation for finite σ_Y^2. We have examined this issue in Sect. 3.7.3 about the Y, ϕ covariances and found that their first-order terms are quite accurate for two- and three-dimensional flows in isotropic aquifers up to values of σ_Y^2 as large as unity. On the other hand, for transport in stratified formations (Chap. 4.5), for which exact solutions are available, it was shown that the first-order approximation is of a more limited range of validity. Hence, it is of considerable interest to determine the degree of accuracy of the main quantity of interest, i.e. the displacement covariances $X_{j\ell}$, or the associated effective dispersion coefficients $D_{j\ell}$ ($j,\ell=1,2,3$), for flow through heterogeneous formations of three-dimensional structures. Since numerical solutions, presumably not limited by the requirement of smallness of σ_Y^2, are not yet available, we shall employ the procedure recalled in Sect. 3.7.3, namely to derive the next higher-order correction $O(\sigma_Y^4)$. Furthermore, we shall assume that the smallness of this higher-order correction in comparison to the first one is indicative of the accuracy of the latter. The procedure presented here follows closely the one of Dagan, 1988.

Again, the starting point is the general Eq. (4.3.23), relating $X_{j\ell}$ to the velocity field, with no restriction placed on the variance. We limit the discussion to steady flow, uniform in the average, along the lines of Chap. 4.6. and we include the effect of pore-scale dispersion in the analysis.

As we have already mentioned in Chap. 4.6, nonlinearity in σ_Y^2 stems from two terms of (4.3.23) : first, from the product of the velocity Fourier transforms and secondly, from the exponential term in $\mathbf{X}'$. Each of these sources of nonlinearity have to be dealt with separately, and we shall start with the second one.

The great difficulty in computing the integrand $\langle \hat{u}_j \hat{u}_\ell \exp(i\mathbf{k}''.\mathbf{X}') \rangle$ stems from the fact that $\mathbf{X}'$, the displacement residual, is the unknown of the problem, so that (4.3.23) is an implicit integro-differential equation for $X_{j\ell}$ (in the first-order approximation, by neglecting this term, we have replaced the actual trajectory by its average). In the literature on turbulent diffusion, a scheme known as Corssin's conjecture, has been suggested in order to overcome this difficulty (see, for instance, Lundgren and Pointin, 1976). It is hypothesized that after a long travel time, the fluctuation $\mathbf{X}'$ becomes weakly correlated with the Fourier components of the velocity, i.e.

$$\langle \hat{u}_j(\mathbf{k}')\hat{u}^*_\ell(\mathbf{k}'')\exp(i\mathbf{k}''\cdot\mathbf{X}')\rangle \simeq \langle \hat{u}_j(\mathbf{k}')\hat{u}^*_\ell(\mathbf{k}'')\rangle\langle\exp(i\mathbf{k}''\cdot\mathbf{X}')\rangle \tag{4.8.1}$$

Furthermore, after a sufficiently long time $\mathbf{X}'$ becomes Gaussian and of constant $D_{j\ell}=(1/2)(dX_{j\ell}/dt)$ (Chap. 4.3). Then, we have explicitly $\langle\exp(i\mathbf{k}.\mathbf{X}')\rangle = \exp(\Sigma\text{-}\Sigma k_j k_\ell X_{j\ell}/2) = \exp(\Sigma\text{-}\Sigma k_j k_\ell D_{j\ell} t)$ (Sect. 1.1.2). Comparison of the solution of (4.3.23), after substituting (4.8.1) in it, with some numerical simulations of turbulent diffusion (Lundgren and Pointin, 1976; Salu and Montgomery, 1977) was quite satisfactory, if the spectrum was smooth enough. We shall adopt this nonlinear correction here as well, i.e. we replace (4.3.23) by

$$\frac{d^2X_{j\ell}(t)}{dt^2} = \frac{2}{(2\pi)^{3/2}}\int \hat{u}_{j\ell}(\mathbf{k})\,\exp[i\mathbf{k}\cdot\mathbf{U}t-(1/2)\sum_{p=1}^{3}\sum_{q=1}^{3}X_{t,pq}k_pk_q]\,d\mathbf{k} \qquad (j,\ell=1,2,3) \tag{4.8.2}$$

For a given spectrum, Eq. (4.8.2) is an integro-differential equation for $X_{j\ell}$. Since it is supposedly valid only for large t, integration over time, with $X_{t,j\ell}=2D_{t,j\ell}t$, $D_{t,j\ell}=D_{j\ell}+D_{d,j\ell}$, in the exponent, gives

$$D_{j\ell} = \frac{1}{(2\pi)^{3/2}}\int \hat{u}_{j\ell}(\mathbf{k})\int_0^\infty \exp(i\mathbf{k}\cdot\mathbf{U}t-\sum_{p=1}^{3}\sum_{q=1}^{3}D_{t,pq}k_pk_q t)\,dt\,d\mathbf{k} =$$
$$\frac{1}{(2\pi)^{3/2}}\int [\,\hat{u}_{j\ell}(\mathbf{k})]/(\sum_{p=1}^{3}\sum_{q=1}^{3}D_{t,pk}k_pk_q-i\mathbf{k}\cdot\mathbf{U})]\,d\mathbf{k} \tag{4.8.3}$$

which is an implicit equation to determine $D_{j\ell}$, $D_{d,j\ell}$ being given. Hence, under the limitations of Corssin's conjecture, we seek the nonlinear correction to the asymptotic $D_{j\ell}$, for t'>>1, rather than to $X_{j\ell}$ at any t.

We have shown in Sect. 4.6.4 that the derivatives of the lateral moments, i.e. dX_{22}/dt and dX_{33}/dt, tend to zero for $t\to\infty$ in the first-order approximation, if convection only is taken into account. Thus, if we seek corrections of order $O(\sigma_Y^4)$ of $D_{j\ell}$, we are entitled to keep in the integrals of (4.8.3) only D_{11}, with $\mathbf{U}$ taken along x_1. Furthermore, the same approximation holds for $D_{d,j\ell}$ in view of the smallness of the transverse pore-scale dispersivity as compared to the longitudinal one (see Chap. 2.10). Hence, the denominator in the integrand becomes $D_{t,11} k_1^2 - iUk_1$. By inverting $\hat{u}_{j\ell}(\mathbf{k})$ in (4.8.3) and by integrating over k_2, k_3, x_2 and x_3, the following result is immediately obtained

$$D_{j\ell} = \frac{1}{(2\pi)^{1/2}} \int_{-\infty}^{\infty} u_{j\ell}(x_1,0,0) \int_{-\infty}^{\infty} \frac{\exp(ik_1x_1)}{D_{t,11}\, k_1^2 - iUk_1}\, dk_1\, dx_1 \tag{4.8.4}$$

An additional integration over k_1 in (4.8.4) yields

$$D_{j\ell} = \frac{1}{U} \int_{0}^{\infty} u_{j\ell}(x_1,0,0)\, [1 + \exp(-x_1U/D_{11})]\, dx_1 \tag{4.8.5}$$

Eq. (4.8.5) constitutes an important starting point to investigate the nonlinearity in σ_Y^2 of the asymptotic, effective, dispersion coefficients. Obviously, the first-order approximation of Chaps. 4.6 and 4.7 are obtained by neglecting the exponential in (4.8.5). Since we seek here the next order correction of $D_{j\ell}$, we may substitute in the integral the first approximation of $D_{11} = \sigma_Y^2\, U\, I_h + \alpha_{d,L}\, U$ (Chap. 4.6). Then, after an integration by parts, we obtain to second order,

$$\alpha_{j\ell} = \frac{1}{U^2} \int_{0}^{\infty} u_{j\ell}(x_1,0,0)\, dx_1 + \frac{u_{j\ell}(0,0,0)\, \sigma_Y^2\, I_h}{U^2} \tag{4.8.6}$$

where $\alpha_{j\ell} = D_{j\ell}/U$ are the effective dispersivities. The second expression in the r.h.s. of (4.8.6) represents the effect of the exponential term of (4.3.23), i.e. of accounting for the actual trajectory rather than its average. Since the variances $u_{j\ell}(0,0,0)$ are positive, it is seen that its net effect is to increase the dispersivity, as one would expect on physical grounds. Furthermore, we can adopt the first-order approximation of $u_{j\ell}(0,0,0)$ in (4.8.6), to obtain the correction $O(\sigma_Y^4)$.

As for the other source of nonlinearity, its emergence is quite clear, since it arises from the integral of (4.8.6), in which $u_{j\ell}(x_1,0,0)$, the velocity covariance along the mean trajectory, has to be evaluated at second-order.

We are now in a position to evaluate separately the two terms $O(\sigma_Y^4)$ of $\alpha_{j\ell}$ (4.8.6), and this is done for the longitudinal effective dispersivity $\alpha_L = \alpha_{11}$ first. The difficult problem now is to evaluate the second-order term of u_{11}, a task which has not yet been undertaken in the literature. To grasp the magnitude of this term, we shall adopt a further approximation motivated by the results of Sect. 4.6.4, namely that the asymptotic longitudinal dispersivity can be obtained by assuming that streamlines are straight and the head gradient is equal to its mean $-J$, which led to the simple

ous equilibrium concentrations respectively. On this basis the inferred retardation factors were calculated with the aid of Eq. (2.12.7), i.e. $R = 1 + \rho_b K_d / n$. The average values and their range (Curtis et al, 1986, Table 6) for the different compounds are also represented in Fig. 4.9.3. Rate studies for the various components under laboratory conditions indicated that around 50% of the total sorption occurred in approximately two hours, followed by a gradually decline which lasted two to three days. These times are much larger than the ones indicated in Chap. 2.12, but can be considered as instantaneous at the time scale of the solute motion in the aquifer. Finally, an analysis of the spatial distribution of the sorption coefficients in the aquifer is presented by Mackay et al (1986a). The large variability is illustrated in Fig. 4.9.4, which shows the change of K_d of the PCE compound with depth for one of the cores. This variation could be correlated neither with the specific area of the solid nor with its organic carbon content, and is attributed to a yet unidentified mineral phase. An interesting finding was that spatial variability of the other compounds was closely related to that of PCE, indicating a common cause. Studies about the spatial correlation scale of K_d and its possible correlation with hydraulic conductivity were under way.

After this brief review of a few findings of interest at the Borden site field test, we shall outline the theoretical approach to transport of reactive solutes, along the lines of the preceding chapters. In contrast with the generalizations of the transport equation in Chap. 2.12, pertaining to homogeneous laboratory columns, the interest here resides in natural formations in which coefficients characterizing the solute transformations are not constant, but spatially variable. Fig. 4.9.4 provides a vivid illustration of the type of irregular variations one may expect, suggesting a statistical approach. Unlike the case of hydraulic conductivity, there are very few field data to support a theoretical model for the p.d.f. of the sorption coefficient or other relevant properties. For this reason, the following developments are of a preliminary nature, awaiting further accumulation of systematical field data. In the same vein, we shall limit the analysis to a few simple models of solute reactions.

(i) Radioactive decay.

One of the simplest types of solute transformations is that of radioactive decay, characterized by the exponential equation (2.12.9). This process may be assumed to be space independent, i.e. the decay coefficent a is a constant. Furthermore, the solute does not influence the flow, i.e. it does not affect significantly the fluid density or viscosity.

Our starting point is the basic equation (4.3.2), describing the concentration field associated with a solute particle of mass ΔM. The latter is now time dependent according to the relationship

$$\Delta M(t) = \Delta M_0 \{1 - \exp[-a(t-t_0)]\} \quad ; \quad \Delta M_0 = C_0\, n_0\, \Delta \mathbf{a} \qquad (4.9.1)$$

where ΔM_0 is the initial mass at $t=t_0$ and $\mathbf{a}$ stands for the initial coordinate of the particle. Hence, Eq. (4.3.3) becomes

$$C(\mathbf{x},t,t_0) = \{1 - \exp[-a(t-t_0)]\}\int_{V_0} \frac{n_0}{n} C_0\, \delta(\mathbf{x}-\mathbf{X}_t)\, da \tag{4.9.2}$$

leading to the simple relationship $C=\{1 - \exp[-a(t-t_0)]\}C_{nr}$, where C_{nr} is the concentration of a nonreactive solute transported in the same formation. Thus, the zero-order spatial moment, i.e. the total mass, is simply given by $M(t)= \int C d\mathbf{x}= M_0\{1-\exp[-a(t-t_0)]\}$. It is easy to ascertain that the other moments, i.e. the trajectory of the center of mass $\overline{\mathbf{X}}$ (4.3.37) and the "moment of inertia" $\overline{X}_{j\ell}$ (4.3.40) are not affected by the radioactive decay. In other words the solute cloud is similar to that of a nonreactive component, except for a scaling down of concentration with time.

(ii) Retardation.

In Chap. 2.12 we have discussed the effect of reversible sorption, which follows a linear isotherm, upon transport. In particular, it was shown that in the case in which the time scale characterizing the concentration changes is much larger than the one required to reach equilibrium between the solution and the matrix, the transport equation (2.12.13) is similar to the one of a nonreactive solute, except for a retardation coefficient $R=1+K_d\rho_s(1-n)/n$. The application of the concept to ion-exchanging contaminants even in the cases in which the linear isotherm is not obeyed is discussed by Valocchi, 1984. We shall assume that this is the case, but K_d, and consequently R, is a space random function (Fig. 4.9.4). Furthermore, we assume that $R(\mathbf{x})$ is stationary, of constant mean $\langle R\rangle$, and of residuals $R'(\mathbf{x})$ of covariance $C_R(\mathbf{r})$. The crux of the matter is whether R is correlated to the hydraulic conductivity, which is the spatially variable formation property affecting the flow. In absence of conclusive field data we assume that this is the case, i.e. the cross-covariance $C_{RY}=\langle R'(\mathbf{x}_1)Y'(\mathbf{x}_2)\rangle$ is different from zero. In particular, the simplest correlation between R' and Y' is a linear one $R'=\beta Y'$, in which case $C_R=\beta^2 C_Y$ and $C_{RY}=\beta C_Y$, i.e. all second moments are expressed with the aid of the logconductivity covariance.

To illustrate the impact of retardation upon transport in heterogeneous formations, we shall evaluate the first two moments of a particle displacement, which are instrumental in determining the concentration expected value (see Chap. 4.3). The basic differential equation (4.3.18) for the convective displacement becomes now

$$\frac{d\mathbf{X}}{dt} = \frac{1}{R}(\mathbf{U}+\mathbf{u}) \tag{4.9.3}$$

where $\mathbf{U}$ is constant. To simplify matters we assume that $\sigma_R/\langle R\rangle<1$ and expand in (4.9.3) as follows $1/R= 1/\langle R\rangle(1-R'/\langle R\rangle+...)$, retaining only first-order terms. Then, integration in (4.9.3) yields

$$\mathbf{X}(t) = \frac{1}{\langle R\rangle}\mathbf{U}t - \frac{\mathbf{U}}{\langle R\rangle^2}\int_0^t R'(\mathbf{X}_t)\, dt' + \langle R\rangle\int_0^t \mathbf{u}(\mathbf{X}_t)\, dt' - \frac{1}{\langle R\rangle^2}\int_0^t R'(\mathbf{X}_t)\, \mathbf{u}(\mathbf{X}_t)\, dt' \tag{4.9.4}$$

where the argument $\mathbf{X}_t$ in the integrands is a function of time, denoted by t′, to emphasize that it is an integration variable. The computation of the statistical moments of $\mathbf{X}$ faces the same basic difficulty as before, namely the unknown function appears implicitly as an argument in the various integrals. We adopt, therefore, the same approximation of small σ_Y^2 and replace $\mathbf{X}_t$ by its zero-order average $Ut/\langle R\rangle$ in the argument of R′ and $\mathbf{u}$. Furthemore, we concentrate on the computation of the longitudinal component and adopt the large travel time approximation of u_1, i.e. $u_1 \simeq U\,Y'$ for $\mathbf{U}(U,0,0)$ (see Sect. 4.8.1). Under these assumptions we get from (4.9.4)

$$\langle X_1\rangle = \frac{1}{\langle R\rangle} U\,t - \frac{U}{\langle R\rangle^2} C_{RY}(0,0,0)\,t = \frac{Ut}{\langle R\rangle}\left(1 - \frac{\beta\sigma_Y^2}{\langle R\rangle^2}\right) \qquad (4.9.5)$$

the last equation being valid if R′ is linearly correlated to Y′. In a similar manner we get for the covariance X_{11}

$$X_{11} = \frac{(2U^2)}{\langle R\rangle^2}\int_0^t (t-t')\left\{\frac{1}{\langle R\rangle^2} C_R[(Ut')/\langle R\rangle,0,0] - \right.$$

$$\left.\frac{2}{\langle R\rangle} C_{RY}[(Ut')/\langle R\rangle,0,0] + C_Y[(Ut')/\langle R\rangle,0,0]\right\} dt' \qquad (4.9.6)$$

which becomes for large t and for the linear R,Y correlation

$$X_{11} = \frac{2}{\langle R\rangle}\left[1 - \frac{\beta}{\langle R\rangle}\right]^2 \sigma_Y^2\, U\, I_{Yh}\; t \qquad (4.9.7)$$

Examination of the results shows first that the average velocity of the particle $d\langle X_1\rangle/dt$ (4.9.5) is reduced by a factor equal to the average retardation coefficient if $\beta=0$, or to a different, but constant, retardation factor if $\beta\neq0$. As for the asymptotic macrodispersion coefficient $D_{11}=(1/2)(dX_{11}/dt)$, it is smaller than the one valid for an inert solute. It is emphasized, however, that the requirement for the asymptotic results to be valid is that the average travel distance $Ut/\langle R\rangle$ is much larger than the correlation scale I_{Yh}. In terms of travel time this means that $t \gg \langle R\rangle I_{Yh}/U$, as compared to the condition $t \gg I_{Yh}/U$ for an inert solute. In other words, the time required for a reactive soil to reach the Fickian regime is larger by a factor $\langle R\rangle$ than the corresponding one for an inert solute.

As we have already emphasized, these are preliminary results which await further experimental confirmation, and primarily the statistical analysis of the spatial variability of R in natural formations.

(iii) Irreversible reaction and retardation.

We consider now the combined effects of mass reduction (Fig. 4.9.1) and of retardation (Fig. 4.9.3), suggested by the findings at the Borden site test for a few compounds. Starting again with transport of a solute particle of mass ΔM, two modifications of the basic Eqs. (4.3.2), (4.3.18) for a nonreactive solute are appropriate

$$\frac{1}{\Delta M_0}\frac{d(\Delta M)}{dt} = -\mu(\mathbf{X}_t) \ ; \ \frac{d\mathbf{X}}{dt} = \frac{\mathbf{V}(\mathbf{X}_t)}{R(\mathbf{X}_t)} \qquad (4.9.8)$$

The first equation in (4.9.8) represents mass decay at the linear rate coefficient μ. The latter is, however, supposed to be spatially variable, i.e. $\mu=\mu(\mathbf{x})$. Hence, a traveling particle experiences a variable rate of decay, depending on its position $\mathbf{x}=\mathbf{X}_t$ at time t. The second equation is identical to (4.9.3) to account for retardation, $\mathbf{V}$ and R being the fluid velocity and the retardation coefficient, respectively, again spatially variable. On the same hypothetical basis μ and R are regarded as stationary random space functions, characterized by $\langle\mu\rangle$, $C_\mu(\mathbf{r})$ and by $\langle R\rangle$, $C_R(\mathbf{r})$. Again, the basic question is whether these variables are correlated among themselves and with the matrix conductivity. We shall illustrate the procedure by assuming the first case only, which is plausible in view of the similarity of the factors which underlie the two processes. Hence, we add the cross-covariance $C_{\mu R}$ to the other second-order statistical moments. Again, the starting point is the basic equation (4.3.2) describing the concentration field associated with a particle, which becomes

$$\Delta C(\mathbf{x},t) = \Delta M(t)\ \delta[\mathbf{x}-\mathbf{X}_t(t)] \ ; \ \Delta M(t) = \Delta M_0\{1-\langle\mu\rangle t-\int_0^t \mu'[\mathbf{X}_t(t')]dt'\} \qquad (4.9.10)$$

whereas $\mathbf{X}$ is given by Eq. (4.9.4). The expected value of ΔM is simply given by $\langle\Delta M\rangle= \Delta M_0(1-\langle\mu\rangle t)$ and the same is true for the mass of a finite solute body, i.e. $m= m_0(1-\langle\mu\rangle t)$, i.e. the mass diminishes at a linear rate. It is of interest to evaluate the variance of ΔM, the starting point being the residual in (4.8.10), i.e. $\Delta M'= -\Delta M_0\int_0^t \mu'(\mathbf{X}_t)dt'$. Again, the computations can be simplified considerably by replacing $\mathbf{X}_t$ by its expected value $\langle\mathbf{X}\rangle= Ut/\langle R\rangle$. This leads to

$$\sigma^2_{\Delta M} = 2(\Delta M_0)^2\ t\int_0^t C_\mu(Ut'/\langle R\rangle)\ dt' \simeq \frac{2(\Delta M_0)^2\langle R\rangle\sigma_\mu^2 I_\mu}{U}\ t \qquad (4.9.11)$$

where I_μ is the linear integral scale of μ in the direction of the mean flow, and the last expression applies asymptotically, for a travel distance which is large compared to I_μ. The main conclusion is that for a spatially variable decay coefficient μ, the coefficient of variation of the particle mass increases also with time. In the realistic case of a finite solute body the coefficient of variation is smaller due to the smoothing effect of the space averaging, depending on the ratio between the size of the body and I_μ.

To further illustrate the stochastic approach, we shall evaluate the expected value of the trajectory of the center of mass of a finite solute body (see Chap. 4.3) of initial concentration and volume C_0 and V_0, respectively. By the definition (4.3.37) and by using (4.9.10) we get

$$\langle\overline{\mathbf{X}}\rangle(t) = \langle(1/M)\int C\mathbf{x}d\mathbf{x}\rangle \simeq \frac{Ut}{\langle R\rangle} - \frac{\int_0^t \langle\mu'(Ut'/\langle R\rangle)\mathbf{X}'(Ut/\langle R\rangle)\rangle dt'}{1-\langle\mu\rangle t} \tag{4.9.12}$$

The last expression in (4.9.12) is underlain by a few assumptions: C_0 is constant, V_0 is large at the correlation scales of μ and R, higher than the second-order terms in μ',R$'$ are neglected and the trajectory $\mathbf{X}_t$ is replaced by its mean $Ut/\langle R\rangle$ in the arguments of μ' and $\mathbf{X}'$. As for the latter, it is given by (4.9.4), from which we have retained only the term in R$'$, after assuming that μ is not correlated to $\mathbf{u}$. Hence

$$\int_0^t \langle\mu'(Ut''/\langle R\rangle)\, \mathbf{X}'(Ut/\langle R\rangle)\rangle dt'' = -\frac{U}{\langle R\rangle^2}\int_0^t\int_0^t \langle\mu'(Ut''/\langle R\rangle)\, R'(Ut'/\langle R\rangle)\rangle dt'dt'' =$$

$$-\frac{2U}{\langle R\rangle^2}\int_0^t (t-t')\, C_{\mu R}(Ut'/\langle R\rangle)dt' \simeq -\frac{2\,U\,\sigma_{\mu R}\,I_{\mu R}\,t}{\langle R\rangle U} \tag{4.9.13}$$

where $\sigma_{\mu R}$ is the correlation coefficient between μ and R and $I_{\mu R}$ is the linear integral scale of $C_{\mu R}$ in the direction of the mean flow. The last expression in (4.9.13) is valid for a travel distance which is large compared to $I_{\mu R}$. Substitution of (4.9.13) in (4.9.12) leads to the final result of the analysis of the motion of the center of mass of the solute body

$$\langle\overline{\mathbf{X}}\rangle(t) = \frac{Ut}{\langle R\rangle} + \frac{2\,U\,\sigma_{\mu R}\,I_{\mu R}\,t}{\langle R\rangle U(1-\langle\mu\rangle t)} \tag{4.9.14}$$

The striking result in (4.9.14) is that due to the second term, the velocity of the center of mass is no more constant , and it decreases or increases with time depending on whether $\sigma_{\mu R}$ is negative or positive. This effect results entirely from the fact that M in the definition of $\overline{\mathbf{X}}$ (4.9.12) is time dependent. Again, this result depends on the soundness of the model of spatial variability of μ and R and on the sign of the correlation between the two, which is probably negative.

4.9.2 The effect of parameters estimation errors

One of the main aims of the theory of transport by groundwater is to provide tools for predicting the concentration field in a given formation. The various formulae presented in the preceding chapters may serve for this purpose. These relationships depend, however, on a few parameters which were assumed to be known. Thus, in the simple case of a conservative solute and of average uniform flow through a heterogenous formation of stationary random conductivity (Chaps. 4.3, 4.6, 4.7), the parameters are $\mathbf{U}$, the average velocity, n, the effective porosity, σ_Y^2 , the log conductivity variance, I_{Yh} and I_{Yv}, the log-conductivity horizontal and vertical integral scales, respectively, and $D_{d,L}$ and $D_{D,T}$, the longitudinal and transverse pore-scale dispersion coefficients, respectively. Furthermore, the velocity is related to additional parameters, i.e. $\mathbf{U}= K_{ef}\mathbf{J}/n$ where at first-order $K_{ef}=K_G=\exp(m_Y)$ and $\mathbf{J}$ is the average head gradient. So far it was tacitly assumed that these parameters are given. In practice, however, they are estimated by using field data which as a rule are quite scarce. The discussion of the methodology of parameter estimation is beyond our scope, although we shall touch briefly the issue in Part 5. The main point is, however, that by using statistical methods, e.g. least squares, maximum likelihood, nonlinear regression, one is able to estimate the parameters from existing data. These estimates are themselves random variables, subject to uncertainty, which carries over to the variables on which they depend. We have addressed this topic briefly in Sects. 3.2.5 and 3.7.7 and we shall apply the concept here in the transport context. With $\boldsymbol{\theta}$ the parameters vector, the statistical inference provides their estimates $\tilde{\boldsymbol{\theta}}$, which for a sufficiently large number of data can be assumed to be normal, efficient and unbiased, and the variance-covariance matrix $\Theta_{j\ell}$, with $j,\ell=1,...,P$ and P the number of parameters. The impact of parameters uncertainty upon a random space function which depends on them has been analyzed recently by Feinerman et al (1986), and the following discussion is along their lines.

Referring now to the transport problem, let us consider first the point value concentration for a solute body, Eqs. (4.3.3, 4.3.4). We shall rewrite the result as $C(\mathbf{x},t\,|\boldsymbol{\theta})$ to emphasize that it is conditioned on the parameters vector $\boldsymbol{\theta}$, which was assumed to be known deterministically so far. Similarly, the ensemble mean $\langle C(\mathbf{x},t|\boldsymbol{\theta})\rangle$ (4.3.25), should be viewed now as carried out for the set of realizations of the random space function $Y(\mathbf{x})$, but conditioned on fixed values of $\boldsymbol{\theta}$. Hence, the estimate of $\langle C(\mathbf{x},t)\rangle$ is given by

$$\tilde{C}(\mathbf{x},t) = \langle C(\mathbf{x},t|\tilde{\boldsymbol{\theta}})\rangle \qquad (4.9.15)$$

which is obtained from the previous results by replacing the parameters by their estimates. Similarly, the variance is estimated like in Sect. 3.2.5 as follows

$$\sigma_C^2(\mathbf{x},t) = \langle[C(\mathbf{x},t) - \langle C(\mathbf{x},t)\rangle]^2\rangle = \langle[C(\mathbf{x},t) - \tilde{C}(\mathbf{x})]^2\rangle + \langle[\tilde{C}(\mathbf{x},t) - \langle C(\mathbf{x},t)\rangle]^2\rangle =$$

$$\simeq \sigma_C^2(\mathbf{x},t|\tilde{\boldsymbol{\theta}}) + \sigma_{\tilde{C}}^2 \qquad (4.9.16)$$

i.e. the total variance is equal to the variance conditioned on the parameters estimates *supplemented* by the variance of estimation of $\langle C \rangle$. The computation of the latter can be simplified along the lines of Sect. 3.2.5 by adopting a Taylor expansion of $\langle C(\mathbf{x},t)\rangle$ around $\tilde{\boldsymbol{\theta}}$, i.e. $\langle C \rangle = \tilde{C} + \Sigma(\theta_i - \tilde{\theta}_i)$ $(\partial\tilde{C}/\partial\tilde{\theta}_i) + ...$ Substitution in the last term of (4.9.15) yields $\sigma^2_{\tilde{C}} \simeq \Sigma\Sigma(\partial\tilde{C}/\partial\tilde{\theta}_i)(\partial\tilde{C}/\partial\tilde{\theta}_j)\,\Theta_{ij}$.

We shall illustrate the application of these concepts to some of the most useful results of Chap. (4.3), namely to the spatial moments of the concentration of a solute body. Thus, the expected value of the displacement of the center of mass is given by (4.3.39), now viewed as conditioned on the parameter $\mathbf{U}$ only. Hence, the best estimate of the unconditional expected value is

$$\langle\tilde{\mathbf{X}}(t)\rangle = \bar{\mathbf{a}} + \tilde{\mathbf{U}}\, t \qquad (4.9.17)$$

where $\tilde{\mathbf{U}}$ the estimate of the mean velocity.

Next, we are going to evaluate the covariance of the displacement of the center of mass (4.3.39), which generally depends on the vector $\boldsymbol{\theta}$ of parameters $\mathbf{U}$, σ^2_Y, I_{Yh}, I_{Yv} (we neglect the impact of estimation of the pore-scale dispersivities). Similarly to (4.9.16) we have

$$\tilde{\bar{\sigma}}_{j\ell}(t) = \bar{\sigma}_{j\ell}(t\,|\tilde{\boldsymbol{\theta}}) + \mathrm{COV}(\tilde{U}_j,\tilde{U}_\ell)\, t^2 \qquad (4.9.18)$$

where $\mathrm{COV}(\tilde{U}_j,\tilde{U}_\ell)$ is the covariance of the estimates of the average velocity components. The last term in (4.9.18) originates from the expression of $\langle\bar{\mathbf{X}}\rangle$ (4.3.39), i.e. from the term $\int[\langle X_j(t\,|\boldsymbol{\theta})\rangle - \langle X_j\rangle]$ $[\langle X_\ell(t\,|\boldsymbol{\theta})\rangle - \langle X_\ell\rangle]\, f(\mathbf{U})\, d\mathbf{U} = [t^2 \int(\tilde{U}_j - U_j)(\tilde{U}_\ell - U_\ell)\, f(\mathbf{U})\, d\mathbf{U}$.

In the discussion of Sect. (4.3.6) it was shown that $\bar{\sigma}_{j\ell}(t\,|\boldsymbol{\theta})$ tends to zero if the transversal dimension of the initial volume V_0 is much larger than the log-conductivity integral scale, and the same is true for the first term of (4.9.18). In contrast, the last term of (4.9.18) does not depend on V_0 and, furthermore, it grows quadratically with time. Hence, the uncertainty of the position of the center of mass may result to a greater extent from the estimation error of the mean velocity, than from the effect of spatial variability of Y, and ergodicity does not apply to estimation errors. This is an important point, to be recalled in any analysis in which the theory plays a predictive role.

It is emphasized that as a rule $\mathbf{U}$ is inferred from the head measurements via the relationship $\mathbf{U} = K_{ef}\mathbf{J}/n$ and the variance of $\tilde{\mathbf{U}}$ depends, therefore, on the estimation error of $\mathbf{J}$, $K_{ef} \simeq K_G = \exp(m_Y)$ and n. The variance can be estimated if $\tilde{\mathbf{J}}$, $\tilde{m}_Y$ and $\tilde{n}$ are assumed to be normal and of given covariance.

Finally, we shall evaluate the expected value of the "moment of inertia" of the solute body $\langle\bar{X}_{t,j\ell}(t\,|\boldsymbol{\theta})\rangle$ (4.3.41). Since the expression $\mathrm{COV}(\tilde{\mathbf{U}})t^2$ appears in the expected values of $X_{j\ell}$ and $\sigma_{j\ell}$ as well, it is filtered out from $\langle\bar{X}_{t,j\ell}\rangle$. Hence, we arrive at the relationship

$$\langle \tilde{\overline{X}}_{(t,j\ell)}(t) \rangle = \langle \overline{X}_{t,j\ell}(t\,|\tilde{\boldsymbol{\theta}}\,) \rangle \tag{4.9.19}$$

This is an important result, since it does not contain the term depending quadratically on time which shows up in (4.9.18). This stems from the fact that in each realization the moment of inertia is *relative* to the center of mass in the same realization, and an error in **U** causes a translation of the entire solute body in space. Hence, (4.9.19) is quite robust, but it is emphasized that in the prediction process the ultimate distribution of concentration depends on the location of the center of mass and on the moment of inertia as well.

The same approach can be applied to any of the concentration moments derived in the preceding sections.

So far we have assumed that the parameters $\boldsymbol{\theta}$ have been estimated by using a sufficiently large number of measurements to warranty the asymptotic properties of estimates. This is not always the case and in some circumstances the information is scarce. Then, we might employ a Bayesian approach, namely to regard parameters as random and characterized by an a-priori p.d.f., e.g. a rectangular distribution in which the minimal and maximal values are presumed. Subsequently, the various statistical moments of the concentration are regarded as conditioned on $\boldsymbol{\theta}$ and their estimates and estimation variances are obtained by using Bayes relationship (Chap. 1.2).

It is important to realize that spatial variability and estimation errors are intertwined in prediction of solute transport. Therefore, a refined model of the effect of spatial variability of Y is of little predictive use if parameters are known with great uncertainty.

We shall return to some aspects of uncertainty reduction of estimated parameters in Part 5.

Exercises

4.9.1 Derive expressions similar to (4.9.12) and (4.9.13) in the case of linear mass reduction (Eq. 4.9.8) and in absence of retardation. Assume that μ is linearly correlated to the log-conductivity Y.

4.9.2 A conservative solute is injected in an aquifer, the initial volume V_0 being a sphere of diameter d. Assume that C_Y is isotropic and that $d \gg I_Y$, so that ergodicity is obeyed with regard to the spatial variability of Y. Assuming that the estimate of U, the component of **U** in the x_1 direction, is constant but uncertain and of coefficient of variation 0.2, determine the concentration variance at the point of coordinate $x_1 = \tilde{U}t$.

ginating from large areas (the so called "non-point sources"). The point is, however, that the modeling of the entire range of space averages, from point to large areas, may need consideration.

The necessity to deal with spatially variable entities of large possible uncertainty has motivated the use and development of specific tools which are not common to other branches of physics and engineering science, namely conditional probability or the "geostatistical" approach (Chap. 1.8). In the preceding parts heterogeneity was characterized by the statistical moments (expected values, covariances) of the medium properties and the ensemble underlying the computations was the one of all possible media of same statistical structures. In contrast, the subensemble of possible formations, for which the measured values at various points of the aquifer are regarded as deterministic and given, underlies the geostatistical approach. The Bayesian or conditional probability is the tool which permits one to incorporate the information provided by measurements in order to reduce uncertainty of prediction, and this powerfull instrument will be exploited in this part.

Another subject of considerable interest related to the last topic, to be addressed here, is that of identification or of solution of the "inverse" problem. In essence it implies the use of measurements of water heads in order to identify the underlying transmissivity distribution in the formation.

These topics, as well as the problem of transport at the regional scale, form the content of this part.

5.2 ANALYSIS OF FIELD DATA AND STATISTICAL CHARACTERIZATION OF HETEROGENEITY

5.2.1 A few field findings

Unlike field measurements of hydraulic properties at the local scale (Sect. 3.2.1), which have been seldom conducted in a comprehensive and systematic manner, there is a large body of field data at the regional scale. Indeed, it is a matter of routine to monitor heads by observation wells and to determine transmissivity and storativity by pumping tests at a few locations, or by simplified procedures like specific well capacity (the relationship between the discharge and the drawdown in the pumping well). Deferring the mathematical definition of these properties, we wish to recall a few field findings from the literature. The main aim is to provide a factual support to the argument set forth in the Introduction, namely that as a rule aquifers are heterogeneous at the regional scale. It is beyond the scope of this book to discuss the geological ground of these findings and we concentrate here on the quantitative representation based on direct field measurements.

Starting with the transmissivity, the point is illustrated vividly by the representation of the log-transmissivity fluctuations at different locations in the Avra Valley aquifer in Arizona (Fig. 5.2.1). This aquifer has been investigated quite extensively in relation with the inverse problem by Clifton

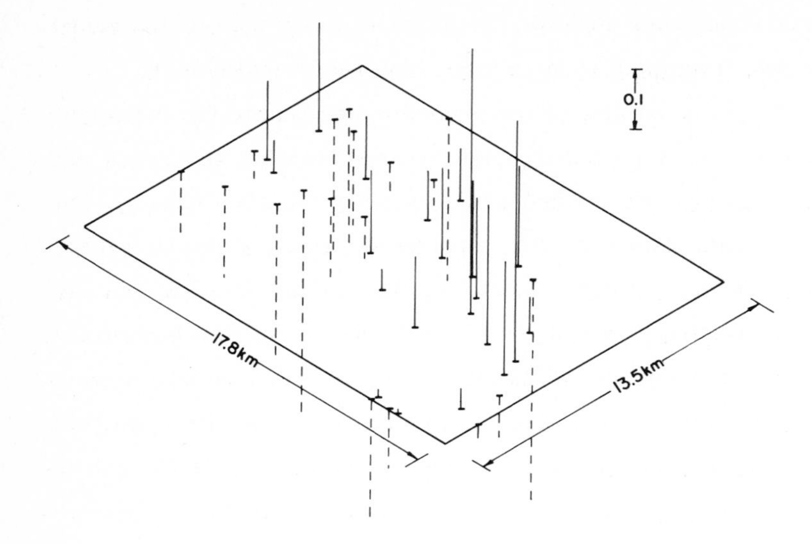

Fig. 5.2.1 **Measured log transmissivity in the Northern Part** of the Avra Valley (data are taken from Clifton, 1981). **In the figure the residual of $\log_{10}T$ for** T in square **feet per** day is represented at different points by vertical segments (**positive deviations**, solid lines, **and** negative deviations, dashed lines). The average value of $\log_{10}T$ is **4.34 and the** scale of log T is represented in the upper right corner (reproduced from Dagan, 1986)

(1981) and by Clifton and Neuman (1982), and has served also as a test case for Rubin and Dagan (1987b). The transmissivity has been measured at a relatively large number of points, 106 altogether, mostly by specific capacity tests. For convenience, the aquifer has been divided into two parts, and Fig. 5.2.1 pertains to the Northern one. The statistical analysis of the data showed that T fits a log-normal distribution, a finding confirmed in other cases as well. The main point illustrated by Fig. 5.2.1 is that $Y=\ln T$ is fluctuating in an irregular manner in the plane, justifying the statistical approach (for the sake of simplicity we employ in this part the same symbol Y, which represented logconductivity in Parts 3 and 4, for the natural logarithm of transmissivity). Clifton and Neuman (1982) have analyzed the correlation structure of Y, and an isotropic semi-spherical covariance (Eq. 3.2.17) has been fitted to the data. Thus, the statistical structure of $Y(\mathbf{x})$ was represented in terms of four parameters: T_G, the transmissivity geometric mean, w, the nugget, σ_Y^2, the variance of log-transmissivity correlated values and I_Y, the integral scale. For North Avra the values were : $T_G=1990$ m²/day, w=0.03, $\sigma_Y^2=0.33$ and $I_Y=3600$ m. We shall return to the discussion of these results in Chap. 5.6, on the inverse problem.

Hoeksema and Kitanidis (1985) have undertaken the task of characterizing in statistical terms the transmissivity and storativity of a large number of aquifers in U.S., based on existing data in the literature. The results, stemming from the solution of the inverse problem (Chap. 5.6), are summarized in Tables 1-4 of op. cit. Without entering into the details of the methodology, which encountered some difficulties, especially in the estimation of the integral scale, we present here a few results.

Both transmissivities and hydraulic conductivities were found to be generally of lognormal distributions. Hoeksema and Kitanidis (1985) fitted an exponential isotropic covariance with a nugget (5.2.2) to measurements of Y and herein are a few of their summarizing results of their Table 6. Thus, for ten aquifers in consolidated materials, σ_Y^2 was quite low, smaller than 0.7, except for two cases in which the values were 1.38 and 5.17, the median for the ten formations being 0.30. In contrast, w varied quite widely in the range 0-16, reflecting probably microregionalizations or large measurement errors. The integral scale varied between 1400 m to 44700 m, with a median of 14800 m, but four of the values were lower bounds compatible with the minimal distance between measurement points. More measurements for aquifers in unconsolidated materials were available, altogether 20 formations. The variance of correlated residuals varied widely among the aquifers, with a maximum of $\sigma_Y^2=6$, and with a median of 0.88. The values of the integral scale were also dispersed, with a median of 6300 m and an average of 16400 m. Much less information was available for the storativity S (8 aquifers), which again was found to be log-normal and displayed generally only a nugget.

Delhomme (1979) presents results of identification of T_G and σ_Y^2 (the total variance) for 13 aquifers, consolidated and unconsolidated, mostly from France. The transmissivity geometric mean T_G varied within a large range of 90 to 12000 m^2/day whereas σ_Y^2 range was found to be 0.69 to 1.9, with one exceptional value of 5.2. Since no attempt was made to separate an uncorrelated component (nugget) from the total variance, it is probable that the variance of correlated residuals is somewhat smaller than the values mentioned above. Delhomme mentions integral scales varying between 10^3 m for alluvial aquifers, to 10^4-2.10^4 m for chalk and limestone.

5.2.2 *Definition of hydraulic properties at the regional scale*

Toward a mathematical characterization of hydraulic properties of porous formations at the regional scale, we have to define first the transmissivity and storativity. In the case of a homogeneous formation at the local scale, for which the hydraulic conductivity K and the specific storativity s do not vary along the vertical, the definition is simply $T = K\,D$ and $S = s\,D$, where D is the saturated thickness, in a direction normal to the formation bottom. However, natural formations are generally heterogeneous at the local scale, and the analysis of this topic has been the content of Part 3. In such a case the appropriate generalization is

$$T = K_{ef}\,D \quad ; \quad S = s_{ef}\,D \tag{5.2.1}$$

where K_{ef} and s_{ef} are the effective hydraulic conductivity and specific storativity, respectively. In Chap. 3.4 K_{ef} has been evaluated for various types of stationary heterogeneity and for steady flow, uniform in the average. The simplest case is the one of a perfectly layered formation (Fig. 3.7a) and flow parallel to bedding, for which $K_{ef} = K_A$, i.e. the arithmetic mean. Assume now that the formation is confined and a pumping test is carried out by a fully penetrating well. The transmissivity determined by using Theis solution (Sect. 3.6.1) is equal in this case with the integrated K over the vertical, i.e. $T = \int_0^D K\, dz$. This last value and (5.2.1) are practically equal provided that D is much larger than the vertical correlation scale (see Sect. 3.2.4), such that the coefficient of variation of T is small. Matters become more complex if heterogeneity is of a three-dimensional nature and if the well is partially penetrating or the aquifer is unconfined. The relationship between T measured by a pumping or by a specific capacity test and (5.2.1) is not a simple one, because of the flow nonuniformity and unsteadiness on one hand (see Chap. 3.4), and space averaging on a limited volume surrounding the well, on the other. In contrast, K_{ef} in (5.2.1) is an ensemble average defined for uniform flow. We shall still assume that these effects are minor and the averaging volume is large enough to allow for exchanging space and ensemble averages. Hence, our picture of heterogeneous formations is a juxtaposition of local heterogenity, which may be regarded as stationary, and characterized in the mean by K_{ef} and S_{ef}, and a slow variation of these properties in the plane. This variation, combined with changes in the saturated thickness D, is modeled in turn as random, with a correlation scale much larger than the thickness. For the regional observer, however, T and S (5.2.1) are point values, and the variations at the local scale are either wiped out or manifest in a nugget present in the variogram of T. Again, this separation is possible due to the disparity between the thickness and the horizontal dimension and correspondingly, between local and regional correlation scales.

5.2.3 *Statistical representation of heterogeneity*

In view of these findings, we regard T and S (5.2.1) as random space functions of the planar coordinate $\mathbf{x}$. Along the lines of our classification in Sect. 3.2.2 and in compliance with the field data of Sect. 5.2.1, $Y_i = \ln T(\mathbf{x}_i)$ is viewed as a multivariate normal (MVN) vector for any set of points $\mathbf{x}_i$ (i=1,...,N). Usually only the assumption of weak stationarity is made about Y, and field data are employed to derive the p.d.f. of Y and its two-point covariance. The stronger assumption implied by the Gaussian representation, is both theoretically convenient and compatible with measurements. Indeed, the unconditional Y is stationary and characterized completely by its expected value m_Y and by its covariance $C_Y(\mathbf{r})$, with $\mathbf{r}$ the planar distance vector between two points, i.e. $\mathbf{r} = \mathbf{x}_1 - \mathbf{x}_2$ and $r = |\mathbf{r}| = [(x_1 - x_2)^2 + (y_1 - y_2)^2]^{1/2}$. We shall recall two types of C_Y from the list of Sect. 3.2.3, namely the isotropic exponential covariance (3.2.14)

$$C_Y(\mathbf{r}) = w[1 - H(\mathbf{r})] + \sigma_Y^2 \exp(-\frac{r}{I_Y}) \tag{5.2.2}$$

and the semi-spherical (3.2.17)

$$C_Y(r) = w[1-H(r)] + \sigma_Y^2 \left(1 - \frac{3}{2}\frac{r}{\ell_Y} + \frac{1}{2}\frac{r^3}{\ell_Y^3}\right) \quad \text{for } \frac{r}{\ell_Y} < 1 \; ; \quad C_Y = 0 \quad \text{for } \frac{r}{\ell_Y} > 1 \qquad (5.2.3)$$

In these equations w is the nugget or the variance of uncorrelated residuals, H is Heaviside step function, i.e. $H(r)=0$ for $r<0$ and $H=1$ for $r>0$, σ_Y^2 is the variance of correlated Y, I_Y is the linear integral scale and ℓ_Y is the range, where by (3.2.18) $I_Y=(3/8)\ell_Y$.

The nugget w incorporates a few possible effects: a microregionalization in the geostatistical terminology, i.e. spatial variations of Y on a scale much smaller than I_Y, the above mentioned effect of local heterogeneity and measurement or modeling errors related to the in-situ determination of T.

It is reminded that the two-point correlation may be also represented with the aid of the variogram $\Gamma_Y = w+\sigma_Y^2-C_Y$ and that the total variance of Y or the sill of the variogram is $w + \sigma_Y^2$. Finally, field data reviewed in Sect. 5.2.1 indicate that σ_Y^2 is generally smaller than unity for sedimentary formations, with possible exceptions, particularly for aquifers of large planar dimensions, while I_Y is of the order of 10^3 m.

The stationarity and ergodicity of Y imply that the horizontal characteristic dimension of the aquifer L is much larger than I_Y. However, if measurements or computations affect only a portion of the formation, Γ_Y may be employed to capture the correlation structure of Y for $r \ll L$ (see Chap. 1.7).

As we have mentioned in Sect. 5.2.1, less field based information is available with regard to the storativity S. In Sect. 3.2.2 we have presented a hypothetical correlation between the specific storativity s and the log-conductivity suggested by Freeze (1975). The application of such a correlation to the regional values does not seem to be warranted and in applications S has to be determined by tests. The data mentioned before suggest a lognormal distribution for S, which is close to a normal one if $\sigma_S^2 \ll 1$. Furthermore, S apparently does not display a spatial correlation structure and may be assumed to have a pure nugget. In absence of specific information, we shall assume that this is generally the case, and regard S as a random variable, rather than a random space function.

We consider now the conditional p.d.f. of Y and the conditional moments. We assume that T has been measured at M points, i.e. $Y_i=\ln T(\mathbf{x}_i)$ $(i=1,...,M)$ are given and they constitute a realization of the MVN vector Y_i. By using methods of statistical inference (see Chap. 5.6), it is assumed that the parameters characterizing the statistical structure of $Y(\mathbf{x})$, namely m_Y, w, σ_Y^2 and I_Y have been identified, after adopting a specific C_Y, say (5.2.2). These parameters are estimates and are subjected to estimation errors, but in this stage we shall disregard these errors (their impact is discussed later, at the end of this section). Consider now a set of arbitrary points $\mathbf{x}_\ell$ $(\ell=1,...,N)$, which includes the measurement points $\mathbf{x}_i$ such that $N>M$, and the MVN vector $Y_\ell=Y(\mathbf{x}_\ell)$. Its unconditional joint p.d.f. is defined with the aid of the expected value m_Y and of the covariance matrix $C_Y(\mathbf{x}_\ell,\mathbf{x}_m) = C_Y(r_{\ell m})$ $(\ell,m=1,...,N)$. The ensemble underlying the random vector Y_ℓ is the one of all possible realizations of the MVN Y_ℓ. In these realizations Y_i $(i=1,...,M)$ have different values compatible with its unconditional p.d.f. In the actual formation, however, Y_i are fixed and given, and only Y_j (j=M+1,...,N) are unknown and random. The conditional joint p.d.f. of Y_j is given by Bayes formula, $f(Y_j|Y_i) = f(Y_j,Y_i)/f(Y_i)$, where the latter are marginal distributions. In Chap. 1.2 we have recalled the classical results of the theory of probability, namely that the conditional Y_j is also multivariate normal, of expected value $\langle Y^c_j \rangle$ and covariance matrix $C^c_Y(\mathbf{x}_j,\mathbf{x}_k)$. In turn these are expressed with the aid of Eqs. (1.2.20-25), which are recalled here

$$\langle Y^c(\mathbf{x}_j)\rangle = m_Y + \sum_{i=1}^{M} \lambda_i(\mathbf{x}_j)(Y_i - m_Y) \tag{5.2.4}$$

$$C^c_Y(\mathbf{x}_j,\mathbf{x}_k) = C_Y(\mathbf{x}_j,\mathbf{x}_k) - \sum_{i=1}^{M} \lambda_i(\mathbf{x}_j) C_Y(\mathbf{x}_k,\mathbf{x}_i) =$$

$$C_Y(\mathbf{x}_j,\mathbf{x}_k) - \sum_{i=1}^{M}\sum_{\ell=1}^{M} \lambda_i(\mathbf{x}_j)\lambda_\ell(\mathbf{x}_k) C_Y(\mathbf{x}_\ell,\mathbf{x}_i) \qquad (j,k=M+1,...,N) \tag{5.2.5}$$

In turn, the coefficients λ_i are the solutions of the linear system

$$\sum_{i=1}^{M} \lambda_i(\mathbf{x}) C_Y(\mathbf{x}_i,\mathbf{x}_n) = C_Y(\mathbf{x},\mathbf{x}_n) \qquad (n=1,...,M) \tag{5.2.6}$$

In particular, from (5.2.5), the conditional variance is given by $\sigma_Y^{2,c}(\mathbf{x}) = C^c_Y(\mathbf{x},\mathbf{x})$. We recall here a few of the basic properties of the conditional moments: (i) even if Y_j is stationary in the strict sense, i.e. m_Y is constant and C_Y is a function of $\mathbf{r}$, the conditional Y^c_j is no more stationary, since $\langle Y^c(\mathbf{x})\rangle$ (5.2.4) depends on the relative position of $\mathbf{x}$ with respect to the measurement points and the same is true for $C^c_Y(\mathbf{x}_j,\mathbf{x}_k)$ (5.2.5) ; (ii) conditioning causes a variance reduction, i.e. $\sigma_Y^{2,c}(\mathbf{x}) \le \sigma_Y^2(\mathbf{x})$. The equality holds for points far from the measurement points, i.e. for $|\mathbf{x}-\mathbf{x}_i| >> I_Y$; (iii) at the measurement points $\langle Y^c(\mathbf{x}_i)\rangle = Y_i$ and $\sigma_Y^{2,c}(\mathbf{x}_i)=0$, i.e. (5.2.4) is an exact interpolator. In the presence of a nugget, however, the variance is equal to zero only at the measurement point and jumps to w at any other one. The latter value is the one to be taken into account, showing that there is very little variance reduction in the presence of a large w; (iv) in the absence of a nugget effect and for a dense distribution of measurement points, i.e. at distances much smaller than the integral scale, the conditional variance tends to zero and $\langle Y^c\rangle$ (5.2.4) becomes practically a deterministic interpolator.

We shall consider a few generalizations of these relationships. The first one refers to the case in which, due to measurements in a restricted area, Y is viewed as a random function of stationary increments. Thus, if $r << I_Y$ it is seen that both C_Y (5.2.2,3) lead to the linear variogram $\Gamma_Y = w[1-H(r)] + a\, r$ where $a=\sigma_Y^2/I_Y$ or $a=3\sigma_Y^2/2\ell_Y$, respectively. In such a case measurements permit one to estimate the constant a only, and in fact σ_Y^2 can be regarded as unbounded. Recalling the developments of Chap. 1.8, C_Y in Eqs. (5.2.4,5) is replaced by $\sigma_Y^2-\Gamma_Y$ and subsequently the limit $\sigma_Y^2\to\infty$ is taken. The immediate result is

$$\langle Y^c(\mathbf{x}_j)\rangle = \sum_{i=1}^{M}\lambda_i(\mathbf{x}_j)\,Y_i \;;\quad C^c_Y(\mathbf{x}_j,\mathbf{x}_k) = -\Gamma_Y(\mathbf{x}_j,\mathbf{x}_k) - \Lambda(\mathbf{x}_j) + \sum_{i=1}^{M}\lambda_i(\mathbf{x}_j)\,\Gamma_Y(\mathbf{x}_j,\mathbf{x}_k) =$$

$$= -\Gamma_Y(\mathbf{x}_j,\mathbf{x}_k) - \Lambda(\mathbf{x}_j) - \Lambda(\mathbf{x}_k) + \sum_{i=1}^{M}\sum_{\ell=1}^{M}\lambda_i(\mathbf{x}_j)\,\lambda_\ell(\mathbf{x}_k)\,\Gamma_Y(\mathbf{x}_i,\mathbf{x}_\ell) \qquad (5.2.7)$$

$$\sum_{i=1}^{M}\lambda_i(\mathbf{x})\,\Gamma_Y(\mathbf{x}_i,\mathbf{x}_k) - \Lambda(\mathbf{x}) = \Gamma_Y(\mathbf{x},\mathbf{x}_k) \;;\quad \sum_{i=1}^{M}\lambda_i(\mathbf{x}) = 1 \qquad (k=1,\ldots,M) \qquad (5.2.8)$$

replacing (5.2.4,6). It is to be reminded that the additional coefficient Λ stems from the $\lim \sigma_Y^2(1-\Sigma\lambda_i)$ for $\sigma_Y^2\to\infty$, which also leads to the last equation in (5.2.8). These relationships are identical with the kriging equations (Chap. 1.8) and $\langle Y^c\rangle$ and $\sigma_Y^{2,c}(\mathbf{x})= C^c_Y(\mathbf{x},\mathbf{x})$ are known as kriged values and kriging variances, respectively. Although the point of departure is different, it is seen that Gaussian conditional probability and stochastic interpolation by kriging lead to the same result for functions of stationary increments. In particular, the expected value is filtered out from (5.2.7) and Y^c is determined entirely with the aid of the variogram and the measured values. We shall adhere here to the terminology of conditional probability, although the results can be converted in most cases into equivalent geostatistical terms.

Another possible generalization of the preceding results refers to the unconditional expected value m_Y. In the strict stationary case m_Y is constant, but Eqs. (5.2.4,6) are valid if m_Y is space dependent as well, i.e. if a trend is present, provided that C_Y or Γ_Y is the covariance or the variogram of the residuals Y', respectively. In the geostatistical approach, however, Γ_Y is the half the variance of the increments of Y, which is identical with those of Y' only for m_Y constant. As we have shown in Chap. 1.8 the need to infer m_Y may be circumvented, even if it has a trend, by operating with generalized increments and generalized covariances (see, for instance, Bras and Rodriguez-Iturbe, 1985), but we shall limit the developments here to the simple case in which m_Y is constant, or may be rendered so by a division of the formation into subdomains.

Next, in some circumstances, e.g. in numerical simulations of flow, we may pursue the space average or the "block" value $\overline{Y}=(1/A)\int_A Y(\mathbf{x})d\mathbf{x}$ over planar elements A. We have discussed the effect of space averaging upon the unconditional moments of $\overline{Y}$ in Sect. (3.2.4), and the subject is investigated extensively in Vanmarcke (1983). Matters were shown to be quite simple for the unconditional moments of the MVN vector $\overline{Y}_j=\overline{Y}(\bar{\mathbf{x}}_j)$, where $\bar{\mathbf{x}}$ is the coordinate of the centroid of the planar elements over which Y is averaged. Indeed, the expected value is m_Y whereas the covariance matrix $C_{\overline{Y}}(\bar{x}_j,\bar{x}_k)$ is obtained from C_Y by the Cauchy algorithm (3.2.34). To obtain similar expressions for the conditional moments the system of equations (5.2.4,6) is generalized as follows (see Dagan, 1985a).

The expected value $<\overline{Y}^c(\bar{x}_j)>$ is given, in terms of m_Y and Y_i, by Eq. (5.2.4) in which λ_i is replaced by a new coefficient $\bar{\lambda}_i$, the latter being the solution of the linear system

$$\sum_{i=1}^{M} \bar{\lambda}_i(\bar{x}_j)\, C_Y(x_i,x_k) = \overline{C}_Y(\bar{x}_j,x_k) \quad (k=1,...,M) \tag{5.2.9}$$

which replaces (5.2.4). The only modification is of the right-hand-side of (5.2.4), which is defined as follows

$$\overline{C}_Y(\bar{x}_j,x_k) = \frac{1}{A}\int C_Y(x',x_k)\, \Omega(x'-\bar{x}_j)\, dx' \tag{5.2.10}$$

where integration is over the entire plane and the unit function Ω is equal to unity for x' within the averaging area A centered at $\bar{x}_j$, and zero otherwise. Finally, the covariance $C^c_{\overline{Y}}(\bar{x}_j,\bar{x}_k)$ is given by an equation similar to the last one in (5.2.5), namely

$$C^c_{\overline{Y}}(\bar{x}_j,\bar{x}_k) = C_{\overline{Y}}(\bar{x}_j,\bar{x}_k) - \sum_{i=1}^{M}\sum_{\ell=1}^{M} \bar{\lambda}_i(\bar{x}_j)\, \bar{\lambda}_\ell(\bar{x}_k)\, C_Y(x_\ell,x_i) \tag{5.2.11}$$

Two extreme limits of the moments of the block values are of particular interest. The first one is for the dimension of A much smaller than the integral scale I_Y. Then, in the absence of w, $\overline{C}_Y$ (5.2.10) differs very little from C_Y, and space averaging has little effect upon the moments of $\overline{Y}$, which become identical to those of Y (5.2.4,5). If a nugget w is present, it is wiped out by space averaging, since its support is zero. If w refelects the presence of measurement errors, this effect is understandable, and the uncertainty of Y shows up, nevertheless, in the estimation of m_Y, which will be discussed in the sequel. The disappearance of w in the case of a microregionalization poses a more delicate problem, since the outcome depends on the relative magnitude of A and of the scale of Y correlation for small distances between measurement points, which is generally not accessible by measurements. This discussion indicates that space averaging on areal elements which are very small compared to I_Y does not offer an advantage over the use of the moments (5.2.4,6) of the point value. The second extreme case is the one of A of a dimension much larger than I_Y. Then, due to division by A in (5.2.10), $\overline{C}_Y$ tends to zero like I^2_Y/A, and the same is true for $\bar{\lambda}$ (5.2.9). Hence, the conditional moments tend to their unconditional expressions, and accounting directly for

$$q = - T\,\nabla\langle\phi\rangle - \int_0^D \mathbf{q}'\,dz \quad ; \quad T = K_{ef}\,D \tag{5.3.4}$$

where K_{ef} is the effective conductivity analyzed in Chap. 3.4. Next, along the discussion of the preceding section, we neglect in (5.3.4) the integral of the specific discharge residual $\mathbf{q}'$ by arguments related to the effect of space averaging (see Sect. 3.7.6), which is an acceptable approximation if D is much larger than the conductivity vertical correlation scale. Otherwise, the neglected term is regarded as part of the random variation of T (see Chap. 5.2). The assumption that $\langle\phi\rangle$ is a slowly varying function of $\mathbf{x}$ at the local scale might be questioned in the case of flow toward wells, but again if the well length is large compared to the local correlation scale, the effective conductivity for uniform flow may be adopted in such cases (see Sect. 3.4.6). Similarly, by the assumption that $\langle\phi\rangle$ is slowly varying in time at the local scale (see Sect. 3.4.7), we may write for the elastic storage term in (5.3.1)

$$\int_0^D s\,\frac{\partial\phi}{\partial t}\,dz \simeq S\,\frac{\partial\langle\phi\rangle}{\partial t} \quad ; \quad S = s_{ef}\,D \tag{5.3.5}$$

where it is reminded that s_{ef} is equal to the arithmetic mean of the specific storativity (Sect. 3.4.7).

To summarize, and with the change of notation $\langle\phi\rangle$=H consistent with the literature, Eqs. (5.3.1), (5.3.4) and (5.3.5) lead to

$$\nabla.(T\,\nabla H) = S\,\frac{\partial H}{\partial t} - R_{ef} \qquad (\mathbf{x}\in\Omega) \tag{5.3.6}$$

for confined or semi-confined formations, where R_{ef} is the sum of a distributed recharge R, due to infiltration or leakage, and of R_w (5.3.2), while Ω is the flow domain. Strictly speaking, in the case of wells, Eq. (5.3.6) is valid only for fully penetrating ones. Otherwise, the shallow water approximation, which underlies (5.3.6), is not valid in the neighborhood of the well (see Sect. 3.6.4) . Still, (5.3.6) will be used in any case, with restriction about its applicability in the close vicinity of partially penetrating wells.

In the case of phreatic aquifers we adopt the Dupuit-Forcheimer approximation for the vertically averaged variables (see Sect. 3.6.4), i.e. $H \simeq \eta$, where η is the mean free-surface elevation. Then, we get

$$\nabla.(T\,\nabla H) = n_{ef}\,\frac{\partial H}{\partial t} - R_{ef} \quad ; \quad T = K_{ef}\,D \tag{5.3.7}$$

where the elastic storage term has been neglected in comparison with the free-surface storage. The main difference between (5.3.7) and (5.3.6) is that D, the saturated thickness, depends on H. Thus, for a horizontal bottom, D=H (Fig. 5.3.1b).

In unsteady flow, H and η are functions of x,y and t, whereas in steady flow the time dependence and the time derivatives in (5.3.6) and (5.3.7) are dropped out.

It is common in the literature to linearize (5.3.7) by a procedure similar to that of Sect. 3.6.2, namely to assume that in the expression of T, D is the distance from the bottom to the average, horizontal, water-table. Hence, $T = K_{ef} D$ is regarded in (5.3.7) as a function of x,y, but not of H. Then, there is no difference between (5.3.6) and (5.3.7), except for the interpretation of H (5.3.6) as the piezometric head in a confined aquifer, while it is the free-surface elevation in (5.3.7). The assumption of small deviations of η from a constant value is not obeyed in a heterogeneous formation in which T varies wildly through the aquifer. This point will be retaken in Chap. 5.5.

Eqs. (5.3.6,7) form the basic relationships satisfied by H, and the direct problem consists in solving them, with appropriate boundary and initial conditions. The boundaries of the formation define the regional hydrologic unit, which is of concern in this part. A typical boundary is one of no flow, for which $\boldsymbol{q}\cdot\boldsymbol{\nu}=0$, i.e. $\partial H/\partial\nu=0$, where $\boldsymbol{\nu}$ is a unit vector normal to the boundary. In applications this is either a surface of contact with an impervious material or a water divide, which is assumed to stay so. Another typical boundary is one of contact with a water body (lake, river, sea) for which $H=H_b$ is given. These are particular cases of the one known as a radiation boundary condition

$$a \frac{\partial H}{\partial \nu} - b\, H = c \qquad (\mathbf{x}\in\partial\Omega) \tag{5.3.8}$$

where a,b,c are constant or functions of time and $\partial\Omega$ stands for the boundary of Ω. Eq. (5.3.8) may pertain, for instance, to a leaky boundary for which the flux depends linearly on the head, while the previous relationships are obtained for a or b equal to zero. Finally, in the simplest form the initial condition is of given head $H(x,y,0)=H_0(x,y)$.

In the traditional, deterministic, approach, the input data for the solution of the direct problem, consist of the functions $T(\mathbf{x})$, $S(\mathbf{x})$, $R_{ef}(\mathbf{x})$ in the flow domain Ω and the boundary and initial conditions for H on $\partial\Omega$. In the case of heterogeneous aquifers and in the probabilistic framework, T, S and R_{ef} are represented as random space functions, and (5.3.6,7) become stochastic differential equations. The solution $H(\mathbf{x},t)$ is, therefore, a random space function itself, which is defined by its statistical moments. At second-order these are $\langle H(\mathbf{x},t)\rangle$ and $C_H(\mathbf{x},\mathbf{x}',t) = \langle[h(\mathbf{x},t)\, h(\mathbf{x}',t)]\rangle$ where $h(\mathbf{x},t)= H(\mathbf{x},t)-\langle H(\mathbf{x},t)\rangle$ is the head residual. If H is normal, these moments define completely its spatial statistical structure at any t, but we shall limit the information to second-order in any case. We state now the direct problem as follows: derive $\langle H\rangle$ and C_H for H satisfying Eq. (5.3.6) or (5.3.7) in the flow domain, for Y, S and R_{ef} random and given, and with given boundary conditions (5.3.8) and appropriate initial conditions. Generally, the coefficients showing up in the boundary conditions are also uncertain, but we shall regard them as fixed and the results for H conditioned on their value. The impact of their randomness can be easily considered a-posteriori by using the basic Bayes formula. To narrow down the scope, we restrict T to a log-normal and stationary p.d.f., characterized by the four parameters of Sect. 5.2.3. As for S and n_{ef}, we assume that they are random variables, with a "nugget" covariance only. Last, R_{ef} is generally made up from two terms: the contribution of wells (5.3.2), which is assumed to be deterministic, and a term R, resulting from natural recharge. The latter is regarded as a random variable, depending on t, but not on $\mathbf{x}$. In other words, (5.3.6,7) can be solved for fixed parameters S and R, with the results for H regarded as conditioned on their values, precisely like for boundary conditions. From this discussion it is apparent that the crux of the matter is the randomness of T and its impact upon H, and most of the developments in the sequel deal with it.

The stochastic solution of the direct problem may be considered for the two possible types of probabilities of Y= ln T, unconditional and conditional. The joint p.d.f. of Y at a set of arbitrary points in both cases has been elaborated in Sect. 5.2.3. The unconditional Y is to be adopted whenever the complete information about measurements of T, value and locations, are missing, or when block averages of H over areas much larger than I_Y^2 are sought (see Sect. 5.2.3). Otherwise, the use of Y^c is preferable, since it takes maximum advantage of the information contained in measurements.

Exercises

5.3.1 An unconfined aquifer overlies a semi-confined one, the two being separated by a semi-pervious thin layer. Formulate the equations satisfied by the head in each of the formation, at the regional scale. Under what conditions is it possible to treat the two aquifers as one unit for solving the direct problem?

5.3.2 Rewrite Eq. (5.3.6) for H in steady flow and for R_{ef}=0, in terms of Y = ℓn T. Derive the equation satisfied by the streamfunction $\Psi(\mathbf{x})$ (Chap. 3.5) and reformulate the boundary conditions in terms of Ψ.

5.4 EFFECTIVE PROPERTIES AND SOLUTIONS OF THE EQUATIONS OF MEAN FLOW

5.4.1 *Effective transmissivity and storativity (confined flow, unconditional probability)*

Along the lines of Part 3, we start with the development of the equations satisfied by the expected values of the specific discharge $\langle \boldsymbol{q} \rangle$ and of the water head $\langle H \rangle$, for heterogeneous formations of *two-dimensional* random structures. The simplest and fundamental case is the one of steady and average uniform flow in a confined formation, pertaining to the configuration of Fig. 5.4.1. Unlike Part 3, we shall limit the discussion to an isotropic covariance C_Y, since analysis of most field data does not indicate the existence of transmissivity anisotropy in the horizontal plane. The effective transmissivity is defined in a manner similar to that of Sect. 3.4.1, by the equation

$$\langle \boldsymbol{q} \rangle = - T_{ef} \nabla \langle H \rangle \tag{5.4.1}$$

where it is reminded that the ensemble underlying the averaging operator is of the heterogeneous formations of stationary multivariate normal $Y = \ln T$, of given m_Y and C_Y. In the case of uniform flow of Fig. 5.4.1 $\nabla\langle H \rangle = -\mathbf{J}$, $J = (H_A - H_B)/L_x$, and on grounds of dimensional analysis $T_{ef} = f(m_Y, \sigma_Y^2, I_Y/L_x, I_Y/L_y)$, where we have assumed that C_Y (e.g. 5.2.2) is represented by the parameters σ_Y^2 and I_Y. The nugget w does not affect T_{ef} directly, since its derivation involves an integration over C_Y. T_{ef} is defined as the limit of the above function for $I_Y/L_x \ll 1$ and $I_Y/L_Y \ll 1$, i.e. for formations obeying the basic requirement of stationarity. Hence, T_{ef} is a function of m_Y and σ_Y^2 solely, being a genuine property of the medium, independent of the geometry of the flow domain or of the flow regime. We shall examine in the sequel the effect of boundaries, flow nonuniformity and unsteadiness upon T_{ef}, remembering that the flow here is a particular, two-dimensional case, of the general case analyzed in Part 3.

The most general results about T_{ef} are its bounds, and according to Sect. 3.4.1 $T_H \le T_{ef} \le T_A$, where $T_H = T_G \exp(-\sigma_Y^2/2)$ is the harmonic mean, $T_A = T_G \exp(\sigma_Y^2/2)$ is the arithmetic mean and $T_G = \exp(m_Y)$ is the geometric mean. However, these absolute bounds are two wide apart, since they apply to arbitrarily anisotropic formations.

We have derived in Sect. 3.4.2. the effective conductivity by a small perturbation approach, valid for $\sigma_Y^2 \ll 1$. The results there contain the ones applying to two-dimensional heterogeneity as particular cases. Thus, Eq. (3.4.18) provides the effective conductivity for an axi-symmetric C_Y, isotropic in the horizontal plane. The two-dimensional *isotropic* case is obtained at the limit $e = I_{Yv}/I_h \to \infty$, where I_{Yv} and I_{Yh} are the vertical and horizontal integral scales, respectively. In other words, the two-dimensional structure is regarded as the limit of a three-dimensional one in which Y is fully correlated in the vertical direction. At this limit ν (3.4.17) tends to unity and by Eq. (3.4.18)

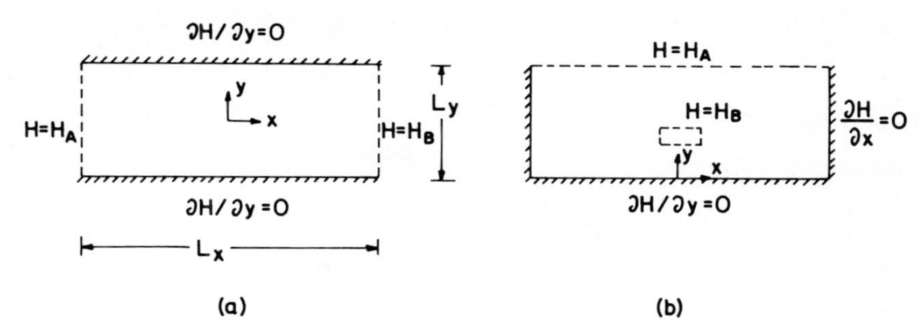

Fig. 5.4.1 Flow domains and boundary conditions for the numerical simulations of Smith and Freeze (1979): (a) uniform average flow and (b) flow to a tunnel.

$T_{ef} \rightarrow T_G$, up to terms $0(\sigma_Y^4)$. This result has been strengthened by the self-consistent analysis of Sect. 3.4.3, which is not limited by the requirement of small σ_Y^2. The result is given by Eq. (3.4.35) and the comment following it, which states that its numerical evaluation has rendered $T_{ef} = T_G$ with a great degree of accuracy for arbitrary σ_Y^2 (Dagan, 1979b), confirming an early result of Matheron (1967).

Smith and Freeze (1979) have solved by Monte Carlo simulations the flow of Fig. 5.4.1. The formation has been partitioned in rectangular blocks of constant T, whose values were generated at random, by an auto-regressive scheme. The T of neighboring blocks were correlated, such that an integral scale larger than the blocks dimensions could be simulated. Due to numerical convergence problems, the largest attained value of σ_Y^2 was 4. The expected value of the specific discharge was computed by integrating first the specific discharge over the outflow face, and subsequently ensemble averaging for the various Monte Carlo simulations. Smith and Freeze (1979) report that for all values tested T_{ef} was equal to the geometric mean T_G. Hence, these different computations led to the same conclusion, namely that for average uniform flow through an isotropic formation in a domain of large extent compared to the log-transmissivity integral scale, the effective transmissivity is equal to the geometric mean $T_G = \exp(m_Y)$.

Along the lines of Chap. 3.4, we examine now the impact of various flow conditions which depart from the above ones.

The first issue is that of the influence of boundaries, e.g. the possible change of $\langle q \rangle$ for points close to the impervious or constant head boundaries of Fig. 5.4.1. The analysis of Sect. 3.4.5, based on a first-order perturbation approximation, led to the conclusion that the effect of the boundaries is negligible. This conclusion is supported indirectly by the numerical results of Smith and Freeze (1979) mentioned above. Indeed, if the boundaries had a significant impact upon $\langle q \rangle$, its integral over L_y would have also been affected, though to a lesser extent.

A more critical issue is that of the influence of the nonuniformity of the average flow, which has been analyzed in Sect. 3.4.6, again by a first-order approximation in σ_Y^2 and for slowly varying flows. The main conclusions were that the additional effect of a polynomial trend of $\langle H \rangle$ would manifest only through the derivatives of third-order of $\langle H\rangle$ in the $\langle q\rangle,\langle H\rangle$ relationship (see Eq. 3.4.48). Furthermore, in the case of radial flow and for an isotropic three-dimensional structure, it was found that the average flow, and correspondingly the effective conductivity, is somewhat reduced as compared to the value for uniform flow (see Eq. 3.4.49). The reduction is proportional to $\sigma_Y^2 I_Y^2/R^2$, where R is the length scale characterizing the average flow nonuniformity. The formalism of Sect. 3.4.6 leads to similar results for two-dimensional flow. Generally, regional flows are slowly varying in space even at the large regional I_Y, e.g. flow caused by uniform recharge. There is, however, a notable exception which deserves attention, namely flow to wells. Indeed, in this case the length scale R is r_w, the well radius. This is generally much smaller than I_Y, the transmissivity regional scale, and the situation is diametrically opposite to that of slowly varying flow. In the case of three-dimensional flows, the smoothing effect of vertical averaging along the well axis was invoked in order to reduce the specific discharge variance, but this argument is not valid in the case of two-dimensional flow (the same issue has been raised by Matheron, 1967 and analyzed by Dagan, 1982a). The problem of the effective transmissivity for radial well flow in a heterogeneous random structure is a difficult one, and its numerical treatment is bound to be complex, due to the need of refining the numerical grid in the neighborhood of the well. It is possible, however, to derive approximately T_{ef} by observing that most of the drawdown occurs in the neighborhood of the well (see Chap. 3.5), over an area smaller than I_Y^2. Hence, practically speaking, the well is submerged in a homogeneous formation with T constant and equal to its value at $\mathbf{x}_w$. This picture agrees with the idea set forth in Chaps. 5.1-5.2, namely that by a pumping test we can determine the point value of T. Under these circumstances, we can employ in the neighborhood of the well the solutions derived in Chap. 5.4 for flow in a homogeneous formation. For instance, in the case of steady flow, the radial head gradient is given by

$$q_r = -T\frac{dH}{dr} = -\frac{Q_w}{2\pi r} \quad ; \quad r = |\mathbf{x}-\mathbf{x}_w| \tag{5.4.2}$$

For a given well discharge Q_w, ensemble averaging of (5.4.2) yields

$$\frac{d\langle H\rangle}{dr} = \frac{Q_w}{2\pi r T_H} = -\frac{\langle q_r\rangle}{T_H} \quad ; \quad T_H = \langle 1/T\rangle^{-1} = T_G \exp(-\sigma_Y^2/2) \tag{5.4.3}$$

Hence, the effective transmissivity is equal to the harmonic mean T_H, which is smaller than T_G by the factor of (5.4.3). Rigorously speaking, this should be regarded as the limit of T_{ef} for $I_Y/r_w \to \infty$, a limit which is practically encountered in actual formations. Besides, as we shall show in Chap. 5.5, H is affected by a large variance in the neighborhood of the well, making the prediction of its expected value of limited use. In practice, however, this difficulty is alleviated by the fact that for existing wells the transmissivity has been measured at its location and it is given.

Hence, the large uncertainty associated with the use of the unconditional p.d.f. of T may be drastically reduced by employing conditional p.d.f., and this will be treated in Chap. 5.5.

Smith and Freeze (1979) has solved numerically, by the methodology discribed above, a few problems of nonuniform flow, the one selected here for illustration being that of flow toward a tunnel (Fig. 5.4.1b). In this case there is a strong flow nonuniformity in the region of convergence of streamlines. The authors have computed the expected value $\langle Q \rangle$ of the total discharge through the formation and compared it with Q_{ef} corresponding to a homogeneous formation of $T=T_G$. For $\sigma_Y^2 = 1$ and for $I_Y/L_y \simeq 0.17$, the reduction of $\langle Q \rangle$ as compared to Q_{ef} was 12%, which is quite moderate. However, the tunnel dimension was not as small, as are the wells diameters, compared to I_Y.

Finally, we shall discuss the effect of flow unsteadiness upon T_{ef}, which has been investigated by Dagan (1982c) under a first-order analysis in σ_Y^2 and for slowly varying flows in time. Two types of unstediness have been considered: transients and periodic. The transient consisted in the passage from rest to a steady state of uniform average flow, and it has been discussed in Sect. 3.4.7. This case is of interest mainly at the local scale, e.g. for the incipient period of pumping by a well. The main result was that the effective conductivity is time dependent and it relaxes from the initial value K_A, the arithmetic mean, to the steady state value K_G, according to the function of time of Fig. 3.4.8. In the case of two-dimensional flows, relevant to transmissivity, the dimensionless relaxation time (Fig.3.4.8, Eq. 3.4.62 with m=2) is given by $tT_G/\langle S \rangle I_Y^2 \simeq 10$. For the large values of I_Y of transmissivity, this time may be quite large. However, time changes at the regional scale are generally slow, being related mainly to seasonal variations of recharge and pumping. Thus, the second analysis of Dagan (1982c), of a periodic variation of the average head gradient, is more relevant to the present discussion, and herein are the main results.

The average head gradient was decomposed in a steady component J_{st} and a periodic one $J_u \exp(i\omega t)$, both slowly varying in space. After solving for the first-order approximation of the head (see Sect. 3.4.7) and assuming that the dimensionless parameter $\mathcal{I}_Y = I_Y (\omega \langle S \rangle / T_G)^{1/2}$ is smaller than unity, the following close form expression was obtained

$$T_{ef} = T_G \left[1 + \frac{J_u}{J_{st}} \sigma_Y^2 \, d_2 e^{i\omega t} + 0(\sigma_Y^4) \right] \quad ; \quad d_2 = -\frac{i\mathcal{I}_Y^2}{2} \left[\ln \mathcal{I}_Y + 1 + \frac{i\pi}{4} + 0(\mathcal{I}_Y^2) \right] \tag{5.4.4}$$

The first important finding is that generally speaking T_{ef} is a function of time and of the gradient, and cannot qualify as an intrinsic property of the formation. Since d_2 (5.4.4) is complex, there is a phase shift between T_{ef} and the head gradient. A meaningful T_{ef}, equal to its steady state value T_G, can be defined only if the second term in (5.4.4) is negligible compared to unity, i.e. $|J_u/J_{st} \sigma_Y^2 d_2| << 1$. Hence, for given formation parameters σ_Y^2, I_Y, T_G and $\langle S \rangle$, and flow parameters J_{st} and ω, the magnitude of the unsteady gradient J_u has to be sufficiently small to warrant a quasi-steady approach. Following the example of Dagan (1982c), namely assuming an annual variation, i.e. $\omega=2\pi/365\ \mathrm{day}^{-1}$, and the values $T_G/\langle S\rangle = 0.5\times10^5\ \mathrm{m^2/day}$ and $I_Y = 10^3$ m, yields $|J_u/J_{st}| \ll 6.4/\sigma_Y^2$. This condition is bound to be obeyed in most conceivable cases, unless I_Y is much larger than the assumed value, other parameters equal. We shall assume that this is the case and consider T_{ef} to be constant and equal to T_G in the sequel. It is emphasized, however, that the problem of unsteady flows in heterogeneous formations of a two-dimensional structure for large σ_Y^2, or for flows which do not obey the above requirement of slow time variation, is an open subject.

In the same study (Dagan, 1982c), the effective storativity S_{ef} has also been investigated, the result being that it is equal to its arithmetic average $\langle S\rangle$ under quite general conditions.

Summarizing this Section, it was found that the effective transmissivity is equal to the geometric mean in steady and uniform flow in a formation of large planar extent compared to I_Y. This value applies locally and instantaneously to nonuniform or unsteady average flows, only under conditions of slow spatial and temporal variations of the average head gradient at the scales I_Y and $I_Y^2\langle S\rangle/T_G$, respectively. These conditions are much more restrictive than in three-dimensional flows. An outstanding exception is that of flow toward a well, which is characterized by a length scale much smaller than I_Y, and for which T_{ef} is markedly different from T_G.

5.4.2 Effective transmissivity (unconfined flow, unconditional probability)

In this section we are going to examine the effect of the presence of a free surface (Fig. 5.3.1b) upon the effective transmissivity (phreatic flow under uncertainty has been investigated for the first time by Gelhar, 1976). The main difference between confined and unconfined flows is that in the latter (see Eq. 5.3.7) the transmissivity $T = K_{ef}H$ depends not only on the effective conductivity, but also on the head. A few solutions of free-surface flow problems in homogeneous aquifers have been examined in Chap. 3.6, while here we consider the effect of heterogeneity. It has not been investigated so far by numerical simulations that may serve as a basis for comparison. To obtain simple results, we adopt from the outset the small perturbation approximation of Sects. 3.3.2, 3.4.2. In the present context, let $\langle\eta\rangle$ be the average saturated thickness, while $\langle\eta\rangle+\eta'$ is the total one (Fig. 5.4.2a), the fluctuation η' being caused by heterogeneity. We can write, therefore, $T = \mathcal{T}(1+\eta'/\langle\eta\rangle)$, where $\mathcal{T} = K_{ef}\langle\eta\rangle$ is the transmissivity corresponding to the average elevation of the free-surface. $\mathcal{T}$ is regarded as a random space function, characterized statistically by the same p.d.f. as in confined flow, i.e. by the unconditional moments m_Y and C_Y. Since under the Dupuit assumption $H = \langle\eta\rangle+\eta'$, we may also write $\nabla H = -\mathbf{J}+\nabla\eta'$, where $\mathbf{J} = -\nabla\langle\eta\rangle$, the gradient of the average free-surface (Fig. 5.4.2a), is taken as constant. This picture would have been exact for a sloping aquifer (Fig. 5.4.2b) under uniform average flow, with J the slope, but the scheme applies to the configuration of Fig. 5.6.2a as well, provided that $\langle\eta\rangle$ is slowly varying at the I_Y scale, which we presume. As a matter of fact the order of magnitude of the variation of $\langle\eta\rangle$ is ΔH, the head drop between boun-

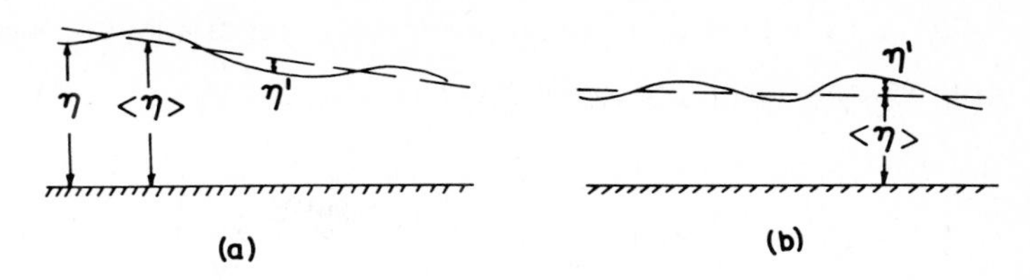

Fig. 5.4.2 Definition sketch for free-surface flow: (a) horizontal impervious bottom and (b) sloping bottom.

daries, which is generally small compared to the average depth $\langle\eta\rangle$. Under these conditions, the equation of steady flow can be rewritten as follows

$$\nabla.[T(1+\frac{\eta'}{\langle\eta\rangle})(-\mathbf{J}+\nabla\eta')] = 0 \quad ; \quad T = T_G \exp(Y') \tag{5.4.5}$$

with Y′ multivariate normal and characterized by σ_Y^2 and I_Y. We linearize (5.4.5) along the lines of Sect. 3.3.2, i.e. for small σ_Y^2, by expanding $\exp(Y') = 1+Y'+(1/2)Y'^2+...$ and retaining in (5.4.5) linear terms in Y′ and η', to obtain

$$\nabla^2\eta' - \frac{1}{\langle\eta\rangle}\mathbf{J}.\nabla\eta' = \mathbf{J}.\nabla Y' \tag{5.4.6}$$

The basic equation (5.4.6) is the starting point for evaluating T_{ef} here, as well as higher-order moments in Chap. 5.5. Comparison with the corresponding equation in a confined formation, e.g. (3.3.9), reveals that they differ by the presence of the second term in the left-hand side of (5.4.6), which originates from the dependence of T upon η. Once the statistical moments of η', solution of (5.4.6) are derived, the expected value of the specific discharge is given by (3.4.9), i.e. in the present case by

$$\langle q\rangle = T_G(\mathbf{J} + \frac{\sigma_Y^2}{2}) - \boldsymbol{\alpha} + O(\sigma_Y^4) \quad ; \quad \boldsymbol{\alpha} = -\langle Y' \nabla\eta'\rangle \tag{5.4.7}$$

Hence, precisely like in Sect. 3.4.2 the evaluation of T_{ef} at first order in σ_Y^2 boils down to the computation of $\boldsymbol{\alpha}$. This can be carried out with the aid of the Green function pertaining to the linear operator of (5.4.6), i.e. with $G(\mathbf{x},\mathbf{x}')$ satisfying

$$\nabla^2 G - \frac{1}{\langle\eta\rangle}\mathbf{J}.\nabla G = -\delta(\mathbf{x}-\mathbf{x}') \tag{5.4.8}$$

which should be compared with (3.4.10) for confined flow. Like in Sect. 3.4.2 we adopt for G its expression in an unbounded domain, in which case it is a function of $\mathbf{r} = \mathbf{x}-\mathbf{x}'$. The expression of the Green function can be found in Carslaw and Jaeger (1962, Sect. 10.7), being analogous to the solution of the heat conduction equation for a moving source. In the present notation, G and its Fourier transform, are given by

$$G(\mathbf{r}) = \frac{1}{2\pi} \exp\left(\frac{\mathbf{J.r}}{2\langle\eta\rangle}\right) K_0\left(\frac{Jr}{2\langle\eta\rangle}\right) \quad ; \quad \hat{G}(\mathbf{k}) = \frac{1}{2\pi} \frac{1}{[k^2 - i(\mathbf{J.k})/\langle\eta\rangle]} \tag{5.4.9}$$

where K_0 is the modified Bessel function of second kind and zero order. For $r \to 0$ or $k \to \infty$, the limit of (5.4.9) is $G \to -(1/2\pi)\ln(r)$ and $\hat{G} \to (1/2\pi)(1/k^2)$, respectively, i.e. the expressions similar to (3.4.10) pertaining to G for Laplace equation. In other words the singular behavior of G is dictated by the operator ∇^2 in (5.4.8). In contrast, for large r, $G \to (\langle\eta\rangle/4\pi Jr)^{1/2} \exp[-Jr/2\langle\eta\rangle + \mathbf{J.r}/2\langle\eta\rangle]$ tends to zero, whereas G (3.4.10) grows logarithmically with r.

The function α is given by Eq. (3.4.2) in terms of G, or by (3.4.13) for $\hat{G}$, and it is rewritten for two-dimensional flows, as follows

$$\alpha = \int \mathbf{k}(\mathbf{J.k})\, \hat{G}(\mathbf{k})\, \hat{C}_Y(\mathbf{k})\, d\mathbf{k} = \frac{1}{2\pi} \int \frac{(\mathbf{J.k})\mathbf{k}\, \hat{C}_Y(\mathbf{k})}{k^2 - i(\mathbf{J.k})/\langle\eta\rangle}\, d\mathbf{k} \tag{5.4.10}$$

The logtransmissivity covariance and its Fourier transform, in the two-dimensional $\mathbf{x}$ and $\mathbf{k}$ spaces, respectively, are given in Sect. 3.2.3, e.g. Eq. (3.2.16), for the exponential C_Y. Its main contribution to the integral in (5.4.10) arises around the value of $k \simeq 1/I_Y$ (see Sect. 3.4.2). For this k, the ratio between the two terms making up the denominator of $\hat{G}$ (5.4.9) is equal to the dimensionless number $\mathcal{I}_Y = JI_Y/\langle\eta\rangle$, which plays an important role in this analysis. Under the present assumptions $\mathcal{I}_Y \ll 1$ since $J \simeq \Delta H/L$ and $\mathcal{I}_Y = (\Delta H/\langle\eta\rangle)(I_Y/L)$, ΔH being the head drop between boundaries and L the horizontal extent of the formation. Both ratios contributing to $\mathcal{I}_Y$ are much smaller than unity and $\mathcal{I}_Y$ is indeed a very small number. Under these circumstances we can compute α in an approximate manner in (5.4.10) by carrying out an asymptotic expansion in $\mathcal{I}_Y$. Toward this aim we switch to polar coordinates in the $\mathbf{k}$ plane, with θ the angle between the vectors $\mathbf{J}$ and $\mathbf{k}$. The result is

$$\alpha = \frac{J}{2\pi} \int_0^{2\pi} \cos^2\theta \int_0^{\infty} \frac{k^3 \hat{C}_Y(k)}{k^2 + J^2\cos^2\theta/\langle\eta\rangle^2}\, dk\, d\theta =$$

$$\frac{J}{2\pi} \int_0^{2\pi} \cos^2\theta \left[\int_0^{\infty} k\hat{C}(k)\, dk - \frac{J^2}{\langle\eta\rangle^2}\cos^2\theta \int_0^{\infty} \frac{k\hat{C}_Y}{k^2 + J^2\cos^2\theta/\langle\eta\rangle^2}\, dk \right] d\theta = \tag{5.4.11}$$

$$J\sigma_Y^2\left[\frac{1}{2} + \frac{1}{16}\mathcal{I}_Y^2 \ln(\mathcal{I}_Y^2) + O(\mathcal{I}_Y^2)\right]$$

Substitution of (5.4.11) in (5.4.7) yields for the effective transmissivity in the equation $q = T_{ef} J$ the final result

$$T_{ef} = T_G \left[1 - \frac{1}{16} \sigma_Y^2 I_Y^2 \ln(I_Y^2) + O(\sigma_Y^2 I_Y^2) \right] \qquad (5.4.12)$$

Eq. (5.4.12) encapsules the main result of the analysis, namely that the effective transmissivity in unconfined flow is equal to T_G, the one valid for confined flow, at a high degree of accuracy, provided that the flow is slowly varying in space, to ensure that $I_Y = I_Y J / \langle \eta \rangle << 1$.

5.4.3 A few solutions of the mean flow equations

The determination of the expected value, unconditional or conditional, of H reduces to solving the flow equations (5.3.6,7) in which T and S are replaced by their effective values. These are deterministic partial differential equations in the plane and a few sophisticated and versatile codes are available in order to solve them numerically, under general geometries and boundary conditions. The review of these finite differences, finite elements or boundary elements methodologies and codes is beyond our scope. Along the lines of Chaps. 3.5 and 3.5 we shall limit the discussion to deriving briefly a few analytical solutions for simple cases, to grasp some salient features of regional flows. Furthermore, these solutions will serve us in the following chapters.

The simplest case is the one of constant unconditional effective properties $T_{ef} = T_G$ and $S_{ef} = S_A$, the geometric mean and the arithmetic mean, respectively. Thus, the equations and solutions are the same as the ones pertaining to a homogeneous aquifer of constant properties. In steady flow, $\langle H \rangle$ satisfies Poisson equation (see 5.3.6)

$$\nabla^2 \langle H \rangle = - R_{ef} / T_{ef} \qquad (5.4.13)$$

Two geometries which lend themselves to simple solutions are those of Fig. 5.4.1a, i.e. a rectangle bounded by two impervious and two constant head boundaries, or Fig. 5.4.3, a circular formation with boundary condition $\langle H \rangle = -\mathbf{J}.\mathbf{x}$ for $r=|\mathbf{x}|=L$. The same boundary condition, of a uniform gradient **J**, applies to Fig. 5.4.1a, with $\mathbf{J}(J,0)$ and $J=(H_A - H_B)/L_x$. The solutions are

$$\langle H \rangle = -Jx - (R_{ef}/T_{ef})\left(\frac{x^2}{2} - \frac{L_x^2}{8}\right) \quad ; \quad \langle H \rangle = -\mathbf{J}.\mathbf{x} - (R_{ef}/T_{ef})\left(\frac{|\mathbf{x}|^2}{4} - \frac{L^2}{4}\right) \qquad (5.4.14)$$

respectively. The head is, therefore, made up from a linear trend, corresponding to uniform flow,

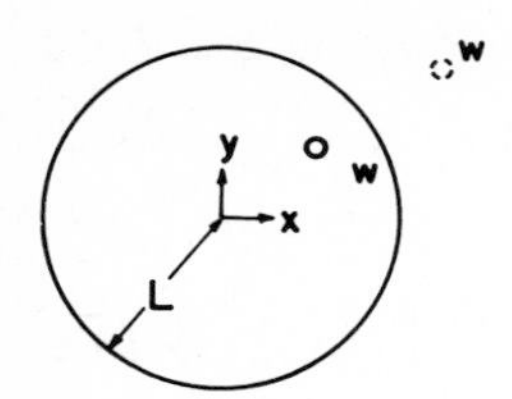

Fig. 5.4.3 A pumping well in a circular formation and its image.

and a quadratic trend, resulting from the existence of the recharge. We consider now a general configuration and $\langle H\rangle$ as regular function of $\mathbf{x}$. A Taylor expension of $\langle H\rangle$ in the neighborhood of a point of regularity $\mathbf{x}_0$, yields

$$\langle H\rangle = H_0 - \sum_{i=1}^{2} c_i(x_i - x_{0i}) - \frac{1}{2}\sum_{i=1}^{2}\sum_{j=1}^{2} c_{ij}(x_i - x_{0i})(x_j - x_{0j}) + \dots \tag{5.4.15}$$

where H_0, c_1, c_2, c_{11}, $c_{12}=c_{21}$ and c_{22} are six constant coefficients. Since $\langle H\rangle$ satisfies identically (5.4.23) we get: $c_i = J_i$, $c_{11}+c_{22} = R_{ef}/T_{ef}$. Hence, up to quadratic terms, we can rewrite the local expression of $\langle H\rangle$ as follows

$$\langle H\rangle = H_0 - \mathbf{J}.\mathbf{x} - (R_{ef}/4T_{ef})|\mathbf{x}-\mathbf{x}_0|^2 + c[(-x_0)^2-(y-y_0)^2] + d(x-x_0)(y-y_0) \tag{5.4.16}$$

in terms of a uniform gradient $\mathbf{J}$, a constant recharge R_{ef} and two additional constants $c = (c_{22}-c_{11})/4$ and $d = -c_{12}$. Hence, the quadratic trend of $\langle H\rangle$ can be attributed to recharge and to an additional term which is harmonic, the latter reflecting the influence of boundaries. For the simple configurations discussed above, $\langle H\rangle$ (5.4.16) is valid everywhere. The expression (5.4.16) will be employed in Chap. 5.6 dealing with the inverse problem.

It is easy to incorporate the presence of pumping or recharging wells in the solution (5.4.14) pertaining to a circle. Indeed, the solution for a well in an unbounded formation is given in Sect. 3.6.1. The presence of the boundary of constant head can be accounted for by supplementing it by a singularity of opposite sign at the image of the well across the boundary (see, e.g. Carslaw and Jeager, 1962). Thus, for a well of discarge Q_w, located at $\mathbf{x}_w$, the solution to be added to the second equation in (5.4.14) is

$$\langle H_w\rangle = \frac{Q_w}{2\pi T}\left[\ln|\mathbf{x}-\mathbf{x}_w| - \ln|\mathbf{x}-\boldsymbol{x}_w|\right] \quad ; \quad \boldsymbol{x}_w = \frac{L^2}{|\mathbf{x}_w^2|}\,\mathbf{x}_w \tag{5.4.17}$$

where L is the radius (Fig. 5.4.3). It is emphasized that in line with our discussion of Sect. 5.4.1, the transmissivity in (5.4.17) is the one pertaining to the point $\mathbf{x}_w$, rather than the effective transmissivity in uniform flow. Eq. (5.4.17) can be generalized for a number of wells by summing up similar expressions, each well being represented in terms of its discharge Q_w, location $\mathbf{x}_w$ and transmissivity $T=T(\mathbf{x}_w)$. The head gradient $\langle\nabla H_w\rangle$ and the related specific discharge $\boldsymbol{q}$ are $O(|\mathbf{x}-\mathbf{x}_w|^{-1})$ for points far from the well. Hence, if the well and the observation point are sufficiently far from the boundary, the effect of the image term in (5.4.17) is negligible, and only the term representing the well in an unbounded domain must be retained. A semi-infinite formation bounded by the constant head boundary, say y=0, is a particular case of (5.4.17), with $\boldsymbol{x}_w$, the image across the boundary, given by $x_w=x_w$, $y_w=-y_w$. Similarly, if the boundary of the formation is one of zero flux, in the solution (5.4.17) the sign of the image term should be reversed, i.e. the well and its image have discharges of same sign. In this case, however, the total well discharges have to be equal to the effective recharge times the area, to satisfy the continuity equation.

The head field around a well in the rectangular formation of Fig. (5.4.1a) has a more complex representation, implying an infinite array of images across the four boundaries. However, the leading term, supplemented eventually by a few close images, may yield an accurate expression of the specific discharge, for points sufficiently far from the boundaries.

These simple results can be extended to unsteady flow in formations of constant effective properties, $\langle H\rangle$ satisfying now (see Eq. 5.3.6)

$$\nabla^2\langle H\rangle = \frac{S_{ef}}{T_{ef}}\frac{\partial\langle H\rangle}{\partial t} - \frac{R_{ef}}{T_{ef}} \tag{5.4.18}$$

Eq. (5.4.18) is the classical equation of heat conduction and an extensive compendium of solutions can be found in Carslaw and Jaeger (1962). Generally speaking, the time variations of $\langle H\rangle$ may be related to transients and to periodic changes. Transients are related to a monotonous change of boundary conditions, i.e. the period following a sudden start of pumping. They are characterized by a certain time scale characterizing the transition from a steady state to a new one. Periodic variations are related to a periodic change of boundary conditions or recharge which lead, after a transient, to a periodic variation of $\langle H\rangle$, the "quasi-steady" solution. We shall illustrate the solution of (5.4.18) by considering first a periodic variation of $\langle H\rangle$ caused by an harmonic time dependence of the recharge. We consider the rectangular domain of Fig. 5.4.1a and seek the solution of $\langle H(x,t)\rangle$ of (5.4.28), with $R_{ef} = \overline{R}_{ef} + \mathcal{R}_{ef}\exp(i\omega t)$, where i is the imaginary unit and it is understood that the real part of $\langle H\rangle$ is considered. The boundary conditions are depicted in Fig. 5.4.1a and we adopt the decomposition $\langle H\rangle = \mathcal{H} - (i/\omega)(\mathcal{R}_{ef}/S_{ef})\exp(i\omega t)$. The solution for $\mathcal{H}$ is provided by Carslaw and Jaeger (1962, Sect. 3.6), and it leads to

$$\langle H(x,t)\rangle = -Jx - \frac{\overline{R}_{ef}}{T_{ef}}\left(\frac{x^2}{2} - \frac{L_x^2}{8}\right) - \frac{i}{\omega S_{ef}}[1 - A(x,t)]\,\mathcal{R}_{ef}\,e^{i\omega t}$$

$$A = \frac{\cosh[\omega' x(1+i)]}{\cosh[\omega'(L_x/2)(1+i)]} \quad ; \quad \omega'=(\omega S_{ef}/2T_{ef})^{1/2} \tag{5.4.19}$$

where, again, the real part of $\langle H \rangle$ is to be considered. It is seen that the solution for $\langle H \rangle$ is made up from the steady-state terms (5.4.14), supplemented by the periodic one. The modulus of the complex function A represents the attenuation of the periodic head change due to the presence of the constant head boundaries at $|x|=L_x/2$. The imaginary part of A represents a phase difference between $\langle H \rangle$ and $-iR_{ef}$. Both $|A|$ and arg(A) are represented as functions of $2x/L_x$, for different values of $\omega' L_x/2$, in Figs. 13 and 14 of Carslaw and Jaeger (1962), respectively. Two values of the parameter $\omega' L_x/2$ are of special significance : for sufficiently low frequencies, such that $\omega' L_x < 1$, the expansion of A (5.4.19) in a power series leads to the steady-state solution (5.4.14) in which R_{ef} is the instantaneous, time-depending, recharge. In other words, the flow adapts itself to the recharge at each instant ; in contrast, for $\omega' L_x > 20$, $|A| \simeq 0$ over most of the formation and the unsteady component of $\langle H \rangle$ is a function of t solely. This case is of particular significance in regional flows, since for usual values of the parameters of interest $\omega' L_x >> 1$. Hence, in this case the right-hand side of (5.4.18) is a function of t solely and $S_{ef}\, \partial\langle H \rangle/\partial t$ may be regarded as part of the effective recharge. This interpretation will be found to be useful in Chap. 5.6. Furthermore, the decomposition (5.4.16) is valid in this case, if H_0 is regarded as a function of time.

Next, we consider the unsteady head field associated with pumping wells. In Sect. 3.6.1 we have given already the solution for a well in an unbounded domain, which start to pump at a constant discharge at t=0. With $\langle s \rangle$ the drawdown caused by the well, we have in the present notation

$$\langle s \rangle = \frac{Q_w}{2\pi T}\, Ei\left(-\left[|\mathbf{x}-\mathbf{x}_w|^2 S_{ef} \frac{)}{4Tt}\right]\right)$$

$$\langle s \rangle \rightarrow -\frac{Q_w}{2\pi T}\, \ell n[|\mathbf{x}-\mathbf{x}_w|\; S_{ef}/2\sqrt{Tt}] \quad \text{for} \quad \frac{|\mathbf{x}-\mathbf{x}_w|^2 S_{ef}}{Tt} \rightarrow 0 \tag{5.4.20}$$

where Ei is the exponential integral and (5.4.30) is known as Theis solution (see Sect. 3.6.1). Again, the presence of a boundary can be accounted for by supplementing $\langle s \rangle$ by the drawdown caused by an appropriate image well across the boundary. However, except in the neighborhood of the boundary, the influence of the image may be quite small, since $\langle s \rangle$ decays exponentially with the distance from the well for a finite t. It is easy to generalize (5.4.20) for a well of varying discharge. Indeed, by using the ready made solution of Carslaw and Jaeger (1962, Sect. 10.4) we have

$$\langle s \rangle = \frac{S_{ef}}{4\pi T} \int_0^t Q_w(t')\, \exp\left[-\frac{|\mathbf{x}-\mathbf{x}_w|^2 S_{ef}}{4T(t-t')}\right] \frac{dt'}{t-t'} \tag{5.4.21}$$

which leads to (5.4.20) as a particular case, for Q_w=const. In practice, Q_w is a sum of time step functions, which can be easily expressed as a sum of solutions of type (5.4.20) (see Exercise 5.4.1). Like in the steady flow regime, the contribution of a few wells located at different $\mathbf{x}_w$, of discharges $Q_w(t)$ and of transmissivity $T = T(\mathbf{x}_w)$, is represented by summation of (5.4.21) for various wells. Finally, the wells drawdown has to be subtracted from $\langle H \rangle$ (5.4.19) associated with recharge and with effects of boundaries.

Exercises.

5.4.1 Write a close form solution for $\langle s \rangle$ caused by a well whose recharge Q_w varies by time steps. Show that the solution may be regarded as a superimposition of (5.4.20) with different starting times.

5.4.2 Draw the mean free-surface profile for flow in a circular aquifer with constant recharge and for a well pumping at constant Q_w for $t>0$ at its center. The head on the boundary is constant. Assume that the effective transmissivity is constant.

5.4.3 Analyze with the aid of (5.4.19) the variation of the head with time at the aquifer center, $x=0$, for various values of the dimensionless frequency $\omega' L_x$.

5.5 *SECOND-ORDER STATISTICAL MOMENTS OF FLOW VARIABLES. THE EFFECT OF CONDITIONING*

5.5.1 *Introduction*

Along the lines of Part 3, we are going to investigate now the second-order statistical moments of the head $H(\mathbf{x},t)$. In the context of the direct problem, H is regarded as a random space function by the reasons given in Chap. 5.3. Indeed, H is the solution of the equations of flow (5.3.6) or (5.3.7) in the domain Ω, subject to boundary and initial conditions (5.3.8). Eq. (5.3.6) is a stochastic partial differential equation, since $T(\mathbf{x})$ is a given random function, while S and R_{ef} are given random variables. In the preceding section we have derived the equations satisfied by the expected value $\langle H \rangle$, for unconditional probability of $Y=\ell nT$, by using the concept of effective properties. In this chapter we examine the second-order moments of H and their dependence upon the statistical properties of Y (the discussion of the impact of the errors of estimation of S and R_{ef}, as well as of the parameters characterizing Y, is deferred to Chap. 5.6). With Y stationary and characterized by m_Y and $C_Y(r)$, which are given, the second-order unconditional moments of H are $C_{YH}(\mathbf{x}_1,\mathbf{x}_2)$ and $C_H(\mathbf{x}_1,\mathbf{x}_2)$ or the associated variogram $\Gamma_H(\mathbf{x}_1,\mathbf{x}_2)$. These moments have been examined in detail in Chap. 3.7 and we recall their definition

$$C_{YH}(\mathbf{x}_1,\mathbf{x}_2) = \langle Y'(\mathbf{x}_1)h(\mathbf{x}_2)\rangle \quad ; \quad \Gamma_H(\mathbf{x}_1,\mathbf{x}_2) = \frac{1}{2}\langle[h(\mathbf{x}_1)-h(\mathbf{x}_2)]^2\rangle \tag{5.5.1}$$

where $Y'=Y-m_Y$ and $h=H-\langle H\rangle$ are, as usual, the residuals. If H is normal, the multivariate normal vector of the values of Y and H at an arbitrary set of points is completely characterized by the expected values m_Y and $\langle H\rangle$ and by the covariances C_Y, C_{YH} and C_H. We shall limit our discussion, however, to these second-order moments even if H is not normal.

The next topic of this chapter is the investigation of the statistical moments of H^c, the head field for Y^c, the log-transmissivity conditioned on measurements, as input variable in (5.3.6). Putting it into words, the unconditional moments of H are based on the solution of (5.3.6) for the set of all possible realizations of the MVN vector of values of Y, of moments m_Y and C_Y. In contrast, H^c is the solution of (5.3.6) for realizations of Y^c, a MVN vector of moments $\langle Y^c\rangle$ (5.2.4) and covariance C^c_Y (5.2.5). Hence, the moments $\langle H^c\rangle$, C^c_{YH} and C^c_H are based on a subset of realizations of Y and H, those for which Y is fixed and equal to the measured values Y_i at the measurement points $\mathbf{x}_i$ (i=1,...,M). This topic is of interest at the regional scale and has not been investigated in Part 3 dealing with the local scale.

As we have mentioned in Sect. 5.2.1, there are many field measurements of water head in aquifers. To illustrate the erratic variation of H encountered in field studies, we have depicted in Fig. 5.5.1 the planar distribution of the head residual, after subtracting a linear trend. The data pertain to the Northern part of the Avra Valley aquifer (Clifton, 1981), the same formation for which the transmissivity is given in Fig. 5.2.1. More detailed maps of measured head planar distributions are given in Chap. 5.6.

The main aim of the remaining sections of this chapter is to derive the theoretical expressions of the second-order moments, unconditional and conditional. We shall review first in this section a few results obtained by numerical simulations by Smith and Freeze (1979), mentioned already in Sect. 5.4.1. The results pertain to the configuration of Fig. 5.4.1a, i.e. for flow driven by a constant head gradient J. By using 300 Monte Carlo simulations (see Sect. 3.3.2) and numerical solutions of the flow equations, Smith and Freeze (1979) were able to evaluate the head variance σ^2_H at the nodes of the grid which discretizes the flow domain. They also derived indirectly the integral scale I_Y for each set of runs. To illustrate their results, we reproduce in Fig. 5.5.2 the distribution of σ_H as a function of x on the central line y=0. The variance is equal to zero at the boundaries of constant head and grows toward the center of the formation. The three curves are for $I_Y/L_x \simeq 0.17$ and for three values of σ^2_Y, namely 0.23, 0.98 and 4.4. As one would expect, σ_H grows with σ_Y, other parameters equal. To assess the dependence of σ^2_H upon σ^2_Y, we have represented in Fig. 5.5.9 its value at the center of the formation, at x=y=0, as a function of σ^2_Y. It is seen that the growth is linear for $\sigma^2_Y \leq 1$ and the nonlinear dependence upon σ^2_Y for the largest value is quite weak. Thus, let us assume that one can write at a fixed x/L_x the relationship $\sigma^2_H = a\,\sigma^2_Y + b\,\sigma^4_Y$, as suggested by the theoretical developments of Sect. 3.7.3. Then, by using the results of Fig. 5.5.9 it is found that $b/a \simeq 0.05$.

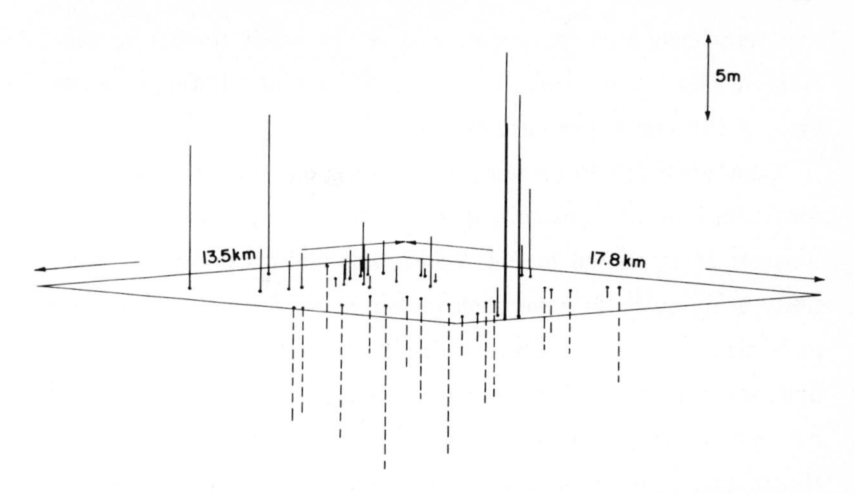

5.5.1 Measured heads in the Northern Part of the Avra Valley (data are taken from Clifton, 1981). In the figure the residual of H for H in meters is represented at different points by vertical segments (positive deviations, solid lines, and negative deviations, dashed lines). A linear head trend has been subtracted from measured values. The scale of H is represented in the upper right corner.

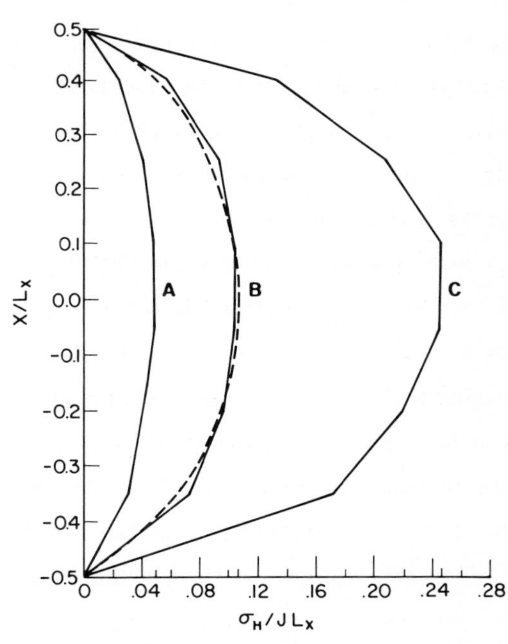

5.5.2 Dependence of σ_H, the standard deviation of the head along the center line y=0, upon x, for flow in a rectangular domain with constant head at $x=\pm L_x/2$ and no flow at $y=\pm L_{y/2}$ (Fig. 5.4.1a). Full line: numerical results of Smith and Freeze (1979) for $L_x/L_y=1/2$ and for (A) $\sigma_Y^2 = 0.23$, $I_Y/L_x=0.18$, (B) $\sigma_Y^2 = 0.98$, $I_Y/L_x=0.175$ and (C) $\sigma_Y^2 = 4.4$, $I_Y/L_x=0.166$. Dashed line: analytical results (Eq. 5.5.33) of Rubin and Dagan (1988) in an infinite domain, for conditioning of head at six points at $x=-L_x/2$, y=-d,0,d and $x=L_x/2$, y=-d,0,d and for $d=2I_Y$.

$$\langle H^c(\mathbf{x}')\rangle = \langle H(\mathbf{x}')\rangle + \sum_{i=1}^{M} \mu_i(\mathbf{x}')(Y_i - m_Y) \qquad (5.5.16)$$

$$C^c_{YH}(\mathbf{x},\mathbf{x}') = C_{YH}(\mathbf{x},\mathbf{x}') - \sum_{i=1}^{M} \mu_i(\mathbf{x})\, C_Y(\mathbf{x}_i,\mathbf{x}') \qquad (5.5.17)$$

where the coefficients μ_i are solutions of the linear system

$$\sum_{i=1}^{M} \mu_i(\mathbf{x})\, C_Y(\mathbf{x}_i,\mathbf{x}_k) = C_{YH}(\mathbf{x}_k,\mathbf{x}) \qquad (k=1,...,M) \qquad (5.5.18)$$

By a similar application of the general relationship (1.2.20-1.2.24) to the MVN vector $H(\mathbf{x})$, $H(\mathbf{x}')$, Y_i we obtain

$$C^c_H(\mathbf{x},\mathbf{x}') = C_H(\mathbf{x},\mathbf{x}') - \sum_{i=1}^{M}\sum_{\ell=1}^{M} \mu_i(\mathbf{x})\, \mu_\ell(\mathbf{x}')\, C_Y(\mathbf{x}_i,\mathbf{x}_\ell) \qquad (5.5.19)$$

The last equation can be rewritten in terms of the head variogram by employing the definition relationship $\Gamma^c_H(\mathbf{x},\mathbf{x}') = (1/2)[C^c(\mathbf{x},\mathbf{x}) + C^c(\mathbf{x}',\mathbf{x}') - 2C^c(\mathbf{x},\mathbf{x}')]$. Substitution of C^c (5.5.19) in Γ^c_H leads to

$$\Gamma^c_H(\mathbf{x},\mathbf{x}') = \Gamma_H(\mathbf{x},\mathbf{x}') - \frac{1}{2}\sum_{i=1}^{M}\sum_{\ell=1}^{M} [\mu_i(\mathbf{x})-\mu_i(\mathbf{x}')][\mu_\ell(\mathbf{x})-\mu_\ell(\mathbf{x}')]\, C_Y(\mathbf{x},\mathbf{x}') \qquad (5.5.20)$$

Summarizing these developments, the computation of the moments of the head conditioned on measurements of logtransmissivity comprises the following steps: (i) for a given m_Y and C_Y, the unconditional direct problem is solved first and $\langle H\rangle$, C_{YH} and Γ_H are derived and (ii) the conditional moments $\langle H^c\rangle$, C^c_{YH} and Γ^c_H are determined by (5.5.16), (5.5.17) and (5.5.20), respectively, in terms of the unconditional moments and of the measurements Y_i (i=1,...,M). This stage requires to solve the linear system (5.5.18).

Herein are a few general observations about conditioning : (i) the procedure is rigorous only if the first-order approximation is adopted, since only then is h (5.5.15) normal; (ii) the procedure here is similar to cokriging (Journel and Hujbregts, 1978). However, the latter does not imply normality; (iii) Eqs. (5.5.16)-(5.5.20) are quite general and can be applied to any type of steady flow and domain, the unconditional H being obtained by a numerical solution of (5.5.2). In Exercise 5.5.3 it is required to prove that the two procedures, solving Eq. (5.5.2) for h with Y replaced by Y^c or solving for H and applying the above steps, are equivalent, provided that consistency in keeping terms of order σ_Y^2 is obeyed; (iv) as we have mentioned already in Chap. 1.8 the expected value $\langle H^c \rangle$ (5.5.16) depends on the actual values of the measured Y_i, whereas the covariance (5.5.19) depends only on the relative position of $\mathbf{x}$ and $\mathbf{x}'$ with respect to the measurement points $\mathbf{x}_i$ and (v) in a finite domain C_H is bounded and Eq. (5.5.19) can be employed in order to compute the head variance $\sigma_H^{2,c}(\mathbf{x}) = C_H^c(\mathbf{x},\mathbf{x})$. It is readily seen that conditioning reduces the unconditional variance σ_H^2, since the sum in (5.5.19) is positive for $\mathbf{x}=\mathbf{x}'$. At the limit of a very dense grid of measurement points covering the entire flow domain, the variance tends to zero and $H^c(\mathbf{x})$ becomes essentially a deterministic function. The effect of heterogeneity manifests then in the large variability of $\langle H^c \rangle$ (5.5.16). Hence, conditioning or the geostatistical approach enjoy the fundamental property mentioned in Chap. 1.8: it ensures a transition from maximum uncertainty and a smooth average, when measurements are not taken into account directly, to a deterministic outcome and a spatially variable average, for abundant measurements. The distinctive feature of conditioning is that it provides in a systematic manner a measure of uncertainty in any intermediate case.

The scheme represented by Eqs. (5.5.16-20) is easy to implement in the case in which analytical expressions are available for the unconditional moments C_Y, C_{YH} and Γ_H. This was the case for an uniform average unconditional head, i.e. $\nabla\langle H\rangle = -\mathbf{J}$, for an unbounded domain and for the exponential C_Y (5.2.2), leading to the close form covariance C_{YH} (3.7.8) and variogram Γ_H (3.7.11). In this case conditioning reduces mainly to solving the linear system (5.5.18). To illustrate the results we have considered first a very simple configurations, namely two measurement points of Y_i (i=1,2) along a line parallel to the average gradient $\mathbf{J}$ at a distance $|\mathbf{x}_1-\mathbf{x}_2|=2\ell$ and the head variance at the middle point between them, i.e. on the same average streamline. In this case the solution of the linear system (5.5.18) is very simple (the derivation is left for Exercise 5.5.6), and the variance reduction is given by

$$\sigma_H^2 - \sigma_H^{2,c} = \frac{2\, C_{YH}^2(\ell,0)}{\sigma_Y^2 - C_Y(2\ell)} \qquad (J_x=J,\ J_y=0) \tag{5.5.21}$$

The dependence of $(\sigma_H^2 - \sigma_H^{2,c})/\sigma_Y^2 J^2 I_Y^2$ upon ℓ/I_Y is shown in Fig. 5.5.5 for the exponential C_Y. It is seen that no variance reduction by conditioning is achieved for $\ell=0$, which is understandable in view of the nature of the covariance C_{YH} (Eq. 3.7.8, Fig. 3.7.3) which is anti-symmetrical. Similarly, the effect is small for large ℓ/I_Y due to the decay of C_{YH} with the distance. The maximal effect is achieved for $\ell/I_Y \simeq 2$, corresponding again to the maximum of C_{YH}. The maximal variance reduction caused by a pair of points is seen to be $0.18\ \sigma_Y^2 I_Y^2 J^2$. Dagan (1982) has carried out a similar calculation for one-dimensional flow and has found that the variance reduction is approximately $2\sigma_Y^2 I_Y^2 J^2$. This disparity is due entirely to the two-dimensional nature of the flow:

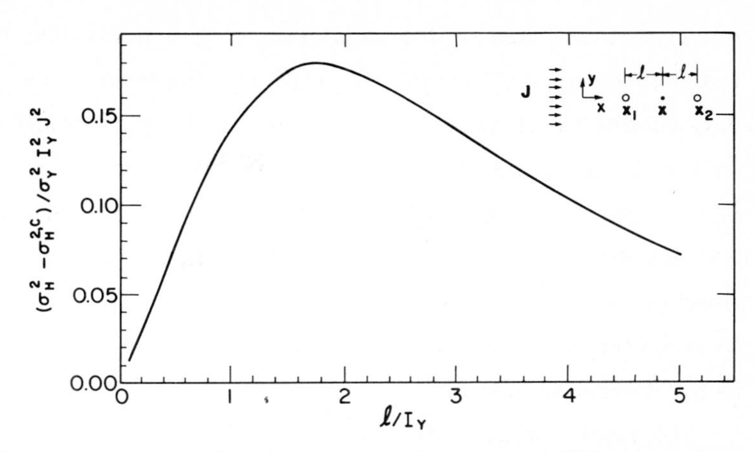

5.5.5 The reduction of the head variance due to conditioning by two log-transmissivity measurements (Eq. 5.5.21).

in 1D flow the conductivity at a point affects the heads in the entire domain (Fig. 3.7.3), whereas in a 2D formation its influence is local. A lesser impact is achieved by measurement points which are not on the same streamline, the extreme case being of points on a line normal to **J**. In the latter case there is no variance reduction, since $C_{YH}=0$. It is emphasized that although both $\sigma_H^{2,c}$ and σ_H^2 are not defined in an unbounded domain, their difference, i.e. the variance reduction due to conditioning is finite. We shall show in Sect. 5.5.4 that the covariances pertaining to an infinite domain can be employed in a bounded one, provided that the two points are not too close to the boundary. Furthermore, a boundary condition of constant head can be accounted for by implanting fictitious measurement points along it. Hence, the result of (5.5.21) is valid in a bounded domain as well, and this property simplifies considerably the investigation of the impact of Y measurements upon the conditional moments of H.

Inspection of (5.5.21) shows that for $\ell > I_Y$ the interaction between the two conditioning points is negligible and their effect is the cumulative one for two independent measurements. Indeed, conditioning by an isolated log-transmissivity measurement, i.e. M=1 in (5.5.18,19), leads to the simple result

$$\sigma_H^2(\mathbf{x}) - \sigma_H^{2,c}(\mathbf{x}) = \frac{C_{YH}^2(\mathbf{x}_1,\mathbf{x})}{\sigma_Y^2} \tag{5.5.22}$$

which leads to (5.5.21) for $|x_1-x|=\ell$, $y=y_1=0$ and $\rho_Y(2\ell)<<1$. Suppose now that measurements of Y are available on an array of points in the entire flow domain, which is a circle of radius $R>>I_Y$, at distance I_Y among them. The variance reduction at the center can be computed by summing up (5.5.22) for all measurements points. Furthermore, for distant points C_{YH} (3.7.8) tends to $-\sigma_Y^2 I_Y(\mathbf{J}.\mathbf{r})/r^2$, where r is the distance from the center to the measurement point. The contribution to the variance reduction by a large number of points can be evaluated approximately by integrating

(5.5.2) over the area of the annulus $I_Y<r<R$ and dividing the result by I_Y^2. The result is easily found to be $\sigma_H^2-\sigma_H^{2,c} \simeq \pi\sigma_Y^2 J^2 I_Y^2 \ell n(R/I_Y)$. Since in a bounded aquifer with constant head boundary conditions (see Sect. 5.5.4) the head grows logarithmically with the distance from the boundary, it is seen that conditioning by head measurements on an array covering the formation at a distance of the order of the integral scale is able to reduce considerably the head uncertainty.

The above results can be extended to the case of nonuniform average flow, with a quadratic trend (5.5.7). Again, the covariance (5.5.8) pertaining to an infinite domain can be applied in order to assess the effect of conditioning at a point by neighboring measurements, except for regions close to the domain boundaries.

The results obtained for the point values of the head H^c can be easily extended for "block" averages, i.e. for $\overline{H}(\mathbf{x}) = (1/A) \int_A H(\mathbf{x})\, d\mathbf{x}$, where $\mathbf{x}$ is the centroid of the area A. The procedure (see Dagan, 1985a) is very similar to the one of Sect. 5.2.3. It consists of replacing the system of equations (5.5.16)-(5.5.20) by a similar one in which H is replaced by $\overline{H}$ and μ_i by $\overline{\mu}_i$. Similarly to (5.2.9), the right-hand side of the linear system (5.5.18) for μ_i is replaced by $\overline{C}_{YH}$, which is defined by the operation (5.2.10) applied to C_{YH}.

5.5.4 The effect of boundaries on head covariances (unconditional probability)

We have shown in the preceding section that it is easy to obtain close form expressions for the head covariances by a first-order approximation, for a linear or quadratic trend of $\langle H\rangle$ and for an unbounded domain. On the other hand, we have emphasized that in applications formations are bounded, but solving the problem in a finite domain complicates the solution considerably and a numerical approach is generally needed. It is therefore of interest to examine the influence of boundaries upon covariances and to assess the applicability of the simple expressions pertaining to the infinite domain to the bounded one. We have mentioned in Sect. 3.7.4 the study of Naff and Vecchia (1986) on the influence of impervious boundaries upon the head variance in three-dimensional flow. We are concerned here with two-dimensional flow and with both covariances C_{YH} and Γ_H.

We start with the discussion of the impact of a boundary of constant head, again by using a first-order approximation in σ_Y^2 and by assuming that the average flow is normal to the boundary (see Fig. 5.5.6a). The logtransmissivity is supposed to be a stationary random function, not influenced by the presence of the boundary. To simplify matters we start with the simplest configuration, namely flow in a half-plane bounded by a constant head boundary y=0 (Fig. 5.5.6a). The mathematical problem for the head becomes now (see Eq. 5.2.2)

$$\langle H\rangle = H_0 - Jy \quad ; \quad \nabla^2 h = J\,\frac{\partial Y'}{\partial y} \quad (\text{for } y>0) \quad ; \quad h(x,0) = 0 \tag{5.5.23}$$

where J is the hydraulic gradient. The solution for h can be written like in (5.5.15) as follows

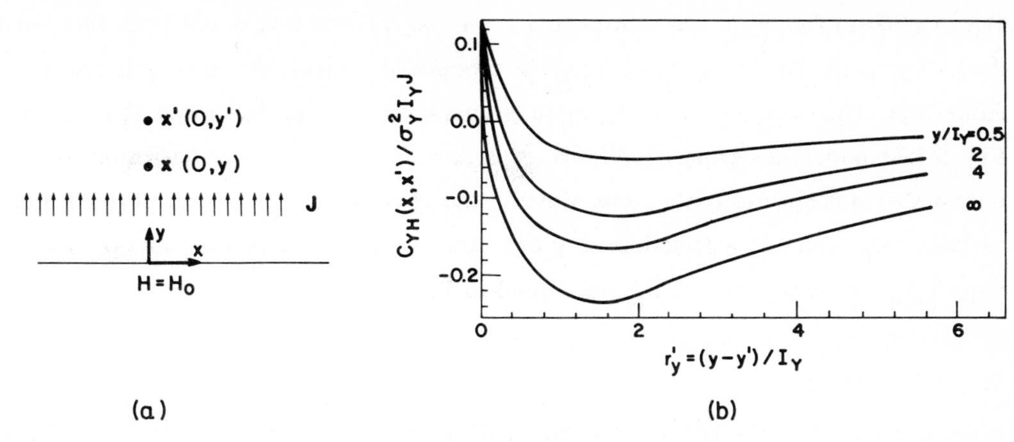

5.5.6 The dependence of (b) the logtransmissivity head covariance on the interval r'_Y and on the distance from the boundary y for flow in (a) a semi-infinite domain y>0 with constant head on y=0 and average head gradient **J** normal to it. Exact analytical results by a first-order approximation in σ_Y^2 and for C_Y (5.5.26) (reproduced from Rubin and Dagan, 1988).

$$h(\mathbf{x}) = J \int_{-\infty}^{\infty}\int_{0}^{\infty} Y'(x',y') \frac{\partial G(x,x',y,y')}{\partial y'} dx' \, dy' \tag{5.5.24}$$

The Green function for the half-plane is given by

$$G(\mathbf{x},\mathbf{x}') = - \frac{1}{4\pi} \ell n[(x-x')^2+(y-y')^2] + \frac{1}{4\pi} \ell n[(x-x')^2+(y+y')^2] \tag{5.5.25}$$

i.e. the Green function for an infinite domain supplemented by its negative image across the boundary.

The covariances C_{YH} and C_H are obtained from (5.5.24) in a formal way by multiplying $h(\mathbf{x}')$ by $Y'(\mathbf{x})$ or by $h(\mathbf{x})$, respectively, and ensemble averaging, leading to expressions which depend on the input covariance C_Y only. The computations become tedious in comparison with the simple calculations based on (3.7.6, 3.7.9), which yield the covariances in an infinite domain. Rubin and Dagan (1988a) have been able to obtain analytical, close form, solutions of (5.5.23) for the exponential separated C_Y (3.2.25), i.e. for

$$C_Y(\mathbf{x},\mathbf{x}') = \sigma_Y^2 \exp(-|x-x'|/I_Y - |y-y'|/I_Y) \tag{5.5.26}$$

which is very close to the isotropic, exponential, C_Y (3.2.13). We shall present here the final results only. The covariances $C_{YH}(\mathbf{x},\mathbf{x}')$ and $C_H(\mathbf{x},\mathbf{x}')$ are stationary with respect to x, i.e. they depend on $r_x = x-x'$, but they are functions of both $r_y = y-y'$ and y. In Fig. 5.5.6b the results of Rubin and achieve a regularization of the unknown function T($\mathbf{x}$) (see Emsellem and De Marsily, 1971), i.e. they imply a certain degree of continuity and smoothness, and on the other they remove high wave-number noise of the input function. Indeed, in a finite element scheme, for instance, it is usually assumed that T varies linearly within elements and is continuous across them. This means that variation of T between streamlines on which T is given is no more arbitrary, but is subjected to some requirements of smoothness and continuity which limits the class of possible solutions. By the same token, the superimposition of a noisy component to H at the nodes of the grid, with vanishing amplitude ϵ, but with nodes kept fixed, is equivalent to letting the product ϵk tend to zero and for such a disturbance, the solution is well behaved. Hence, most existing methods to solve the identification problem are limiting the solution to classes of functions, which enjoy a certain degree of regularity. The troubles mentioned above manifest then either in a flat response surface, i.e. insensitivity of the objective function to changes in T and S, or in different results for the identified T at the same point, depending on the selected parametrization. These difficulties have been coined under different terms, e.g. overparametrization or unidentifiability. Of course, the difficulties depend on a large extent on the number, location and degree of smoothness of measured data. Thus, synthetic examples which are often selected for illustration purposes, may succeed or fail depending on the particular choice of the fictitious data.

A different approach to the identification problem which has emerged in the last few years is the statistical one. The outlook differs in principle from the deterministic one, since one gives up any attempt to determine the precise values of T or S or their space averages. Instead one seeks to employ data and the equations of flow in order to estimate the expected values of the parameters and their variance or some related measure of their error. Thus, in the context of modeling uncertainty by regarding the transmissivity as a random space function adopted in this book, the main objective of the solution of the inverse problem is to identify the statistical structure of the parameter of interest, say the logtransmissivity Y. This is achieved by making some assumptions with regard to stationarity and again by a parametrization of the p.d.f. characterizing Y. Thus, if Y is assumed to be stationary and normal and of covariances (5.2.2) or (5.2.3), its unconditional statistics is exhausted by the four parameters m_Y, w, σ_Y^2 and I_Y. The basic assumption of the stochastic approach is that although data may vary in an erratic fashion (see Figs. 5.2.1, 5.5.1), these parameters may be identified in an unique and stable manner. In principle, the probabilistic methodology takes care of the two troublesome problems faced by the deterministic procedures in the following manner. First, by the assumption of stationarity the arbitrariness of the Y values on streamlines on which it is not measured is reduced, in a statistical sense, by identifying σ_Y+w, which is a measure of the variability of Y around m_Y. Second, a propagation of values across streamlines is achieved through the isotropic correlation scale I_Y, which expresses precisely the connection between Y at two different points. Hence, we expect that in regions remote from the measurement points, Y is subjected to the largest degree of uncertainty, as delimited by m_Y and σ_Y^2+w , whereas the variance is reduced in areas of extent I_Y around such points, the minimum being w, which stems from mial or by other known functional representations for m_Y, i.e. of a trend of Y. Any such generalization implies increasing the number of unknown parameters, which might not be justified in view of scarcity of data. On the other hand, in the case of abundant data, the formation may be partitioned into zones of constant but different m_Y, as suggested for instance by Clifton and Neuman (1982) for the Avra Valley aquifer (see Chap. 5.2), and each zone is treated separately. Under these conditions, the head H, which satisfies Eq. (5.3.6), is itself a random space function, due to its dependence upon T and upon other sources of uncertainty. Finally, a considerable simplification of the computations is achieved by the first-order approximation in σ_Y^2, discussed at length in Part 3 and in the preceding sections. Thus, with $H(\mathbf{x},t) = \langle H(\mathbf{x},t)\rangle + h(\mathbf{x},t)$, the linearized, first-order version of (5.3.6) is as follows (see Eq. 5.5.2)

$$T_G \; \nabla^2\langle H\rangle = S \,\frac{\partial\langle H\rangle}{\partial t} - R_{ef} \tag{5.6.6}$$

$$\nabla^2 h = \frac{S}{T_G}\,\frac{\partial h}{\partial t} - \nabla.[Y' \,\nabla\langle H\rangle] \tag{5.6.7}$$

Hoeksema and Kitanidis (1984, 1985) and Dagan (1985a) have considered steady flow in absence of recharge, in which case the right-hand side term of (5.6.6) is equal to zero. Rubin and Dagan (1987a,b) have investigated steady flow, but in presence of a distributed, constant, natural recharge $R_{ef} = R$. They have assumed that R is a random variable and solved the problem for fixed R, as if it were deterministic, and accounted for its uncertainty in the final results. Finally, Dagan and Rubin (1988) have investigated the unsteady problem, with $R_{ef}=R + R_w$, where R_w (5.3.2) is the contribution of recharging or pumping wells. Again, S was taken as constant, but random, whereas R_w (5.3.2) was deterministic and given.

It is to be reminded that under the linearization of (5.6.7) h is also normal, since it can be represented (see Eq. 5.5.11) as a linear, deterministic, integral operator applied to Y'. Furthermore, it is not influenced by an uncorrelated, or "nugget" component, of Y', which is wiped out by integration.

All the above assumptions are well known and have been already discussed in the frame of the studies of the direct problem. The main contribution of Kitanidis and Vomvoris (1983) is to devise a two stage approach to solving the inverse problem, which is exposed herein for steady flow.

The first stage consists in selecting first a structural statistical model for Y, e.g. the one mentioned above. Furthermore, the unconditional structure of Y is represented with the aid of the components of a parameters vector $\boldsymbol{\theta}$. Thus, with the exponential covariance (5.2.2) adopted in most of the cited articles, the parameters (see Sect. 5.2.3) are: $\theta_1=m_Y$, $\theta_2=w$, $\theta_3=\sigma_Y^2$ and $\theta_4=I_Y$. Next, the unconditional covariances $C_{YH}(\mathbf{x}_1,\mathbf{x}_2)$ and $C_H(\mathbf{x}_1,\mathbf{x}_2)$ are evaluated by solving the direct problem, i.e. by using Eqs. (5.6.6,7), either analytically or numerically. It is precisely at this step that advantage is taken of the linearization in σ_Y^2, which simplifies tremendously the computation of the covariances. It is emphasized that the expected values and covariances of Y and H depend on the vector

θ. At this point, the information contained in the data, namely measurements of $Y_j = Y(\mathbf{x}_j)$ (j=1,...,M) and of $H_j = H(\mathbf{x}_j)$ (j=M+1,...,N) for steady flow, is employed. The formation and the flow are regarded as a realization belonging to the ensemble of formations of same statistical structure and boundary conditions. An MLP (maximum likelihood procedure) is applied to the measurements vector Z_j, with $Z_j = Y_j$ (j=1,...,M) and $Z_j = H_j$ (j=M+1,...,N) (see, e.g. Schweppe, 1973). This consists in estimating the parameters θ by minimizing the negative log likelihood function L

$$L(Z|\theta) = -\ell n[p(Z|\theta) = \frac{N}{2}\ell n(2\pi) + \frac{1}{2}\ell n|Q| + \frac{1}{2}\sum_{i=1}^{N}\sum_{j=1}^{N}(Z_i - \langle Z_i\rangle)(Z_j - \langle Z_j\rangle)C^{-1}_{Z,ij} \qquad (5.6.8)$$

where $C_{Z,ij} = C_Z(\mathbf{x}_i,\mathbf{x}_j)$ is made up from C_Y (i,j≤M), C_{YH} (i≤M, j>M) and C_H (i,j>M) and both $\langle Z\rangle$ and C_Z depend on the vector θ. The ML method provides estimates $\tilde{\theta}$, as well as their covariance matrix Θ_{mn} (m,n=1,...,4), as mentioned in Sect. 5.2.3. Low estimation variances Θ_{ii} are indicative of the appropriateness of the model and of sufficiency of data. It is emphasized that exactly the same procedure could be employed by using only Y measurements, as mentioned in Sect. 5.2.3, and then C_Z reduces to C_Y. However, in this case the problem becomes one of straightforward statistical inference, no use being made of the measured heads and of the equations of flow. In the typical inverse problem the situation is different: Y measurements are scarcer than H. By the same token, the ML procedure can be applied to head measurements solely, i.e. with C_Z reducing to C_H. In this case, however, the expected value $\theta_1 = m_Y$ remains undetermined, since it does not show up in the likelihood function L (5.6.8). This agrees with the results of the investigation of the inverse problem in the preceding section in which it was shown that in absence of measurements of Y one can establish only the relationship between its values at two points on the same streamline, but not the values themselves. The uncorrelated component of Y, the nugget, remains also unidentified, since it is filtered out from the head covariance.

The ML procedure is a particular case of nonlinear regression, which takes into account the correlation structure of the variables, and has some definite advantages (see Sect. 5.3.2). It reduces to the common least squares procedure, if Z components are uncorrelated and have same variance.

The second stage of the methodology devised by Kitanidis and Vomvoris (1983) is the one of cokriging, or the conditional stage in the terminology of Dagan (1985a). In this stage, the presence of data is taken into account by restricting the ensemble of formations and flows to a subensemble in which the values taken by Y_j and H_j at the measurement points are fixed. The solution of the identification problem for the random Y^c consists of the conditional expected value $\langle Y^c(\mathbf{x})\rangle$ and the conditional covariance $C^c_Y(\mathbf{x}_1,\mathbf{x}_2)$, and in particular of the variance $\sigma^{c,2}_Y(\mathbf{x}) = C^c_Y(\mathbf{x},\mathbf{x})$. In this stage the unconditional covariances are known, θ being replaced by its estimate $\tilde{\theta}$, the outcome of the first stage. In both cokriging and Gaussian conditioning, the conditional expected values are minimum variance estimators, compatible with the data and the statistical structure. Obviously, in the case in which both Y and H measurements are available, the advantage of incorporating the H data and solving the inverse problem, as opposed to statistical inference of Y^c from Y measurements solely, is measured by the reduction of $\sigma_Y^{c,2}(\mathbf{x})$ as compared to σ_Y^2, in part or in the entire domain. That this is the case has been demonstrated in a few applications to be discussed later. The conditioning stage may be regarded as an application of a Bayesian approach in which the unconditional distribution serves as prior.

As we have mentioned already, uncorrelated errors, like the ones associated with measurements, are part of the nugget w, which constitutes a lower bound of the variance of Y. Sometimes, a few isolated measurements are affected by abnormally large errors and Kitanidis and Vomvoris (1983) suggest methods to detect and to eliminate such outliers.

It is realized that the selection of the model, i.e. the drift and covariance, is somewhat arbitrary. However, along the process there are some diagnostic checks and new models can be tested and compared in terms of their performance.

Kitanidis and Vomvoris (1983) have applied the above two stage procedure to steady one-dimensional flow, which is easy to simulate by analytical means. They have demonstrated the capability of the method and its robustness, i.e. good agreement with the input transmissivity even for a model which is quite different from the correct structure. This robustness stems from the strong influence of conditioning by measurements.

Since field applications are of a two-dimensional nature, we shall review in the sequel the later developments of the methodology.

(ii) Analytical solution, steady average uniform flow (Dagan, 1985a).

In order to investigate the impact of head data upon identification of transmissivity in a simple and general manner, Dagan (1985a) has investigated the second stage of the procedure described above. Toward this aim it has been assumed that the flow domain is unbounded and the average head has a linear trend, i.e. $\langle H \rangle = H_0 - J_x x - J_y y$.

The assumption of unboundedness simplifies considerably the analytical solution of the direct problem, i.e. the computation of C_{YH} and C_H for a given C_Y, as shown in Part 3 and in Sect. 5.5.2. The accuracy of this simplification has been discussed in Sect. 5.5.4, in which it was shown that C_{YH} and C_H differ from their expressions in an unbounded domain in a strip of the order of the integral scale along the boundary. Hence, when conditioning Y at a point inside the flow domain by measurements which are not close to the boundary, the effect of this approximation is negligible.

The assumption of average uniform flow is also simplifying the solution of the direct problem, as shown repeatedly in Parts 2 and 3. This is a more restrictive assumption, which has been relaxed in the later studies discussed below. In absence of recharge and wells, the assumption of linear $\langle H \rangle$ may be justified, at least in a restricted zone of the formation. Again, in the conditioning stage the largest impact on $Y^c(\mathbf{x})$ is that of measurements points close to $\mathbf{x}$, such that a local linear approximation for $\langle H \rangle$ in the area around $\mathbf{x}$ is justified.

Under these conditions, Dagan (1985a) selected the exponential C_Y (5.2.2) with w=0 as representing the logtransmissivity structure, reducing, therefore, the parameters characterizing the head and transmissivity to six: $\theta_1=m_Y$, $\theta_2=\sigma_Y^2$, $\theta_3=I_Y$, $\theta_4=H_0$, $\theta_5=J_x$ and $\theta_6=J_y$. These parameters were supposed to be known from the first stage, and the task was to investigate the presence of measurements of Y and H upon the identification of Y in the neighborhood of the measurement points. Toward this aim, one needs to construct the conditioning system for $Y^c(\mathbf{x})$ and $Y^c(\mathbf{x}')$, where $\mathbf{x}$ and $\mathbf{x}'$ are two arbitrary points, for given $Y_j=Y(\mathbf{x}_j)$ (j=1,...,M) and $H_j=H(\mathbf{x}_j)$ (j=M+1,...,N). This is simply done by applying the general relationships of Chap. 1.2 to the MVN vector of Y, Y_j and H_j, with covariance matrices C_Y, C_{YH} and C_H. We have carried out already twice a similar exercise, in Sects. 5.2.3 and 5.5.4, and again, special attention has to be paid to C_H, which is not defined in an unbounded domain. As in Sect. 5.5.3, it is replaced formally by $\sigma_Y^2-\Gamma_Y$ and the limit $\sigma_Y^2\to\infty$ is taken. The conditional expected value and covariance of Y^c, similar to results of cokriging, are as follows

$$\langle Y^c(\mathbf{x})\rangle = m_Y + \sum_{j=1}^{M}\lambda_i(\mathbf{x})\, Y'_j + \sum_{j=M+1}^{N}\mu_j(\mathbf{x})\, h_j$$

$$C_Y^c(\mathbf{x},\mathbf{x}') = C_Y(\mathbf{x},\mathbf{x}') - \sum_{j=1}^{M}\lambda_j(\mathbf{x})\, C_Y(\mathbf{x}',\mathbf{x}_j) - \sum_{j=M+1}^{N}\mu_j(\mathbf{x})\, C_{YH}(\mathbf{x}',\mathbf{x}_j) \qquad (5.6.9)$$

where as usual $Y' = Y - m_Y$. The coefficient vectors $\boldsymbol{\lambda}(\mathbf{x})$ and $\boldsymbol{\mu}(\mathbf{x})$ are solutions of the following linear system

$$\sum_{j=1}^{M}\lambda_j\, C_Y(\mathbf{x}_j,\mathbf{x}_k) + \sum_{j=M+1}^{N}\mu_j\, C_{HY}(\mathbf{x}_j,\mathbf{x}_k) = C_Y(\mathbf{x},\mathbf{x}_k) \qquad (k=1,...,M)$$

$$\sum_{j=1}^{M}\lambda_j\, C_{YH}(\mathbf{x}_j,\mathbf{x}_k) - \sum_{j=M+1}^{N}\Gamma_H(\mathbf{x}_j,\mathbf{x}_k) + \Lambda = C_{YH}(\mathbf{x},\mathbf{x}_k) \qquad (k=M+1,...,N) \qquad (5.6.10)$$

$$\sum_{j=M+1}^{N}\mu_j = 0$$

It is easy to ascertain from the expressions (5.5.9) and (5.5.12) of the various covariances that the coefficients $\boldsymbol{\lambda}$ and $\boldsymbol{\mu}/JI_Y$ are functions of the normalized coordinates $\boldsymbol{x}=\mathbf{x}/I_Y$ solely, whereas $\sigma_Y^{2,c}(\mathbf{x})/\sigma_Y^2 = C_Y^c(\mathbf{x},\mathbf{x})/\sigma_Y^2$ depends only on $\boldsymbol{x}$, $\boldsymbol{x}'$ and $\boldsymbol{x}_j$. Hence, while the actual values of the

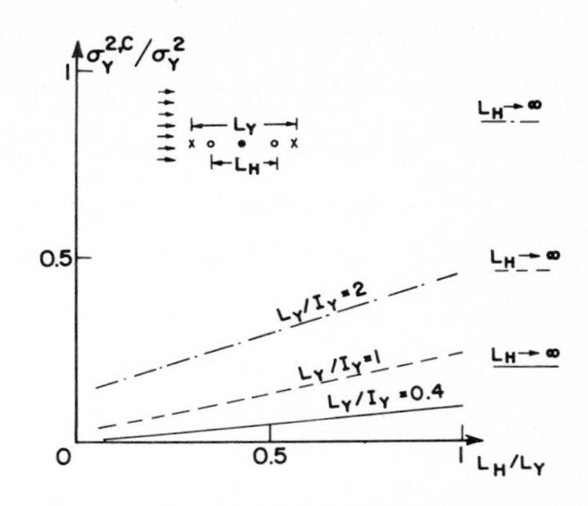

5.6.1 The reduction of the logtransmissivity variance due to the combined effect of two Y measurement points at distance L_Y and two H measurement points at L_H (reproduced from Dagan, 1985a).

parameters and of measurements are needed in order to compute $\langle Y^c \rangle$ (5.6.9), the variance reduction due to conditioning, expressed by $\sigma_Y^{2,c}/\sigma_Y^2$, depends only on the relative position of the observation point $\boldsymbol{x}'$ with respect to the measurement points $\boldsymbol{x}'_j$. Thus, the reduction of variance, which is one of the major objectives of solving the inverse problem, can be studied independently of the actual values of the parameters. One of the results of interest is represented in Fig. 5.6.1 which displays the variance reduction at a point, due to the presence of two transmissivity measurements at distance L_Y and two head measurements at distance L_H. The measurement and observation points are on the same average streamline, i.e. a straight line parallel to **J**, and are located symmetrically, so that the conditioning system (5.6.10) reduces to three independent coefficients $\lambda_1=\lambda_2$, $\mu_1=-\mu_2$ and Λ. The variance reduction is represented as function of the ratio L_H/L_Y for a few values of L_Y/I_Y. Inspection of Fig. 5.6.1 reveals that in absence of head measurements ($L_H \to \infty$), the Y measurements have a variance reduction effect if $L_Y/I_Y<2$, which is in agreement with the results of Sect. 5.5.3. The striking result, however, is that the presence of head measurements, has a dramatic effect upon the reduction of the variance, justifying their incorporation in the identification process. It must be said, however, that head measurements have their maximal impact when they are located on the same average streamline as the observation point, as one can easily find by inspecting the covariance C_{YH} (5.5.9) and Fig. 3.7.8. In the extreme case in which the head measurement points are on a line passing through the observation point and normal to the average gradient, $C_{YH}\equiv 0$ and the coefficients $\mu=0$, i.e. there is no variance reduction. In actual field studies, head measurement points are generally spread in an irregular pattern, but their impact upon identification of Y is maximal if they are on the average streamline passing through the point. This is also in agreement with the general analysis of Sect. 5.6.2. Finally, it is seen that for a fixed L_Y/I_Y, the conditional variance does not tend to zero when L_H/L_Y tends to zero. A couple of measurements at small L_H is equivalent to measuring the actual head gradient at the observation point, which is not enough to identify Y exactly. In contrast, one or two close transmissivity measurements can reduce the conditional

H^* is a pure nugget. When the measurements are projected onto the nodes, V_H needs not to be diagonal, since an interpolation scheme is used in order to related H^* to neighboring nodes. Somewhat inconsistent with the previous assumption, kriging is used for interpolation purposes, after identifying a raw head variogram with the aid of measurements. However, kriging serves only as an interpolator, which does not reflect an underlying spatial structure of the errors. In unsteady flow, an additional simplification is that $V_{H,ijk\ell}=X_{ij}\rho_{k\ell}$, where the indices stand for errors at a point $\mathbf{x}_i$ and time t_k, and point $\mathbf{x}_j$ and time t_ℓ, respectively. In other words, the spatial and time covariances are separated, and furthermore, ρ is assumed to represent a simple auto-regressive process which depends on an unique parameter ρ in case the sampling time intervals are equal. Under these assumptions, and with measurement errors assumed to be normal, their entire statistical structure depends on the two parameters σ_H^2 and ρ (only the first for steady flow).

As for C_p it is first assumed that the prior errors affecting the various parameters, $p_1=Y=\ell n\ T$, $p_2=S$ and $p_3=H_b$,..., are statistically independent. Furthermore, the prior errors of each parameter have a covariance matrix $C_i = \sigma_i^2 V_i$ (i=1,2,...,P), where σ_i^2 is either known or unknown. Whenever data are available, V_i is obtained by kriging, after identifying the variogram or the auto-correlation from measurements (it is reminded that the kriging coefficients in Eq. 5.2.6 depend only on the auto-correlation function and not on the variance).

The key to the estimation of the values of the vector **Z**, comprising **H** and **p**, in Carrera and Neuman (1986a) is by an ML procedure, i.e. by minimizing the negative log likelihood function (5.6.8) in which the expected values $\langle \mathbf{Z} \rangle$ are replaced by the sought estimates, whereas **Z** is replaced by the $\mathbf{Z}^*$, the vector of measurements in the case of **H** or by the prior estimates, based on measurements, for **p**. Finally, the covariance C_Z is made up from the prior errors covariances C_H and C_p discussed above. Thus, after a few transformations, the loglikelihood criterion is written explicitly as follows (Carrera and Neuman, 1987a)

$$L = J_H/\sigma_H^2 + \sum_{k=1}^{P} J_k/\sigma_k^2 + N_x \ell n|\rho| + \sum_{k=1}^{P} \ell n|V_k| + N_H \ell n\ \sigma_H^2 + \sum_{k=1}^{P} N_k\ \ell n\ \sigma_k^2 + N\ \ell n(2\pi) \qquad (5.6.15)$$

In the above expressions J_H is the head residual criterion $J_H = \sum_m \sum_k V^{-1}_{H,km}(H^*_m - H_m)(H^*_k - H_k)$, J_k is the *k*th parameter "penalty criterion" of a similar form, N_H is the total number of head measurements, N_i is the number of prior data for parameter *i* and $N=N_H + \sum_k N_k$ is the total number of prior data. If σ_H^2, σ_k^2 and ρ are fixed, the minimization of L is equivalent to minimizing the generalized least squares criterion $J = J_H + \sum_k \lambda_k J_k$, where $\lambda_k = \sigma_H^2/\sigma_k^2$ are relative weights assigned to the parameters penalty criteria. This was the starting point of Neuman and Yakowitz (1979).

The minimization of L (5.6.15) is carried out by Carrera and Neuman (1986a,b,c) in an iterative manner, in which the estimates of **p** are updated and at each step **H** is determined by solving numerically the equations of flow (5.6.14). If the process converges, estimates of the parameters values are obtained, which are different and smoother than measurements, a regularization being thus achieved.

The next topic discussed by Carrera and Neuman (1986a) is that of the derivation of the parameters estimation errors, by a procedure similar to that described in Sect. 5.6.3. Additional important subjects are: model identification criteria (Carrera and Neuman, 1986a), uniqueness and stability and development of solution algorithms (Carrera and Neuman, 1986b). Finally, the application of the methodology is illustrated for a synthetic example and an actual field problem in Carrera and Neuman (1986c).

The methodology developed by Carrera and Neuman (1986a,b,c) seems to differ in principle from that of Hoeksema and Kitanidis (1984) (see, for instance, the classification proposed by Loaiciga and Marino, 1987). As a matter of fact, there is a great deal of common ground, as it is apparent from the formulation in terms of the likelihood function. Both methodologies characterize parameters in statistical terms, the estimates of Carrera and Neuman (1986a,b,c) being equivalent to the conditional expected values of Sect. 5.6.3. In order to achieve the ambitious goal of Carrera and Neuman (1986a,b,c), namely to determine transmissivity and other parameters for arbitrary errors variances, one has to undertake extensive numerical computations, both by solving the equations of flow (5.6.14) and by minimizing L (5.6.15), which depends on a large number of unknowns. The limitations of numerical schemes, e.g. inability to cope with large spatial variations and smoothing over elements, have been already discussed in previous sections. In this respect, the synthetic example selected by Carrera and Neuman (1986c) does not submit the methodology to the ultimate test. Thus, by prescribing the fluxes on the boundary (Carrera and Neuman, 1986c, Fig. 1), and with relatively abundant head data, the problem of transmissivity nonuniqueness is avoided to a large extent. Similarly, the selected "true" transmissivity values are constant over relatively large areas, the entire flow domain being covered by nine elements of constant, but different, transmissivities. In turn, each of these areas is covered by eight finite elements, achieving a level of discretization which eliminates the smoothing effect of discretization mentioned before. In actual applications, these favorable conditions are encountered quite seldom. Besides differences in computational aspects, it is worthwhile to mention that in the methodology of Carrera and Neuman (1986a,b,c) the parameter prior estimates $\mathbf{p}^*$ are set at the beginning of the iterative process, e.g. by kriging the logtransmissivity values, after determining the variogram from T measurements. In particular, the integral scale I_Y, is estimated from transmissivity measurements only. In contrast, in the procedure of Kitanidis and Vomvoris (1983) and its developments, the identification of C_Y parameters is part of the process in which both head and transmissivities data are employed. Thus, the weight coefficients of the Y measurements in the likelihood function is reflected by both σ_1^2 and V_1, while the latter is taken as fixed in the implementation of the methodology of Carrera and Neuman (1986a,b,c).

The methods of solution of the identification problem reviewed in the last two sections are too recent to permit one to draw definite conclusions about their ability to overcome the difficulties mentioned in Sect. 5.6.2 or about selecting a preferable avenue. Nevertheless, the application of the statistical approach so far has yielded encouraging results which are a good omen for its wide application.

Exercises

5.6.1 Assuming that H and Y are multivariate normal and characterized by stationary two-point covariances C_Y and C_{YH} and by a variogram Γ_H, derive the conditional expected value and the variance of H, for given measurements of H and Y (see Dagan, 1984). Determine explicitly the weight coefficients λ and μ for one or two heads and one transmissivity measurements.

5.6.2 In the identification procedure described in Sect. 5.6.3 (iii) it was assumed that the head variogram is not affected by a nugget effect, since generally heads are measured quite accurately. Generalize the suggested scheme if such an effect is present and write down the expression of the corresponding head variogram.

5.6.3 Determine explicitly the coefficients of Θ_{ij} in (5.2.15) for the impact of estimation errors of m_Y and $\mathbf{J}$, for one transmissivity and two heads measurements. Assume that C_Y is exponential.

5.7 *TRANSPORT AT THE REGIONAL SCALE*

5.7.1 *General*

At present we consider a solute body or a plume which move for a considerable period in a porous formation in the sense that the travel distance of the body or the plume length, respectively, are comparable to or larger than the log-transmissivity integral scale I_Y. In such a case, the regional scale heterogeneity is going to influence the motion of the solute.

In line with the general approach to model the flow, we average the concentration C along the vertical over the thickness of the solute body and eventually over a horizontal area of much smaller scale than I_Y. Then, the space averaged $\overline{C}$ becomes a function of the horizontal coordinates x,y and the time t. This is the "point value" at the regional scale and we shall denote it by C(x,y,t) for the sake of simplicity. Our aim is to investigate the impact of the large scale heterogeneity upon C.

In Part 4, dealing with transport at the local scale, we have regarded the pore-scale dispersion as a "Brownian motion" type of transport, due to the large disparity between the two scales, of local heterogeneity and of pores. By the same token, it is justified to regard here the local heterogeneity effect as a diffusive one, due to the smallness of the local heterogeneity scale compared to that of the transmissivity. Furthermore, we assume that the solute body or the plume have transversal dimensions which are large compared to the local scale, such that ergodicity prevails. Under these conditions, the concentration of a conservative solute satisfies the transport equation

$$\frac{\partial C}{\partial t} + \mathbf{V}.\nabla C = \nabla.(\boldsymbol{D}_{\ell}\nabla C) \tag{5.7.1}$$

where $\mathbf{V} = \boldsymbol{q}/n_{ef}$, a function of x,y and t, is the groundwater velocity at the regional scale, i.e. $\boldsymbol{q} = -T\nabla H$. The tensor $\boldsymbol{D}_{\ell}$, of components $D_{\ell,pq}$ (p,q=1,2) represents the effective dispersion at the local scale in the plane, whose derivation was one of the main objectives of Part 4. In line with the general approach of this part, **V** is modeled as a random space function, to reflect uncertainty associated with transmissivity spatial variability. Hence, C is also an RSF, and has to be represented in terms of its statistical moments, similarly to the representation adopted at the local scale. However, the contrast in the heterogeneity scales is reflected in qualitative differences between the transport phenomena. This can be better understood by referring to two categories of solute input zones: "non-point" and "point" sources.

Non-point sources are defined in the environmental literature as sources of pollutants of a large areal extent. A typical example is the one of pesticides or fertilizers which are applied over large agricultural fields and are recharged to aquifers by infiltration. Another case which may be regarded as of the same type is of a large repository or a battery of numerous injecting wells. In the present context, non-point sources are defined as the ones for which the dimension L_0, transverse to the direction of mean flow, is much larger than the logtransmissivity integral scale (it is reminded that according to the field findings of Chap. 5.2, the latter is of the order 10^2-10^4 meters). Furthermore, let us assume that one is interested in the space average of the concentration $\overline{C}$ at the outflow zone of a similar dimension, e.g. a river serving as an outlet for the solute body or the plume over its entire width, or a battery of pumping wells distributed over a large area. If these two conditions are fulfilled, ergodicity can be invoked and $\overline{C}$ can be assumed to be approximately equal to its ensemble average $\langle\overline{C}\rangle$. In such a case, the theory developed in Part 4 at the local scale for two-dimensional transport can be applied to the regional scale as well, provided we replace σ_Y^2 by the variance of correlated logtransmissivity residuals and I_Y by its integral scale. In particular, for average uniform flow and sufficiently small σ_Y^2 , the results of Figs. 4.6.1-4 for $e\to\infty$ are applicable. Then, $\overline{C}$ satisfies a convective-dispersion equation in which **U** is constant, whereas the "megadispersion" coefficients are given by $D_{t,jk} = (1/2)(dX_{jk}/dt + dX_{\ell,jk}/dt)$ (j,k=1,2). X_{jk} stands now for the displacements covariance related to regional heterogeneity, whereas $X_{\ell,jk}$ stems from the local heterogeneity effect. Thus, from a theoretical standpoint very little has to be added to the developments of Part 4. It is emphasized, however, that D_{jk} may be very large, due to the magnitude of I_Y. Furthermore, Fickian regime for longitudinal transport is seldom reached, since this implies a distance between the source and the output zone of tens of logtransmissivity integral scale. Hence, accounting for the travel time dependence of the "megadispersivity" may be essential in applications. This picture is based on theoretical considerations and has yet to be supported by systematic and reliable large scale field experiments. At any rate, the largest values of dispersivities encountered in Fig. 4.2.1 can be attributed to regional scale heterogeneity, but the results should be regarded with caution because of their low reliability.

Point sources are usually defined as input zones of small dimensions, of the order of tens of meters, e.g. a few neighboring injecting wells or a repository. From our perspective, the transverse dimension L_0 is supposed to be small compared to the logtransmissivity integral scale. Most of the waste disposal sites obey this requirement, in view of the usual magnitude of I_Y (see Chap. 5.2). Generally, in such a case one is interested in predicting the point value of the concentration, or its spatial average over the width of the plume or the average concentration spatial moments of a solute body. Due to the small transverse dimension of the plume compared to I_Y, the large scale transmissivity variation manifests mainly in convection of the whole solute body along a slowly winding path, rather than in a dispersive effect. The prediction of the ensemble average of the concentration is of little use, since in the particular realization of interest the concentration may differ considerably from its average, i.e. ergodicity is not obeyed. In a more systematic framework, this state of affairs is reflected by the exceedingly large concentration variance. This can be seen from the simple Eq. (4.3.35), expressing in an approximate manner the variance by $\sigma_C^2/\langle C\rangle^2=C_0/\langle C\rangle-1$, which is an upper bound. Because of the extremely large regional scale dispersivity, $\langle C\rangle$ is much smaller than the initial concentration C_0 for a point source and $\sigma_C^2/\langle C\rangle^2$ becomes very large. Under these circumstances, the direct application of the results of Part 4 to transport from point sources at the regional scale is not useful. The simplest approach to overcome this difficulty is to regard the convective velocity **V** in (5.7.1) as deterministic and given, which is possible if **V** is known deterministically. In practice this may be the case if transmissivities are measured at a dense grid of points, relative to I_Y, along the trajectory of the plume. Then, a numerical solution of the equations of flow may provide **V**(x,y,t) with little uncertainty, and C can be determined from (5.7.1), with $\boldsymbol{D}_\ell$ based on local heterogeneity effect. However, this is an ideal situation and in applications the transmissivities are measured sparsely, precluding their accurate mapping along the plume trajectory. This dilemma may be solved by using the concept of conditional probability (or the geostatistical approach), which was employed extensively in the previous chapters of this part. The basic idea is to model the velocity **V** as an RSF conditioned on measurements of transmissivity and heads in the formation. The latter can be accounted indirectly in evaluating the conditional logtransmissivity field (see Chap. 5.6) or directly, if prediction has to be carried out for the same flow conditions. As we have already mentioned in this part, one of the main advantages of the conditional probability approach is that it encompasses the two extreme cases above as particular ones. Indeed, if measurements in the transport area are not available, **V** is based on the unconditional statistics of Y and H, precisely like in Part 4. In contrast, for abundant measurements, the variance of estimation of **V** becomes small and its prediction is practically deterministic. The great advantage of conditional probability is that it accounts for presence of measurements in intermediate cases as well, permitting one to assess their impact upon concentration uncertainty. Most of this chapter will be devoted to exploring this line of attack, by following the developments of Dagan (1984).

5.7.2 Transport from "non-point sources"

This is the category of problems in which the area A_0 of solute input to groundwater has a transverse dimension L_0 much larger than the transmissivity heterogeneity scale. Furthermore, the solute body travels a distance which is much larger than the heterogeneity local scale $I_{\ell,Y}$. As mentioned above, this case is amenable to the framework elaborated in Part 4. Indeed, we shall first assume that the transmissivity is lognormal and stationary, along the lines of Sect. 5.2.3. Its unconditional statistical structure is defined in terms of the parameters m_Y, w, σ_Y^2 and I_Y, for a covariance C_Y given by (5.2.2) or (5.2.3). Furthermore, in the simplest conceivable case we assume that the flow is uniform in the average, i.e. $\langle \mathbf{V} \rangle = U = \text{const}$, and adopt a first-order approximation in σ_Y^2, along the lines of Sect. 5.5.2. Under these conditions the expected value of the conductivity $\langle C(x_1,x_2,t) \rangle$ satisfies the transport equation similar to (4.3.29), namely

$$\frac{\partial \langle C \rangle}{\partial t} + U \frac{\partial \langle C \rangle}{\partial x_1} = (D_{11} + D_{\ell,11}) \frac{\partial^2 \langle C \rangle}{\partial x_1^2} + (D_{22} + D_{\ell,22}) \frac{\partial^2 \langle C \rangle}{\partial x_2^2} \qquad (5.7.2)$$

where, for the sake of simplicity, we have taken the axis $x_1 = x$ parallel to the average velocity vector. We shall discuss now the various components of the dispersion tensor appearing in (5.7.2).

$D_{\ell,11} = D_{\ell,L}$ is the longitudinal dispersion coefficient associated with the heterogeneity local effect. As we have shown in Part 4, it is dependent on the travel time, but it tends to a constant value after a travel distance of a few tens of local heterogeneity scales. Thus, under the present circumstances it can be taken as approximately constant and equal to $\sigma_{\ell,Y}^2 I_{\ell,Y} U$ (see Fig. 4.6.4). $D_{\ell,22} = D_{\ell,T}$ is the transverse component, which was shown in Part 4 to be much smaller than $D_{\ell,L}$ and can be neglected in a first approximation. In the case of confinement by thin impervious layers, the approximation of Chap. 4.7 may be adopted, with a logarithmic growth of $X_{\ell,22}$. Altogether, the local components are small compared to the regional heterogeneity related terms, to be discussed next.

$D_{11} = (1/2)(dX_{11}/dt)$ is the longitudinal dispersion coefficient for two-dimensional transport. The displacement covariance X_{11} is given explicitly as a function of time by (4.7.2) and is represented in Fig. 4.6.1 for $e = \infty$. In these expressions σ_Y^2 stands now for the variance of correlated log-transmissivity residuals, whereas I_Y is its integral scale. The asymptotic limit $D_{11} = \sigma_Y^2 U I_Y$, i.e. $\alpha_{ef,L} = \sigma_Y^2 I_Y$ is quite large, due to the magnitude of the integral scale. However, for the same reason it is attained only after a considerable travel distance. Hence, the time dependent nature of D_{11} has to be preserved. By the same token $D_{22}=(1/2)(dX_{22}/dt)$ is given by Eq. (4.7.3) and X_{22} is represented in Fig. 4.6.2 for $e=\infty$, with same meaning of σ_Y^2 and I_Y. Again, it is essential to preserve the time dependence of D_{22} in (5.7.2) in any case.

In the case of constant U considered here, a closed form solution of (5.7.2) for a parcel of solute of mass Δm, originating at $\mathbf{x}=\mathbf{a}$ for $t=t_0$ has been given in Eq. (4.3.13). It can be rewritten explicitly as follows

$$\Delta\langle C\rangle = \Delta M(a_1,a_2,t_0)\, f(x_1-a_1,x_2-a_2,t-t_0)$$

$$f = \frac{1}{2\pi X^{1/2}}\exp\left\{-\frac{1}{2}\frac{[x_1-a_1-U(t-t_0)]^2 X_{t,11}}{X} - \frac{1}{2}\frac{(x_2-a_2)^2 X_{t,22}}{X}\right\} \tag{5.7.3}$$

$$X_{t,11} = X_{11}+X_{\ell,11}\ ;\ X_{t,22}=X_{22}+X_{\ell,22}\ ;\ X=X_{t,11}^2 - X_{t,22}^2$$

The expected value of the concentration of a solute body inserted at $t_0=o$ over an area A_0 and at concentration $C_0(\mathbf{a})$ is obtained by substituting $t_0=0$, $\Delta M=C_0 d\mathbf{a}$ in (5.7.3) and integrating over $\mathbf{a}$ within A_0. Similarly, for a plume originating within A_0 at a rate $\dot{M}(\mathbf{a},t_0)$ (mass of solute per unit time and unit area), one has to replace ΔM by $\dot{M}\, d\mathbf{a}\, dt_0$ and to integrate over both $\mathbf{a}$ and t_0. The rate $\dot{M}$ can be also converted into an initial concentration of the plume by the approximate relationship $C_0(\mathbf{a},t_0)\simeq \dot{M}/U$.

The computation of the concentration variance for a solute body inserted over A_0, with the concentration space averaged over an area A, is given by the general formula (4.3.33), which has to be applied to the horizontal plane rather than the space. The key to computing the Gaussian function $f(\mathbf{x}'-\mathbf{a},\mathbf{x}''-\mathbf{b},t)$ in (4.3.33) is the evaluation of the two particle displacement covariance Z_{jk} $(j,k=1,2)$, which is given by (4.3.31). The computaion is simplified if we neglect the local dispersive effect in (4.3.31), leading to the simple formula

$$Z_{jk}(\mathbf{a}-\mathbf{b},t) = 2\int_0^t (t-t')\, u_{jk}(a_1-b_1+Ut',a_2-b_2)\, dt' \tag{5.7.4}$$

Dagan (1984) has evaluated Z_{11} and Z_{22} for u_{jk} (4.6.5) and for an exponential C_Y. The results are presented in Dagan (1984, Fig. 3). The striking feature is that the longitudinal displacements of two particles lying initially on the same average streamline, i.e. for $a_2=b_2$, are fully correlated for initial separations a_1-b_1 as large as three integral scales. In contrast, Z_{11} drops to zero for two particles separated transversally to the direction of mean flow by $|a_2-b_2|\simeq 3I_Y$. This finding supports the assertion made a few times in the preceding parts, namely that space averaging has a variance reducing effect only if the *transverse* dimension of A_0 is large with respect to I_Y.

As we have indicated in Sect. 4.3.5, the computation of the concentration variance is simplified tremendously if $\overline{C}$ is averaged over an area A which is small compared to I_Y^2 and if the local dispersive effect is neglected, leading to $\sigma_{\overline{C}}^2 \simeq \langle C\rangle(C_0-\langle C\rangle)$. This is also an upper bound to the variance and it leads to large coefficient of variation $\sigma_{\overline{C}}/\langle C\rangle$, especially at the fringe of the solute body or the plume. In application this will be the case of measurement or pumping by a well. Again, in the opposite case of a very wide plume at the I_Y scale and for C space averaged over a similarly large area, the variance tends to zero. As we have emphasized in the preceding section, this implies very large dimensions of the plume and of the averaging area, e.g. the average concentration of the effluent of an aquifer into a river crossing it. In more realistic cases the concentration coefficient of variation will be larger than zero and bounded by the above limit. Its computation requires to evaluate the covariances Z_{jk} first, and subsequent four or six times integration for a solute body or a plume, respectively.

The other subjects developed in Part 4, e.g. calculation of spatial moments and incorporation of effects of flow nonuniformity and unsteadiness, can be applied to two-dimensional transport from nonpoint sources without change.

5.7.3 Transport from "point sources"

(i) Unconditional probability.

We turn now to the case of a solute body whose initial dimension L_0 is much smaller than the logtransmissivity correlation scale I_Y, yet much larger than the local scale $I_{\ell,Y}$. Furthermore, we assume that the local heterogeneity dispersive effect is sufficiently small to ensure the first condition at any time, i.e. $I_Y^2/tD_{\ell,11} \simeq I_Y^2/L\alpha_{\ell,L} >> 1$, with L=Ut the travel distance and with $\alpha_{\ell,L}$ the local longitudinal dispersivity. For the large values of I_Y usually encountered at the regional scale, this condition is bound to be obeyed.

Under these conditions the analysis of the transport process is greatly simplified. Indeed, we refer to the spatial moments of Sect. 4.3.6, and particularly to the fundamental Eq. (4.3.41) for the expected value of the "moment of inertia" of the solute body. As explained already in the comment following (4.3.41), for a small solute body at the I_Y scale, the second and fourth terms in the right-hand-side of (4.3.41) cancel out and we get

$$\langle \overline{X}_{t,jk}\rangle = \frac{1}{M}\langle \int (x_j-\overline{X}_{t,j})(x_k-\overline{X}_{t,k})\, n\, C(\mathbf{x},t)\, d\mathbf{x}\rangle \to \overline{X}_{jk}(0) + X_{\ell,jk} \quad (j,k=1,2) \qquad (5.7.5)$$

Putting it into words, this result reads: the expected value of the second spatial moment of the solute body with respect to its centroid $\overline{\mathbf{X}}_t$, is equal to its initial value augmented by the local dispersive effect. Hence, at the regional scale the solute body behaves like an infinitesimal particle concentrated in the center of mass. Once its motion is determined, the spread of the tracer with respect to the center of mass can be determined by adding the effect of local heterogeneity, which is ergodic and Gaussian.

Concentrating now on the motion of the centroid $\overline{\mathbf{X}}$, the large uncertainty associated with its position is reflected by its covariance $\bar{\sigma}_{jk} \simeq X_{jk}$ (4.3.39), which is precisely the displacement covariances discussed at length in Part 4. Although mathematically identical, it has a profoundly different meaning in the two cases under investigation, i.e. for transport from nonpoint and point sources. While in the first case and under the conditions of ergodicity discussed in the previous section, it represents the dispersive effect of the regional scale heterogeneity, in the second one it is a measure of uncertainty of the position of the solute body. The p.d.f. $f(\overline{\mathbf{X}})$ is characterized completely by $\langle\overline{\mathbf{X}}\rangle$ and by $\overline{X}_{jk}$ if it is Gaussian. This is the case for the first-order analysis in σ_Y^2. Furthermore, for a uniform average velocity $\mathbf{U}(U,0)$ and for the exponential C_Y (5.2.2), $\langle\overline{\mathbf{X}}\rangle = \mathbf{a} + \mathbf{U}t$, while X_{11} and X_{22} are given in a close form by Eqs. (4.7.2) and (4.7.3), respectively, and by Figs. 4.6.1-2. Again, for the large values of I_Y encountered in applications, the travel time dependency of X_{jk} has to be taken into account, and the asymptotic results, valid for $UT/I_Y \gg 1$, are not applicable. Hence, transport from point sources is both affected by uncertainty and transient effects, and the prediction is subjected to large possible errors.

These results can be extended to the case of a plume, which is seen by the regional observer as a winding line in the plane, representing its axis. Its equation is given by

$$\overline{\mathbf{X}}(\tau) = \mathbf{a} + \int_0^{\tau} \mathbf{V}[\mathbf{X}(t')]\, dt' \quad ; \quad \mathbf{V}(\mathbf{x}) = \mathbf{U} + \mathbf{u}(\mathbf{x}) \quad ; \quad 0<\tau<t \tag{5.7.6}$$

where τ is the "running" time and t is the actual total travel time.

Again, under the conditions of Chap. 4.6, $\overline{\mathbf{X}}$ is multivariate normal, of expected value $\mathbf{a} + \mathbf{U}t$ and covariance X_{jk}. Thus, the "leading edge" of the plume, given by (5.7.6) for τ=t, is completely characterized by the same moments as the ones pertaining to a solute body. However, if one is interested in the random location of the entire plume, additional information is needed. Indeed, let us assume that the plume is represented by an array of discrete points $\overline{\mathbf{X}}^{(m)} = \overline{\mathbf{X}}(\tau_m)$ $(m=0,\dots,N\ ;\ \tau_0=0,\dots,\tau_N=t)$. The statistics of the array is exhausted by the joint p.d.f. of $\overline{\mathbf{X}}^{(m)}$. Under the assumption that it is multivariate normal, it is characterized in turn by the expected values $\langle\overline{\mathbf{X}}^{(m)}\rangle = \mathbf{a} + \mathbf{U}\tau_m$ and the covariance matrix $Z_{jk}^{(mn)} = \langle\overline{X}_j'^{(m)}\ \overline{X}_k'^{(n)}\rangle$. The latter can be evaluated in a simple manner with the aid of the velocity covariance by using the first-order approximation in σ_Y^2 of Chap. 4.6. Thus, the appropriate generalization of (4.6.8) is

$$Z_{jk}^{(mn)} = \int_0^{\tau_m}\int_0^{\tau_n} u_{jk}[U(t'-t'')]\, dt'\, dt'' \quad (j,k=1,2) \tag{5.7.7}$$

After a proper change of variables, $Z_{jk}^{(mn)}$ (5.7.7) can be expressed with the aid of the one particle covariances $X_{jk}(t)$ as follows

$$Z_{jk}^{(mn)} = \frac{1}{2}X_{jk}(\tau_m) + \frac{1}{2}X_{jk}(\tau_n) - \frac{1}{2}X_{jk}(|\tau_n-\tau_m|) \quad (j,k=1,2) \tag{5.7.8}$$

Again, like in the case of a solute body, the location of the plume axis is subjected to a large degree of uncertainty and furthermore, the transient part of X_{jk} has to be employed in evaluating $Z_{jk}^{(mn)}$ at any t. Finally, to obtain the actual concentration distribution we have to supplement $\overline{\mathbf{X}}$ by the solution of the diffusion equation (5.7.1) for a plume for the effect of local heterogeneity.

A reduction of the uncertainty of the solute body or plume location can be achieved by employing the actual measurements of head and transmissivity and this is the next topic to be examined.

(ii) Conditional probability.

We consider now the case in which transmissivity and head measurements are available. These measurements were supposedly employed to identify the statistical structures of $Y=\ell nT$ and H by the procedures outlined in Chap. 5.6. We may face now two prediction scenarios: transport at the regional scale has to be analyzed for same flow conditions or for different ones. In the first case we can use the statistical information about Y and H, while in the second one H has to be computed for the future flow conditions, with no head measurements available. Since the first scenario is more comprehensive, we shall concentrate on it.

The basic idea is to employ the *conditional* joint p.d.f. of Y^c and H^c, conditioned on their measurements, rather than the unconditional ones. The derivation of the spatial moments which characterize these variables, namely $\langle Y^c(\mathbf{x})\rangle$, $\langle H^c(\mathbf{x})\rangle$, $C_Y^c(\mathbf{x},\mathbf{x}')$, $C_{YH}^c(\mathbf{x},\mathbf{x}')$ and $C_H^c(\mathbf{x},\mathbf{x}')$ (or Γ_H^c) have been discussed in the preceding sections of this part (for details see also Dagan, 1984, and Dagan and Rubin, 1988). Assuming that they are available, the task is to derive the moments of particles displacements $\mathbf{X}^c(t,\mathbf{a})$, the latter being defined as the coordinate of a particle at time t which was at t=0 at $\mathbf{x}=\mathbf{a}$, and which is convected by the random velocity field $\mathbf{V}^c(\mathbf{x},t)$ stemming from the heterogeneous structure characterized by Y^c and H^c. The basic equation defining $\mathbf{X}^c$ is given by (4.3.16), namely

$$\frac{d\mathbf{X}^c}{dt} = \mathbf{V}^c(\mathbf{X}^c) \quad ; \quad \mathbf{X}^c = \mathbf{a} \text{ for } t=0 \tag{5.7.9}$$

In turn, the velocity is related to the transmissivity and head fields by (5.3.4), i.e.

$$\mathbf{V}^c = \frac{q^c}{n_{ef}} = -\frac{1}{n_{ef}} \mathrm{T}^c \, \nabla \mathrm{H}^c \tag{5.7.10}$$

where n_{ef} is the effective porosity, at the regional scale, and $\mathrm{T}^c = \exp(\mathrm{Y}^c)$.

The two Eqs. (5.7.9) and (5.7.10) may serve as starting points for numerical, Monte-Carlo, simulations, similar to the ones described in Chap. 4.4. The only difference is that the generation of the transmissivity field is carried out from the lognormal population Y^c rather than Y. Smith and Schwartz (1986b) have indeed considered the impact of fixing a few conductivity values in their simulations. Along the lines of the preceding parts, we shall pursue here simplified, approximate, solutions which lend themselves to more general interpretations.

The major assumption is the one based on the first-order approximation in σ_Y^2, leading first to the replacement of $\mathbf{X}^c$ in the argument of $\mathbf{V}^c$ in (5.7.9) by the average displacement. With $\mathbf{V}^c$ = $\mathbf{U}^c + \mathbf{u}^c$, where $\mathbf{U}^c = \langle \mathbf{V}^c \rangle$, we get from (5.7.9) under this approximation

$$\frac{d\langle \mathbf{X}^c \rangle}{dt} = \mathbf{U}^c(\langle \mathbf{X}^c \rangle) \; ; \; \frac{d\mathbf{X}'^c}{dt} = \mathbf{u}^c(\langle \mathbf{X}^c \rangle)$$

i.e. (5.7.11)

$$X^c_{jk}(t,\mathbf{a}) = \int_0^t \int_0^t u^c_{jk}\,[\langle \mathbf{X}^c(t',\mathbf{a})\rangle, \langle \mathbf{X}^c(t'',\mathbf{a})\rangle]\, dt'\, dt'' \qquad (j,k=1,2)$$

The first equation in (5.7.11) permits one to determine the mean trajectory if the mean velocity field is known. This has to be done generally by a numerical quadrature due to the complex dependence of $\mathbf{U}^c$ upon coordinates, as shown below. Similarly, the last equation leads to the displacements covariance by two quadratures, if the conditional velocity covariance is given, and after determining $\langle \mathbf{X}^c \rangle$. These numerical calculations are considerably simpler than the Monte Carlo simulations employed for solving the exact version (5.7.9). The approximation is bound to hold if the trajectories in various realizations do not depart too much, in a statistical sense, from their average.

Next, we take advantage of the first-order approximation in order to simplify the derivation of the velocity covariance. The procedure is similar to the one employed for average uniform flow, which yields $\mathbf{q}^{(1)}$ (3.4.7) and q_{jk} (3.7.26). Indeed, with $\mathrm{Y}^c = \langle \mathrm{Y}^c \rangle + \mathrm{Y}'^c$ and $\mathrm{H}^c = \langle \mathrm{H}^c \rangle + \mathrm{h}^c$ in (5.7.10), we obtain at first order

$$\mathbf{U}^c(\mathbf{x}) = \frac{1}{n_{ef}} \langle \mathrm{T}^c(\mathbf{x}) \rangle \, \mathbf{J}^c(\mathbf{x}) \; ; \; \mathbf{u}^c = \frac{1}{n_{ef}} \langle \mathrm{T}^c \rangle (\mathbf{J}^c \mathrm{Y}'^c - \nabla \mathrm{h}^c)$$

leading to

$$u^c_{jk}(\mathbf{x},\mathbf{x}') = \frac{\langle T^c(\mathbf{x})\rangle\langle T^c(\mathbf{x}')\rangle}{n_{ef}^2}[J^c_j(\mathbf{x})\, J^c_k(\mathbf{x}')\, C^c_Y(\mathbf{x},\mathbf{x}') - J^c_j(\mathbf{x})\frac{\partial C^c_{YH}(\mathbf{x},\mathbf{x}')}{\partial x'_k}$$

$$- J^c_k(\mathbf{x}')\frac{\partial C^c_{YH}(\mathbf{x}',\mathbf{x})}{\partial x_k} + \frac{\partial^2 C^c_H(\mathbf{x},\mathbf{x}')}{\partial x_j\, \partial x'_k}] \qquad (5.7.12)$$

where we have taken n_{ef} constant and have used the symbols $\mathbf{J}^c=-\nabla\langle H^c\rangle$ and $\langle T^c\rangle=\exp(\langle Y^c\rangle)$. For a constant **J** and in absence of conditioning, (5.7.12) degenerates into (3.7.26). Hence, once the various statistical moments of Y^c and H^c are determined, (5.7.12) permits one to evaluate the velocity covariance. It is emphasized that measured values appear explicitly in the expressions of T^c and $\mathbf{J}^c$, whereas the conditional covariances depend only on the location of measurement points. The developments of the previous chapters and Eqs. (5.7.11) and (5.7.12) permit one to derive systematically the motion of a solute body or a plume, starting from T and H measurements, through the identification of the unconditional Y, H moments, computation of conditional moments by solving the appropriate linear systems and ending with displacement moments. It is worthwhile to mention a few points of interest: (i) under the present scheme we need to evaluate $\mathbf{U}^c$ and u^c_{jk} only for points along the mean trajectory, and not on a grid covering the entire flow domain; (ii) the procedure outlined by Rubin and Dagan (1987) permits one to incorporate in u^c_{jk} both uncertainty related to spatial variability of Y and to the estimation errors of the various parameters $\boldsymbol{\theta}$ which characterize the Y, H moments. This is an important point, since estimation errors might be quite large; (iii) the procedure can be applied to unsteady flows, at least for the quasi-steady approximation of Dagan and Rubin (1988); (iv) if transport is predicted for flow conditions different from the ones prevailing at the time of measurement, one has to employ the Y^c moments, with H based on the solution of the direct flow problem for given Y^c (see Sect. 5.5.3); (v) if transport has to be predicted for the future motion of an existing solute body or a plume, for which past measurements are available, the future displacements can be conditioned on the past ones, reducing further the uncertainty and (vi) if measurements are available on a very dense grid the conditional covariances in (5.7.12) tend to zero and the same is true for X^c_{jk} . Then, the problem becomes essentially a deterministic one, of predicting the mean trajectory $\langle \mathbf{X}^c\rangle$ by solving the first equation in (5.7.12).

The program outlined above has not yet been carried out for synthetic or actual plumes. However, some indications about the effect of conditioning upon the reduction of X_{jk} have been obtained by Dagan (1984) under further assumptions. Assuming that at the points of measurement affecting the solute body motion, the measured Y and H are equal to their unconditional means, one can substitute in (5.7.12) $\langle T^c\rangle=T_G$ and $\mathbf{J}^c=\mathbf{J}$, both constant. Hence, $\mathbf{U}^c=\mathbf{U}$ is constant and the average displacement is a straight line. In other words, the influence of conditioning is left only in covariances and it manifests through the coordinates of the measurement points only. Furthermore, if there are only a few such points and for uniform flow and exponential C_Y, closed form solutions

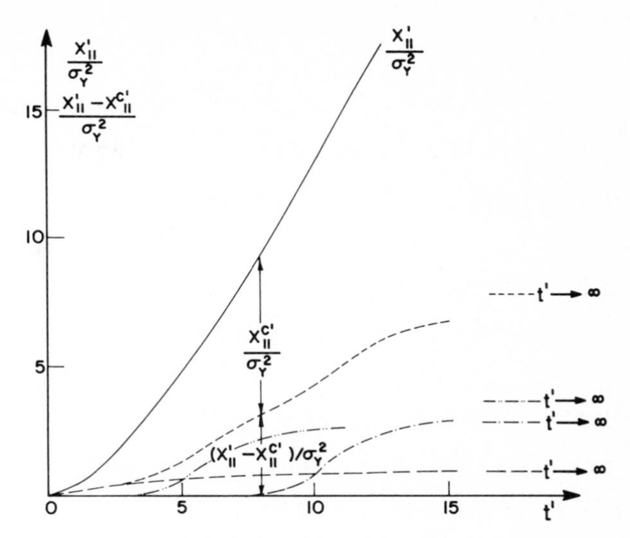

5.7.1 The unconditional X_{11} (Eq. 4.7.2) and the conditional X^c_{11} displacement longitudinal covariances for two-dimensional flow. Conditioning by transmissivity at a point of coordinates $x_1, y_1=0$: ———unconditional, — — — $x_1=0$, -··-·· $x_1/I_Y=5$, -·-·- $x_1/I_Y=10$, ---- cumulative effect for the previous three (reproduced from Dagan, 1984). The dimensionless variables are $X'_{11}=X_{11}/I_Y^2$, $t'=tU/I_Y$.

can be obtained for X^c_{jk}. To illustrate the results, we have reproduced in Fig. 5.7.1 the graphs of Dagan (1984, Fig. 4), depicting the dependence of the longitudinal covariance X_{11} upon the dimensionless travel time $t'=tU/I_Y$. Five cases are shown: unconditional, as given in Eq. (4.7.2), conditioned on a transmissivity measurement on the average streamline at three different distances from the source, and by the three measurements altogether, assuming they are additive. The striking feature is the considerable reduction of the covariance achieved by these measurements. This is a good omen for prediction of transport at regional scale, if appropriate measurements are available. A much lesser impact is achieved by isolated transmissivity measurements in reducing X_{22} or by head measurements (see Figs. 5,6 and 7 in Dagan, 1984) upon both X_{11} and X_{22}.

As we have mentioned at the beginning of this chapter, the impact of regional scale heterogeneity and measurements has not yet been thoroughly assessed, either by systematic field measurements or by modeling. The theoretical developments presented here may serve in helping to design and analyze such field tests and for prediction purposes.

Exercises

5.7.1 Rewrite Eq. (5.7.1) in the case in which effective recharge R(t) is present, such that $\nabla . q=R$. Show that by a suitable transformation of type $C^*(x,y,t) = C(x,y,t).f(t)$, the transport equation for C^* is identical to (5.7.1). Derive the function of time f(t).

5.7.2 Derive an expression similar to (4.3.33) for the concentration variance in the case of a plume, ensuing from an area A_0 at constant concentration C_0.

5.7.3 Derive the analytical expression of X^c_{11} for two transmissivity measurements along the mean streamline. The average unconditional velocity is constant, C_Y is assumed to be exponential and $Y=m_Y$ at the measurement points.

5.8 MODELING TRANSPORT BY THE TRAVEL TIME APPROACH

5.8.1 General

In Parts 2 and 4, as well as in Chap. 5.7, transport has been modeled in terms of the macroscopic concentration $C(\mathbf{x},t)$, the basic physical entity to represent the space and time distribution of the solute in the fluid. However, our approach to derive C was to regard the solute as made up from distinct particles which move in space, their mass distribution being related to concentration. This point of view has underlain Saffman's derivation of the pore-scale dispersion coefficients (Sect. 2.10.3) and has constituted the starting point of the study of transport at the local and regional scales (Parts 4 and 5, respectively). The approach was to seek the distribution of the particles in space at any given time with the aid of the function $\mathbf{x} = \mathbf{X}_t(\mathbf{a},t,t_0)$ (Eq. 4.3.1), the coordinate vector of a particle at time t which was at $\mathbf{x}=\mathbf{a}$ at t_0 ($t_0<t$). The probability density function of one particle displacement $f(\mathbf{X}_t;t,\mathbf{a},t_0)$, as well as the joint p.d.f. for a few particles, were employed in order to describe the spatial distribution of the concentration. A different, but related approach, is to consider a fixed surface S in space and the time τ of travel of a particle from the input zone to S. This approach, representing transport in terms of travel (or of the related terminology of residence, occupancy or life) time has been explored in different branches of physics and quite recently in the context of subsurface transport. In the latter area it has been used primarily in order to represent one-dimensional transport, the main thrust being provided by a series of articles by Jury and Sposito (Jury, 1982, Jury et al, 1986, White et al, 1986, Sposito et al, 1986). Earlier studies were related to column flow (Todorovic, 1970, Simmons, 1982), whereas aspects related to surface hydrology have

been investigated by Rinaldo and Marani (1987). The use of travel time in a deterministic context has been suggested by Nelson (1978) for two-dimensional flows. Finally, the extension of travel time approach to two- and three-dimensional flows in a stochastic frame, and comparison with concentration representation, have been explored recently by Dagan and Nguyen (1989). The same topic, namely, the travel time approach and its relation with the concentration, is the object of the present chapter.

5.8.2 One-dimensional transport

We discuss first the case of one-dimensional transport, i.e. the concentration $C(x,t)$ is a function of one spatial coordinate only. In dealing with transport at the local (Part 4) or regional (Chap. 5.7) scales, we have refrained from discussing this class of flows, since it is of little relevance to natural heterogeneous porous formations. Still, it is of interest to explore the travel time approach in one-dimension for a few reasons: first, the method is particularly advantageous in this case; second, we shall show in the next section that results are applicable to two- or three-dimensional flow and last, as mentioned before, most previous work pertains to this configuration.

We begin with the simplest conditions, namely column flow of an incompressible fluid in a rigid matrix and transport of an inert solute. By the equation of continuity (2.5.16), the specific discharge q is independent of x, i.e. $q(t)$. The transport equation (2.12.11) becomes in its complete version

$$\frac{\partial C}{\partial t} + V\frac{\partial C}{\partial x} = \frac{1}{n}\frac{\partial}{\partial x}\left[n(D_m + D_L)\frac{\partial C}{\partial x}\right] \quad ; \quad V = q/n \qquad (5.8.1)$$

where the effective porosity $n(x)$ is assumed to be spatially variable in order to illustrate the impact of heterogeneity in this case. As usual D_m and D_L stand for the coefficients of macroscopic diffusion and longitudinal dispersion, respectively. At present q is regarded as deterministic and C is conditioned on its value, but we shall also investigate later the possibility of its random variation. We regard n as a stationary random space function of x, with $\langle n\rangle$=const its expected value and with $C_n(x'-x'')$ its two-point covariance. Thus $C(x,t|q)$, solution of (5.8.1) with appropriate initial and boundary conditions, is a random space function as well. The derivation of its statistical moments in terms of those of V has constituted the main object of Part 4.

Along the lines of our basic approach in Part 4, we consider a solute particle of mass $\Delta M_0=\Delta M(a,t_0)$ introduced at $x=a$ and $t=t_0$ in the porous column. By particle we refer to a pulse across the entire cross-section and over a distance Δa which is small compared to the column length, but large at the pore-scale. Hence, M is here in units of mass per unit area. The particle moves in the positive x direction due to convection and pore-scale dispersion. We cast now the transport problem in a form different from that of Part 4, by considering the travel time $t=\tau$ for which the

Polubarinova Kotchina, P. Ya., *Theory of Groundwater Movement*, translation from Russian, Princeton Univ. Press., 1962.

Poreh, M., The dispersivity tensor in isotropic and axisymmetric mediums, Journ. Geophys. Res., 70, 3909-3913, 1965.

Raudkivi, A.J. and R.A. Callander, *Analysis of Groundwater Flow*, Edward Arnold, London, 1976.

Rinaldo, A., and A. Marani, Basin scale model of solute transport, Water Resour. Res., 23, 2107-2118, 1987.

Risken, H., *The Focker-Planck Equation*, Springer-Verlag, Berlin Heidelberg, 1984.

Roberts, P.V., M.N. Goltz and D.M. Mackay, A natural gradient experiment in a sand aquifer, 3. retardation estimates and mass balances for organic solutes, Water Resour. Res., 22, 2047-2058, 1986.

Robertson, H.P., The invariant theory of isotropic turbulence, Proc. Phil. Soc., 36, 209-223, 1940.

Rose, H.E., On the resistance coefficient-Reynolds number relationship for fluid flow througha bed of granular material, Proc. Inst. Mech. Eng., 153, 141-161, 1945.

Rubin, Y. and G. Dagan, Stochastic identification of transmissivity and effective recharge in steady groundwater flow : 1. Theory, Water Resour. Res., 23, 1185-1192, 1987a.

Rubin, Y. and G. Dagan, Stochastic identification of transmissivity and effective recharge in steady groundwater flow : 1. Case study, Water Resour. Res., 23, 1193-1200, 1987b.

Saffman, P.G., A theory of dispersion in a porous medium, Journ. Fluid Mech., 3, 321-349, 1959.

Saffman, P.G., Dispersion due to molecular diffusion and macroscopic mixing in flow through a network of capillaries, Journ. Fluid Mech., 2, 194- 208, 1960.

Saffman, P.G., On the boundary conditions at the surface of a porous medium, Stud. Appl. Math., I, 93-101, 1971.

Sagar, B., Galerkin finite element procedure for analyzing flow through random media, Water Resour. Res., 14, 1035-1044, 1978.

Salu, Y., and D. Montgomery, Turbulent diffusion from a quasi-kinematical point of view, Phys. Fluids, 20, 1-3, 1977.

Sanchez-Palencia, E., *Non-homogeneous Media and Vibration Theory*, Lecture Notes in Physics, Springer-Verlag, Berlin Heidelberg New York, 1980.

Schweppe, F.C., *Uncertain Dynamic Systems*, Prentice-Hall, Englewood Cliffs, N.J., 1973.

Schwydler, M.I., Flow in heterogeneous media (in Russian), Izv. Akad. Nauk SSSR Mekh. Zhidk. Gaza, 3, 185, 1962.

Simmons, C.S., A stochastic-convective transport represention of dispersive in one-dimensional porous media systems, Water Resour. Res., 18, 1193-1214, 1982.

Slattery, J.C., *Momentum, Energy, and Mass Transfer in Continua*, McGraw Hill, New-York, 1972.

Smith, L., and R.A. Freeze, Stochastic analysis of steady state groundwater flow in bounded domain, , Two-dimensional simulations, Water Resour. Res., 15, 1543-1559, 1979.

Smith, L., and F.W. Schwartz, Mass transport, 1. Stochastic analysis of macrodispersion, Water Resour. Res., 16, 303-313, 1980.

Smith, L., and F.W. Schwartz, Mass transport 2. Analysis of uncertainty in prediction,, Water Resour. Res., 17, 351-369, 1981a.

Smith, L., and F.W. Schwartz, Mass transport 3. Role of hydraulic conductivity data in prediction, Water Resour. Res., 17, 1463-1479, 1981b.

Sposito, G.W., W.A. Jury, and V.K. Gupta, Fundamental problems in the stochastic convection-dispersion model of solute transport in aquifers and field soils, Water Resour. Res., 22, 77-88, 1986.

Sposito, R.E. White, P.R. Darrah, and W.A. Jury, A transfer function model of solute transport through soil. 3, The convection-dispersion equation, Water Resour. Res., 22, 255-262, 1986.

Stoker, UJ.J., *Water Waves*, Interscience Publishers, 1957.

Stroud, D., Generalized effective-medium approach to the conductivity of an inhomogeneous material, Phys. Rev. B, 12, 3368-3373, 1975.

Sudicky, E.A., A natural-gradient experiment on solute transport in a sand aquifer: spatial variability of hydraulic conductivity and its role in the dispersion process, Water Resour. Res., 22, 2069-2082, 1986.

Taylor, G.I., Diffusion by continuous movements, Proc. London Math. Soc., A20, 196-211, 1921.

Taylor, G.I., Dispersion of soluble matter in solvent flowing slowly through a tube, Proc. Royal Soc. London, A 219, 186-203, 1953.

Taylor, G.I., Conditions under which dispersion of a solute in a stream of solvent can be used to measure molecular diffusion, Proc. Royal Soc. london, A225, 473-477, 1954.

Terzaghi, K., *Erdbaumechanik auf Bodenphysikalischer Grundlage*, Franz Deuticke, Vienna, 1925.

Terzaghi, K., *Theoretical Soil Mechanics*, John Wiley, New York, 1943.

Theis, C.V., The relation between the lowering of the piezometric surface and the rate and duration of discharge of a well using ground-water storage, EOS Trans. AGU, 16, 519-524, 1935.

Todorovic, P., A stochastic model of longitudinal dispersion in porous media, Water Resour. Res., 6, 211-222, 1970.

Vachaud, G., Contribution a l'etude des problemes d'ecoulement en milieux poreux nonsaturees, Ph. D. Thesis, Grenoble, France, 1968.

Vallochi, A.J., Describing the transport of ion-exchanging contaminants using an effective Kd approach, Water Resour. Res., 20, 499-503, 1984

Van Brakel, J., Pore space models for transport phenomena in porous media, Powder Technology, 11, 205-236, 1975.

Van Dyke, M., *Perturbation Methods in Fluid Mechanics*, Academic, New York, 1964

Vanmarcke, E., *Random Fields: Analysis and Synthesis*, MIT Press, Cambridge, Mass., 1983.

Verrujt, A., Elastic storage of aquifers, in *Flow Through Porous Media* edited by R. J. M. DeWiest, 331-376, Academic, New York, 1969.

Verrujt, A., *Theory of Groundwater Flow*, Mac Millan, 1970.

Warren, J.E., and H.S. Price, Flow in heterogeneous porous media, Soc. Petrol. Eng. Journ., I, 153-169, 1961.

White, R.E., J.S. Dyson, R.A. Haigh, W.A. Jury and G. Sposito, A transfer function model of solute transport through soil. 2, Illustrative applications, Water Resour. Res., 22, 248-254, 1986.

Wilson, J., P. Kitanidis, and M. Dettinger, State and parameter estimation in groundwater models, in *Applications of Kalman Filter to Hydrology, Hydraulics, and Water Resources*, edited by C-L. Chiu, 657-679, Univ. of Pittsburgh, Pittsburgh, Pa., 1980.

Winter, C.L., C.M. Newman, and S. P. Neuman, A perturbation expansion for diffusion in a random velocity field, SIAM J. Appl. Math., 44, 411-424, 1984

Yaglom, A., *An Introduction to Mathematical Stationary Random Functions*, Prentice-Hall, Englewood Cliffs, 1962

Yakowitz, S., and L. Duckstein, Instability in aquifer identification: theory and case studies, Water Resour. Res., 16, 1054-1064, 1980u.

Yeh, T.-C. J., Comment on "Modeling of scale-dependent dispersion in hydrogeologic systems" by J.F. Pickens and G.E. Grisak, Water Resour. Res., 23, 522-523, 1987.

Yeh, W. W-G., and Y.S. Yoon, Aquifer parameter identification with optimum dimension in parametrization, Water Resour. Res., 17, 664-672, 1981.

Yeh, W. W-G., Review of parameter identification procedures in groundwater hydrology: the inverse problem, Water Resour. Res., 22, 95-108, 1986.

Zick, A.A., and G.M. Homsy, Stokes flow through periodic arrays of spheres, Journ. Fluid Mech., 115, 13-26, 1982

Zinszner, B., and Ch. Meynot, Visualisation des proprietes capillaires des roches reservoir, Revue de L'Inst. Franc. du Petrole, 37, 337-361, 1982.

SUBJECT INDEX